ASSEMBLING CRITICAL COMPONENTS

A FRAMEWORK FOR SUSTAINING TECHNICAL AND PROFESSIONAL COMMUNICATION

Foundations and Innovations in Technical and Professional Communication

Series Editor: Lisa Melonçon

Series Associate Editors: Kristin Marie Bivens and Sherena Huntsman

The Foundations and Innovations in Technical and Professional Communication series publishes work that is necessary as a base for the field of technical and professional communication (TPC), addresses areas of central importance within the field, and engages with innovative ideas and approaches to TPC. The series focuses on presenting the intersection of theory and application/practice within TPC and is intended to include both monographs and co-authored works, edited collections, digitally enhanced work, and innovative works that may not fit traditional formats (such as works that are longer than a journal article but shorter than a book).

The WAC Clearinghouse and University Press of Colorado are collaborating so that these books will be widely available through free digital distribution and low-cost print editions. The publishers and the series editors are committed to the principle that knowledge should freely circulate and have embraced the use of technology to support open access to scholarly work.

Other Books in the Series

Michael J. Klein (Ed.), *Effective Teaching of Technical Communication: Theory, Practice, and Application* (2021)

ASSEMBLING CRITICAL COMPONENTS
A FRAMEWORK FOR SUSTAINING TECHNICAL AND PROFESSIONAL COMMUNICATION

Edited by Joanna Schreiber and Lisa Meloncon

The WAC Clearinghouse
wac.colostate.edu
Fort Collins, Colorado

University Press of Colorado
upcolorado.com
Denver, Colorado

The WAC Clearinghouse, Fort Collins, Colorado 80523

University Press of Colorado, Denver, Colorado 80202

© 2021 by Joanna Schreiber and Lisa Melonçon. This work is licensed under a Creative Commons Attribution-NonCommercial-NoDerivatives 4.0 International.

ISBN 978-1-64215-138-1 (PDF) | 978-1-64215-139-8 (ePub) | 978-1-64642-269-2 (pbk.)

DOI 10.37514/TPC-B.2022.1381

Library of Congress Cataloging-in-Publication Data

Title: Assembling critical components : a framework for sustaining technical and professional communication / edited by Joanna Schreiber and Lisa Melonçon.
Description: Fort Collins, Colorado : The WAC Clearinghouse ; Boulder, Colorado : University Press of Colorado, [2021] | Series: Foundations and innovations in technical and professional communication | Includes bibliographical references.
Identifiers: LCCN 2021043906 (print) | LCCN 2021043907 (ebook) | ISBN 9781646422692 (pbk.) | ISBN 9781642151381 (PDF) | ISBN 9781642151398 (ePub)
Subjects: LCSH: Communication of technical information.
Classification: LCC T10.5 .A 83 2021 (print) | LCC T10.5 (ebook) | DDC 601/.4—dc23
LC record available at https://lccn.loc.gov/2021043906
LC ebook record available at https://lccn.loc.gov/2021043907

Copyeditor: Meg Vezzu
Designer: Mike Palmquist
Series Editor: Lisa Melonçon
Series Associate Editors: Kristin Marie Bivens and Sherena Huntsman

The WAC Clearinghouse supports teachers of writing across the disciplines. Hosted by Colorado State University, and supported by the Colorado State University Open Press, it brings together scholarly journals and book series as well as resources for teachers who use writing in their courses. This book is available in digital formats for free download at wac.colostate.edu.

Founded in 1965, the University Press of Colorado is a nonprofit cooperative publishing enterprise supported, in part, by Adams State University, Colorado State University, Fort Lewis College, Metropolitan State University of Denver, University of Alaska Fairbanks, University of Colorado, University of Northern Colorado, University of Wyoming, Utah State University, and Western Colorado University. For more information, visit upcolorado.com.

Contents

Introduction: Promoting a Sustainable Collective Identity for
Technical and Professional Communication 3
    Lisa Melonçon and Joanna Schreiber

PART ONE: EXIGENCY FOR A SUSTAINABLE IDENTITY

Chapter 1: What Are We Really Teaching? Revisiting Technical
and Professional Communication's Pedagogical Training 19
    Sara Doan

Chapter 2: The Ship of Theseus: Change Over Time in Topics of
Technical Communication Research Abstracts 39
    Stephen Carradini

Chapter 3: Mapping Technical Communication as a Field:
A Co-Citation Network Analysis of Graduate-Level Syllabi 69
    Michael J. Faris and Greg Wilson

PART TWO: REFLECTION AND MAINTENANCE OF MAJOR CONCEPTS

Chapter 4: "Visualize a Triangle." What's Professional About
Professional Communication? 119
    Brenton Faber

Chapter 5: Procedural Knowledge and Discourse in Technical
Communication: Easy as 1, 2, 3? 137
    Marjorie Rush Hovde

Chapter 6: Technical Communication Reimagined Through a
Socio-Technical Problem-Solving Lens 169
    Michael J. Albers

Chapter 7: Applied Rhetoric as Disciplinary Umbrella: Community,
Connections, and Identity 199
    Jennifer R. Veltsos, Matthew R. Sharp, Jacob D. Rawlins,
    Ashley Patriarca, and Rebecca Pope-Ruark

PART THREE: REASSEMBLING WITH EMERGING RELATIONSHIPS

Chapter 8: New Ways of Reading: Making Sense of Complex
Biomedical Writing Using Existing Guidelines 221
    Lisa DeTora

Chapter 9: A Critique of Disability and Accessibility Research
in Technical Communication Through the Models of Emancipatory
Disability Research Paradigm and Participatory Scholarship 243
    Sushil K. Oswal and Zsuzsanna B. Palmer

Chapter 10: Localize, Adapt, Reflect: A Review of Recent Research
in Transnational and Intercultural TPC 269
    Nancy Small

Contributors 297

ASSEMBLING CRITICAL COMPONENTS

A FRAMEWORK FOR SUSTAINING TECHNICAL AND PROFESSIONAL COMMUNICATION

Introduction: Promoting a Sustainable Collective Identity for Technical and Professional Communication

Lisa Meloncon
UNIVERSITY OF SOUTH FLORIDA

Joanna Schreiber
GEORGIA SOUTHERN UNIVERSITY

Building from its early history and connection to engineering, computer science, and scientific fields, technical and professional communication (TPC) now addresses a range of industries, organizations, sites, and locations including everything from technology to healthcare to nonprofits. TPC practices are central to facilitating complex communication concerns, with increasingly specialized subject matter, delivered and circulated through sophisticated emerging technologies. These ongoing changes are matched by the field's long-standing commitment to building flexible and ethical communication knowledge and practices. TPC is both a growing range of career opportunities and a thriving academic field represented by a growing number of degree programs and teacher-scholars across the country.

This range of interests and stakeholders is both a strength and a challenge for our field. Some 20 years ago, Johndan Johnson-Eilola and Stuart Selber (2001) cautioned,

> The diversity of perspectives found in and across technical communication contexts enriches the field in important ways. But as the field matures, the distance between these positions increases, then the tensions among different perspectives threaten to divide rather than reward us. (p. 432)

What Johnson-Eilola and Selber observed reminds us that what makes TPC dynamic is also what makes it difficult to delineate and to describe. TPC does not have clear boundaries and pathways found in other fields, such as engineering disciplines, which leads to different types of frustrations. New scholars and students often struggle with trying to find a satisfying definition of TPC. On the other hand, more experienced scholars know all too well that defining the field has been an ongoing challenge and have in some ways accepted the uncertainty of a definitional stance. Brenton Faber and Johndan Johnson-Eilola (2002) observed the dangers of an ongoing "fragmented field." They observed that "technical communication . . . is not yet capable of addressing in a systematic way the question of our collective identity" (p.140). It is our intention in this introduction to engage with the idea of TPC's "collective identity" by focusing on *component parts* that make up this collective identity.

Instead of advocating for a new handy definition of the field, we are arguing that the field is comprised of various components that must be reflected upon from time to time in order to maintain a sustainable and flexible identity. We propose a method of reflection and maintenance for the field's identity. It is in that work that new components, e.g., UX and content strategy, are brought in, and established ones, like procedural knowledge, are reimagined. Our chapters include emerging topics like biomedical writing, a chapter that reimagines the rhetorical situation as socio-technical situation, a chapter that focuses on a framework for transnational work, and a chapter that revisits the role of professionalism in professional writing. The chapters are not intended to pinpoint or bracket every aspect of technical communication, but to illustrate a range of knowledges and practices that comprise important components of the field. The method we share here, we think, both creates space for new knowledges and approaches as well as establishes a "collective identity" that moves us beyond fragmentation so that the various aspects of the field may work and grow together.

In what follows, we provide an overview of some key scholarship devoted to definitions of the field and illustrate the limitations of those definitions. We then move to theorizing TPC's collective identity by discussing how a collective identity functions through its component parts by drawing on assemblage theory. We then introduce each entry in this volume as an instantiation of a component of the field's collective identity. We conclude by describing how this move to a collective identity made up of component parts can perform the reflection and maintenance work for a sustainable field.

■ The Challenges and Limitations of Definitions

TPC has a long tradition with definitions, and there is no shortage of essays devoted to the practice of defining the field or to advocating for a particular definition. From nearly every major collection and a list of classic articles (e.g., Allen, 1990; Dobrin, 1983; Harris, 1978; Lay, 1991; Sullivan, 1990), TPC has consistently tried to define itself. This need for flexible definitions is what has led to the wide diversity of approaches that have included defining TPC as humanistic (e.g., Miller, 1979), as instrumental (e.g., Moore, 1996), and as rhetorical (e.g., Salvo, 2002). We've defined TPC according to the courses offered in its programs (e.g., Melonçon, 2020; Schreiber et al., 2018b) and according to competencies required in industry (e.g., Blythe et al., 2014; Brumberger & Lauer, 2015; Carliner & Chen, 2018; Henschel & Melonçon, 2014; Stanton, 2017), as well as trying to define itself by the research that TPC does (e.g., Carradini, 2020; Friess & Boettger, 2020; Melonçon & St.Amant, 2019; Rude, 2009, 2014; St.Amant & Graham, 2019). The field's ongoing attempts at definitions bring forth the ever-present tensions described by Johnson-Eilola and Faber(2002). These tensions include relations between industry and the academy, a range of industry stakeholders with overlapping and sometimes competing interests, and tensions within the academy itself.

As Jo Allen (1990) points out, definitions in the field have largely been either too broad to offer the field a sense of structure or too narrow to allow for diverse perspectives and emerging practices. Faber's (2002) critique of professionalism illustrates how a term can be applied so broadly that it becomes meaningless. Miles Kimball's recent attempt to scope the field seems to fall into this trap. Kimball (2017) describes TPC as "an activity that manages technological action through communication technologies, including writing itself, in a particular setting and for a particular purpose" (p. 346). The problem with a more ambiguous definition, such as Kimball's, is that in an attempt to provide much needed flexibility it becomes too broad to be helpful at all.

Previous research tells us that there was a brief era, in the early 2000s, where scholars described TPC as having an identity crisis. In at least three well-read and cited collections (Kynell-Hunt & Savage, 2003, 2004; Mirel & Spilka, 2002; Scott et al., 2006), authors and editors brought attention to the ongoing need to create a common identity for the field. The two-volume *Power and Legitimacy* (Kynell-Hunt & Savage, 2003, 2004) argues for strategies for gaining recognition from other disciplines as well as tensions between industry and the academy, and *Critical Power Tools* (Scott et al., 2006) seeks to expand the theoretical frameworks from which TPC traditionally has drawn. By drawing connections between rhetorical traditions and cultural studies concepts and frameworks, *Critical Power Tools* embraces the important role academic research and scholarship has to play in critically engaging TPC disciplines and artifacts. Barbara Mirel and Rachel Spilka (2002) address changing technologies and workplace practices as something with which TPC must contend, and emphasized the connection between academia and the workplace. While the three collections came from three distinct perspectives, they all described TPC as having an identity crisis.

More recently, scholarship has taken a different approach. As Kirk St.Amant and Lisa Melonçon (2016) described it, TPC has "yet to adequately define ourselves in a way that has brought satisfaction to the field in general.... As a result, any sustained attempt to engage in dialogic conversations around definition has been essentially nonexistent in recent years" (p. 269). Rather than outright saying the field is having an identity crisis, these newer collections acknowledge the necessity of identities by calling for what they feel the field's identity should be. For example, Godwin Agboka and Natalia Matveeva (2018) built a collection around their claim that TPC needs to undertake advocacy work in all its various forms.

TPC scholars have begun to address the field's identity from research and programmatic perspectives. Joanna Schreiber and Lisa Melonçon (2019) proposed a continuous improvement model to encourage administrators and faculty to consider programmatic and curricular identity in sustainable ways. By continuously examining programs and the multitude of parts of programs (such as the faculty, administrative constraints, community partners, and the courses themselves), Schreiber and Melonçon's model offers a flexible approach to account for shifts in programmatic identity. In much the same way Schreiber and Melonçon provided a programmatic identity model, we want to broaden that work and offer a field-wide identity model that can flexibly account for the different scholarly areas of TPC, the changing nature of work, and the dynamic contexts in which technical and professional communicators work.

So if we choose not to think of TPC identity through definitions, then the question becomes how can we consider it? We argue that we need to move to questions and concerns of identities that in turn will provide more opportunities for sustainability. We use *identity* traditionally, that is, as the identification of belonging to a specific group based on shared qualities and understanding of the group's beliefs and foundational principles. Thinking in terms of an identity gives TPC a way in which scholars with diverse and varying research and teaching interests can still feel as though they share a common goal. An "identity is a person's knowledge that he or she belongs to a social category or group.... A social group is a set of individuals who hold a common social identification or view themselves as a member of the same social category" (Stets & Burke, 2000, p. 225). However, considering the diversity and "fragmentation" of the field, thinking in terms of a collective identity that has space for a range of diverse concepts, practices, methods, and theories that contribute to a sustainable identity.

We are not aligning identity with structural or political positionality because to do so conflates and collapses two distinct positions that do not further an understanding of identity or of structural/political causes. Does this mean that we do not take structural or political issues seriously? Absolutely not. But it does help us focus an argument specifically on the issue of professional and field identity. Establishing a sustainable identity as a framework for collecting and unifying various parts of the field, we hope, will create a sense of shared understanding across the field and the workplace that will help support the important structural and political work the field and profession need to engage.

While we think these efforts have produced several useful definitions, we do not think any of them effectively provide a way to address the fragmentation of the field and its identity crisis in a sustainable way. Further, we also don't think that definitions, as genres respondent to situations, can effectively address identity. Instead, we came to the realization that a definition, or even a series of definitions, was limiting TPC's capacity of a professional and academic entity. We agree with Jo Allen (1990) that we need an extensive and flexible approach to consider the field and the work it does, but we are insistent for the need to move away from definitions. We suggest that a sustainable approach to TPC identity might be better achieved by critically reflecting on what these fragmentations mean *collectively as an identity*. In the next section, we begin to do this work by theorizing an identity of TPC through articulation and assemblage.

Component Parts: Envisioning a Collective Identity of TPC

What we wish to bring forth in this collection is a dynamic and adaptive identity for the field that is sustainable. Recent work (Johnson et al., 2018; Melonçon & Schreiber, 2018; Melonçon & St.Amant, 2019) has emphasized the necessity of thinking about the field's present and future in terms of sustainability. Extending this work, we turn to Robert Johnson's (2004) advocacy for sustainability in program development. He argued that sustainability was an apt metaphor because it "suggests growth/life but it also invokes the inevitable problem of limits" (Johnson, 2004, p. 102). Johnson's balancing of growth with an appreciation for limits brings a cautious vitality to merging sustainability with the field's need for a more flexible identity. Moreover, Johnson (2004) argued that ongoing reflection and maintenance are keys to sustainability. Thus, in considering TPC's collective identity, we have chosen a flexible framework that promotes this sort of "reflection and maintenance" Johnson suggests for sustainability.

To get TPC to consider issues of sustainability in more deliberate ways, we wanted to think through a collective identity that provides an over-arching framework that can be reflected and maintained. Reflection and maintenance are Johnson's two steps of sustainable practices. Acknowledging a more complex notion of identity as one that is articulated means TPC can embrace the dynamic nature of communication work broadly construed, which is vital for sustainability. Identity has a unifying factor: "Identities are thus contingent; they are dependent on particular elements that could change, thereby changing the composition of the identity" (Slack & Wise, 2015, p. 152).

Our turn to assemblage theory (below) was a deliberate move to work through the following question: How can TPC understand its identity to account for the past, present, and future demands of always in flux communication work? Jennifer Slack (2003) calls for a "cartography of the affective terrain of techcom" (p. 205), which allows for a different reading of the role(s) of technical communicator. More importantly, the technical communication assemblage calls for scholars and practitioners "to understand, though not resolve, the complex work and status of the technical communicator" (Slack, 2003, p. 205). Slack's work, like that of many scholars who have used assemblage theory, is indebted to that of Gilles Deleuze and Félix Guattari (1987). Their work on assemblage was part of a much larger and complex philosophical project.

We want to focus more narrowly by using part of Deleuze and Guattari's (1987) work that explains assemblage as an ontological framework. Ontology, of course, invokes a consistent and ongoing state of becoming or coming into being, which is necessary for a flexible, but stable, collective identity. Collective identity is a form of assemblage that is continually coming into being—stabilizing and re-stabilizing—as the field shifts and changes, offering a body of knowledge from which to draw definitions and make claims. Expanding on Deleuze and

Guattari's (1987) ideas, Manuel DeLanda (2016) makes clear that emergence, this idea of becoming, is essential to the assemblage. With an aspect of emergence, the assemblage can never be reduced simply to its parts. Rather, since it is always in a state of becoming—a state of emergence—the "whole depend[s] on the interactions between its parts [to] ensur[e] that these are not taken to be either necessary or transcendent" (DeLanda, 2016, p. 12). This ontological and emergent emphasis also aligns with the need for TPC to recognize the different nodes of its identity and how at any given moment different facets of identity may need to be emphasized over others.

DeLanda's emphasis on the interactions of the parts and the combination of these parts is not in situ, but contingent. The emphasis on contingent should remind those in TPC of the field's focus on the context. Assemblage theory's strength is that it emphasizes emergence and multiplicity while simultaneously emphasizing the relationship between the parts. Assemblage theory's relational approach has the potential to make room for the multiple identities and shifting identities of TPC scholars and workplace practitioners. The interactions between concepts and theories and practices can be selected at different times based on different contexts for different stakeholders. For example, a technical and professional communicator may need to draw on visual skills to solve a visualization problem one day and make a cogent argument to explain the need for greater attention to equity the next day. Within the TPC assemblage, different components would be necessary to respond to these distinct and different situations, that is, different parts of TPC's identity. Also, both scenarios illustrate the dynamic, contingent, relational, and emergent nature of a collective identity and how it can be used.

Assemblage allows articulated identities, like TPC, to connect to other identities. It territorializes identities. For instance, there is a lot of overlap among the identities of technical and professional communication, business communication, rhetoric of health and medicine (RHM), and user experience (UX). Assemblage allows for this necessary overlap and distinction among various identities; it allows identities themselves to have contexts and relations. As we discussed, definitions are often too static, which limits the force of TPC, but at the same time, the field does need a unifying identity that can bring the diverse parts of the field together. DeLanda explains this phenomenon of components and their use in forming identities:

> One and the same assemblage can have components working to stabilize its identity as well as components forcing it to change or even transforming it into a different assemblage. In fact, one and the same component may participate in both processes by exercising different sets of capacities. (DeLanda, 2006, p. 12)

In other words, at any given time, components of TPC's collective identity may be working to stabilize the field's identity for a specific situation, while simultaneously creating a space for another identity to emerge. A recent example of transforming can be seen in TPC embracing issues of social justice where the social justice has now become a key component of the field's assembled identity (see Jones, 2016).

One of the strengths of assemblage theory, particularly for the way we are using it here as a collective identity, is that "a component part of an assemblage may be detached from it and plugged into a different assemblage in which its interactions are different" (DeLanda, 2006, p. 10). In thinking about TPC identity, knowing that pieces can be moved or brought to the forefront at different times is important and continues the historical trajectory of TPC being diverse with the ability to step into a variety of communication situations and draw on a multiplicity of skills and knowledge. We want to leave this more theoretical section with a concrete example of the necessity for a collective identity with multiple component parts. Consider this scenario:

> Imagine you are at a university majors fair. An associate dean of liberal arts stops by your table and complains that engineering students aren't required to take any humanities courses. You remind him that engineering programs do require a technical communication course. He scoffs that technical communication isn't really humanities and that it is basically a writing in the disciplines (WID) course. At this point, you describe the humanistic qualities of technical communication and value of the assignments and concepts you teach in the class. At this same majors fair, you also encounter parents and prospective students. Parents ask you questions about career options related to the degree and your response favors the practical aspects of technical communication and what students can do with the degree in the world.

This example explicitly shows that neither of the definitions used is right or wrong, but they each highlight a different aspect of the field. We specifically position definitions as genres that tailor knowledge *drawn from a larger identity* for particular audiences and situations. Definitions are situated, and in this case, the situation warranted two distinct approaches to allow two different stakeholders a better understanding of TPC as a field. Like other genres, definitions are malleable and respondent to particular situations. They are used to frame knowledge effectively for audience and situation. As teacher-scholars, we know that we need to consider different audiences and contexts, and when considering how to discuss TPC as a field, a collective identity with multiple component parts gives scholars, teachers, and practitioners a way to invoke different components depending on the situation.

The collective identity can simultaneously stabilize and diversify knowledge making and practice, while emphasizing the micro and the macro and trying to make sense of how they fit together. Most crucially, different components can be invoked, moved, changed, and altered at any time, making and re-making a flexible and adaptable collective identity for the field.

Building a Sustainable Collective Identity: Reflection and Maintenance

The method we propose here is sustainable collective identity through reflection and maintenance of components of TPC. This reflection and maintenance method are the cornerstones of sustainability, and this process involves recognizing stable features of the assemblage as well as emerging in order to build and to maintain a sustainable and rich TPC collective identity. The chapters we present here represent both stable and emerging components of the field. They represent component parts of a collective identity.

Part One: Exigency for a Sustainable Identity

The first chapters of this collection illustrate the exigency of both building an effective and stable identity as well as being flexible enough to bring in new ideas. They illustrate a range of ways to reflect on the field, the ways it is growing and changing, and the consequences of not effectively addressing identity. This book begins by illustrating where we are and what issues we (continue to) face.

Sara Doan's chapter foregrounds underlying tensions and challenges to TPC's identity when it is collapsed with other disciplines like composition. In "What Are We Really Teaching? Revisiting Technical and Professional Communication's Pedagogical Training," Doan reminds us that TPC courses, particularly service courses, have a range of stakeholders that include industry and disciplines like engineering. Doan deftly explains that understanding industry as a stakeholder does not mean pandering to industry, and that TPC has a long history of situating industry needs ethically and rhetorically. She has carefully culled data from syllabi, learning outcomes, and assignments to compare pedagogical practices and concepts in the TPC service course and first-year writing. Doan compares the different aims and scopes and histories of TPC and composition, focusing on issues such as audience and genre. She argues that composition approaches are unsuitable for the TPC service course and advocates for pedagogical training specific to TPC.

Looking to the future, Stephen Carradini presents a meta-research study of stable and emerging concepts in the field. In "The Ship of Theseus: Change Over Time in Topics of Technical Communication Research Abstracts," Carradini conducts a keyword analysis of abstracts from technical and professional communication journals. His study, grounded in previous attempts to understand the field through keywords, seeks to answer questions about prominent, stable, declining, and emerging topics in the field. It illustrates how much and how quickly the field is changing as well as the need for a sustainable identity for the field. After dividing his 2000 to 2017 corpus into three main eras, Carradini identifies keywords that are increasing and decreasing in usage, pointing to possible emerging trends. Keywords with some of the sharpest declines include *ethics*, *rhetoric*, and *scientific* from the first era to the second and from the second to the third. *Content, experience, projects*, and *social* are among those with the strongest increase from the first era to the second and from the second to the third. *Justice, UX*, and *entrepreneurs* emerged as keywords in the third era. Carradini's keyword analysis illustrates shifts in the boundaries of TPC as well as changes in disciplinary values.

In "Mapping Technical Communication as a Field: A Co-Citation Network Analysis of Graduate-Level Syllabi," Michael Faris and Greg Wilson present a systematic analysis of graduate course syllabi for courses purporting to provide foundational knowledge in technical and professional communication. Building from previous analyses, they argue that two major scholarly trends continue to heavily influence the reading list: a focus on the value of practitioners in the workplace and a focus on the value of the discipline in the academy. Their map of citations from 60 syllabi illustrates heavily cited core texts and some emerging texts as well as the frequency of texts being assigned together. Using community detection algorithms, they observe emerging communities of texts strongly linked to anthologies and argue that texts that were not identified as core texts in earlier studies have likely become core texts because of edited collections. Faris and Wilson argue that data overall shows that the field has gained a level of coherence.

▌ Part Two: Reflection and Maintenance of Major Concepts

Next, we turn to major concepts and knowledges that constitute our field. The entries here conduct the necessary reflection and maintenance to move the field toward a sustainable identity. The entries in the second section illustrate an internal reflection on the individual elements (joints and nodes) within technical and professional communication.

Brenton Faber's chapter continues to remind us of the issues that arise with the development and mishandling of broad definitions and labels. In "'Visualize a Triangle': What's Professional About Professional Communication?" Brenton Faber revisits his foundational 2002 work, "Professional Identities: What is Professional About Professional Communication?" Faber's groundbreaking argument that the concept of professionalism has been applied so widely that the term becomes meaningless remains relevant. In this entry, he provides some additional guidance for the field of technical and professional communication to better define professional communication practices by distinguishing various domains of professional communication as well as explaining why professional communication and business communication ought not be considered interchangeable terms. Faber explicitly narrows the purview of professional communication and argues that, properly understood, it provides important strategic checks and balances.

In "Procedural Knowledge and Discourse in Technical Communication: Easy as 1, 2, 3?" Marjorie Rush Hovde revisits the foundational and enduring importance of procedural knowledge to the field. From a historical perspective, Hovde reviews and situates important concepts related to procedural knowledge, including system, task, and user orientations. Providing a range of examples, she illustrates effective documentation practices over time. In doing so, she both provides an important literature review and pushes procedural knowledge as essential to building effective and ethical practices in a changing landscape.

Michael Albers' "Technical Communication Reimagined Through a Socio-Technical Problem-Solving Lens" asks us to rethink one of the most common theoretical frameworks—the rhetorical situation—from the perspective of complex situations. Using examples from service design to software, he invites readers to rethink relatively mundane features of technical and professional communication—writing, communication, and audience—through socio-technical theory. Albers provides a framework for problem-solving and decision-making for increasingly complex environments. Drawing from concrete examples and historically situating concepts, he provides recommendations for both pedagogy and research.

In "Applied Rhetoric as Disciplinary Umbrella: Community, Connections, and Identity," Jennifer Veltsos, Matthew Sharpe, Jacob Rawlins, Ashely Patriarca, and Rebecca Pope-Ruark theorize increasingly disparate TPC components as applied rhetoric. Using several examples, the authors illustrate ways in which applied rhetoric productively brings together sub-disciplines like business communication and science communication, without collapsing their aims and scopes, into a praxis approach that actively engages subjects beyond critique. Using concrete examples, the authors illustrate how these relationships will help TPC address complex issues and build more effective practices.

■ Part Three: Reassembling with Emerging Relationships

The chapters in the final section of this collection illustrate the range of assemblages and identities with which TPC needs to interact. Part of TPC's identity needs to allow for such interaction and development of effective practices across identities. These entries are looking outward, across identities and assemblages. They illustrate emerging relationships and practices and new roles for technical and professional communication.

As scientific and technological disciplines and practices become increasingly specialized, TPC needs to be able to deftly build practices across disciplines and specializations. Lisa DeTora's "New Ways of Reading: Making Sense of Complex Biomedical Writing Using Existing Guidelines" examines how scientific discourses have shifted over time, requiring updated methods. Using biomedicine as an example, she advocates a new approach to critically engaging scientific discourses and appropriately incorporating existing professional genres and guidelines as they affect knowledge and authorship.

TPC is the place to effectively incorporate disability studies, user experience, and accessibility. Sushil Oswal and Zsuzsanna Palmer argue that TPC scholars need to proactively incorporate disability studies scholarship into usability and accessibility methods and analyses. Through critical analysis of recent scholarship, they illustrate how TPC has neglected to effectively incorporate scholarship and methods from disability studies. Advocating for a participatory action model, Oswal and Palmer specifically connect how disability studies can help build more effective and ethical practices, and they provide strategies for better integrating disability studies into TPC research and practice. Using examples from the classroom, they illustrate how to center disabled users in both technical and professional communication research and pedagogy.

TPC needs to be able to both develop and critique effective methods in global contexts. In "Localize, Adapt, Reflect: A Review of Recent Research in Transnational and Intercultural TPC," Nancy Small presents an integrative literature review of intercultural and transnational communication practices in the field. From a corpus of 143 articles, Small draws out issues of conflated terminology and localization practices, including intercultural, transnationalism, localization, and culture. Additionally, she draws from several specific examples to advocate for an ethical model to address transnational work in a more effective and responsible way.

Assembling and Sustaining TPC

A collective identity will always remain despite the changing of the component parts. These parts include concepts, technologies, practices, workplaces, social issues, ethics, industry changes, genres, audiences, and methods. TPC is always changing, as it should, in response to changes in workplace places and innovations in technology, as well as shifts culturally, socially, or politically. Critical reflection that considers the component parts allows TPC to acknowledge the shifting and changing of its collective identity over time. This is the work of sustainability, as argued by Johnson (2004), to reflect and maintain the components of technical and professional communication's collective identity. We hope the readers of this volume will consider their research as part of a larger TPC collective identity—unique but still connected to larger goals and aims.

References

Agboka, Godwin & Matveeva, Natalia. (Eds.). (2018). *Citizenship and advocacy in technical communication: Scholarly pedagogical perspectives*. Routledge.

Allen, Jo. (1990). The case against defining technical writing. *Journal of Business and Technical Communication*, 4(2), 68–77.

Blythe, Stuart Lauer, Claire & Curran, Patrick G. (2014). Professional and technical communication in a web 2.0 world. *Technical Communication Quarterly*, *23*(4), 265–287.

Boettger, Ryan K. & Friess, Erin. (2020). Content and authorship patterns in technical communication journals (1996–2017): A quantitative content analysis. *Technical Communication*, *67*(3), 4–24.

Brumberger, Eva & Lauer, Claire. (2015). The evolution of technical communication: An analysis of industry job postings. *Technical Communication*, *62*(4), 224–243.

Carliner, Saul & Chen, Y. (2018). Who technical communicators are: A summary of demographics, backgrounds, and employment. *Intercom*, *65*(8), 8–12.

Carradini, Stephen. (2020). A comparison of research topics associated with technical communication, business communication, and professional communication, 1963–2017. *Transactions on Professional Communication (IEEE)*, *63*(2), 118–138.

DeLanda, Manuel. (2006). *A new philosophy of society: Assemblage theory and social complexity*. Continuum.

DeLanda, Manual. (2016). *Assemblage Theory*. Edinburgh University Press.

Deleuze, Gilles & Guattari, Félix. (1987). *A thousand plateaus: Capitalism and schizophrenia* (B. Massumi, Trans.). University of Minnesota Press.

Dobrin, David. (1983). What's technical about technical writing? In P. V. Anderson, R. J. Brockmann & C. R. Miller (Eds.), *New essays in technical and scientific communication: Research, theory, practice* (pp. 227–250). Baywood.

Johnson-Eilola, Johndan & Selber, Stuart. (2001). Sketching a Framework for Graduate Education in Technical Communication. *Technical Communication Quarterly*, *10*(4), 403–436.

Faber, Brenton. (2002). Professional identities: What is professional about professional communication? *Journal of Business and Technical Communication*, *16*(3), 306–337.

Faber, Brenton & Johnson-Eilola, Johndan. (2002). Migrations: Strategic thinking about the future(s) of technical communication. In B. Mirel & R. Spilka (Eds.), *Reshaping technical communication: New directions and challenges for the 21st century* (pp. 135–148). Lawrence Erlbaum Associates.

Harris, John S. (1978). On expanding the definition of technical writing. *Journal of Technical Writing and Communication*, *8*(2), 133–138.

Henschel, Sally & Melonçon, Lisa. (2014). Of horsemen and layered literacies: Assessment instruments for aligning technical and professional communication undergraduate curricula with professional expectations. *Programmatic Perspectives*, *6*(1), 3–26.

Johnson, Meredith, Simmons, W. Michele & Sullivan, Patricia. (2018). *Lean technical communication: Toward sustainable program innovation*. Routledge.

Johnson, Robert. (2004). (Deeply) sustainable programs, sustainable cultures, sustainable selves: Essaying growth in technical communication. In T. Kynell-Hunt & G. J. Savage (Eds.), *Power and legitimacy in technical communication Volume II: Strategies for professional status* (pp. 101–119). Baywood.

Johnson-Eilola, Johndan & Selber, Stuart A. (2001). Sketching a framework for graduate education in technical communication. *Technical Communication Quarterly*, *10*(4), 403–437.

Jones, Natasha N. (2016). The technical communicator as advocate: Integrating a social justice approach in technical communication. *Journal of Technical Writing and Communication*, *46*(3), 342–361.

Kimball, Miles A. (2017). The golden age of technical communication. *Journal of Technical Writing and Communication*, *47*(3), 330–358.

Kynell-Hunt, Teresa & Savage, Gerald J. (Eds.). (2003). *Power and legitimacy in technical communication, Volume 1: The historical and contemporary struggle for professional status*. Baywood.

Kynell-Hunt, Teresa & Savage, Gerald J. (Eds.). (2004). *Power and legitimacy in technical communication, Volume 2: Strategies for professional status*. Baywood.

Lay, Mary M. (1991). Feminist theory and the redefinition of technical communication. *Journal of Business and Technical Communication, 5*(4), 348–370.

Melonçon, Lisa. (2018). Critical postscript: On the future of the service course in technical and professional communication. *Programmatic Perspectives, 10*(1), 202–230.

Melonçon, Lisa. (2020). TechComm Programmatic central. [unpublished raw data.]

Melonçon, Lisa & Schreiber, Joanna. (2018). Advocating for sustainability: A report on and critique of the undergraduate capstone course. *Technical Communication Quarterly, 27*(4), 322–335.

Melonçon, Lisa & St.Amant, Kirk. (2019). Empirical research in technical and professional communication: A five-year examination of research methods and a call for research sustainability. *Journal of Technical Writing and Communication, 49*(2), 128–155.

Miller, Carolyn. (1979). A humanistic rationale for technical writing. *College English, 40*(6), 610–617.

Mirel, Barbara & Spilka, Rachel. (Eds.). (2002). *Reshaping technical communication: New directions and challenges for the 21st century*. Lawrence Erlbaum Associates.

Moore, Patrick. (1996). Instrumental discourse is humanistic as rhetoric. *Journal of Business and Technical Communication, 10*(1), 100–118.

Rude, Carolyn D. (2009). Mapping the research questions in technical communication. *Journal of Business and Technical Communication, 23*(2), 174–215. https://doi.org/10.1177/1050651908329562.

Rude, Carolyn D. (2015). Building identity and community through research. *Journal of Technical Writing and Communication, 45*(4), 366–380. https://doi.org/10.1177/00472816 15585753.

Salvo, Michael J. (2002). Critical engagement with technology in the computer classroom. *Technical Communication Quarterly, 11*(3), 317–337.

Schreiber, Joanna, Carrion, Melissa & Lauer, Jessica. (Eds.). (2018). Guest editors' introduction: Revisiting the service course to map out the future of the field. *Programmatic Perspectives, 10*(1), 1–11.

Schreiber, Joanna & Melonçon, Lisa. (2019). Creating a continuous improvement model for sustaining programs in technical and professional communication. *Journal of Technical Writing & Communication, 49*(3), 252–278.

Scott, J. Blake, Longo, Bernadette & Wills, Katherine V. (Eds.). (2006). *Critical power tools: Technical communication and cultural studies*. State University of New York Press.

Slack, Jennifer D. (2003). Technical communicator as author? A critical postscript. In T. Kynell-Hunt & G. J. Savage (Eds.), *Power and legitimacy in technical communication Volume 1: The historical and contemporary struggle for professional status* (pp. 193–207). Baywood.

Slack, Jennifer D. & Wise, James M. (2015). *Culture and technology: A primer* (2nd ed.). Peter Lang.

St.Amant, Kirk & Graham, S. Scott. (2019). Research that resonates: A perspective on durable and portable approaches to scholarship in technical communication and rhetoric of science. *Technical Communication Quarterly, 28*(2), 99–111.

St.Amant, Kirk & Melonçon, Lisa. (2016). Addressing the incommensurable: A Research-based perspective for considering issues of power and legitimacy in the field. *Journal of Technical Writing and Communication, 46*(3), 267–283.

Stanton, Rhonda. (2017). Do technical/professional writing (TPC) programs offer what students need for their start in the workplace? A comparison of requirements in program curricula and job ads in industry. *Technical Communication, 64,* 226–236.

Stets, Jan E. & Burke, Peter J. (2000). Identity theory and social identity theory. *Social Psychology Quarterly, 63*(3), 224–237.

Sullivan, Dale L. (1990). Political-ethical implications of defining technical communication as practice. *JAC: A journal of rhetoric, culture & politics, 10*(2), 375–386.

Part One: Exigency for a Sustainable Identity

Chapter 1: What Are We Really Teaching? Revisiting Technical and Professional Communication's Pedagogical Training

Sara Doan
Kennesaw State University, Marietta, GA

Abstract: In technical and professional communication (TPC), a return to researching the service course provides an opportunity to reflect on current instructor training. I contrast the current approaches centered around genre theory with a theoretical orientation that came from this study: workplace phronesis taught through genre ecologies. Based on results from ten instructor interviews and a content analysis of their syllabi, assignment sheets, and feedback on students' writing, this chapter contrasts instructors' genre-based approaches to teaching the TPC service course with two experienced instructors' use of practical wisdom derived from their own workplace practices. Implications include recommendations for connecting the service course with TPC's content areas, revising the TPC instructor practicum, and encouraging instructors to comment on students' writing through a content-centric, rather than a genre-centric, lens.

Keywords: technical communication, pedagogy, pedagogical goals, phronesis, workplace writing

As technical and professional communication (TPC) continues to flourish in the 21st century, the field should continually be reflective, reconsidering and refining itself, both in external relation to other fields and in internal definition of who we are and what we do. The TPC service, or introductory, course acts as a barometer for the field's trends and pedagogical methods (Knieval, 2007; Melonçon, 2018a) and has recently received renewed attention in TPC research (see for example, Schreiber et al., 2018). The research in this chapter presents a problem: some TPC instruction is framed through a compositionist genre lens, when a lens of workplace phronesis would better teach students to communicate expertise, exercise ethical reasoning, and write and think to grapple with wicked problems. I write this chapter not to disparage composition or compositionists; rather, I wish to point out that these fields have different histories, pedagogies, and epistemologies. This chapter builds the idea of phronesis-based genre ecologies to help students learn to communicate expertise and to re-integrate TPC's content areas into the service course. TPC has great opportunities to reflect on and commit to future research on its history, pedagogical methods, and content areas.

The history of technical communication is bound up in writing to share definitions, knowledge, and processes and problem-solving. Joanna Schreiber, Melissa Carrion, and Jessica Lauer (2018) describe the history of technical communication during the industrial revolution and information age as developed through engineering communication at state universities (Malone, 2011). In the history of technical communication, communicating expertise has been a through line; using the service course to teach students to communicate their expertise is intertwined with the history of the field (Russell, 1993). At their best, TPC's pedagogical values prepare students to engage with the phronesis, or practical wisdom toward the role of problem-solving, of writing in workplaces and organizational spaces (Doan, 2021). Here, I add workplace to phronesis, but this definition encompasses any physical or digital space for communicating expertise, including writing to solve problems. Solving problems leads students to a greater grasp of organizational decision-making: very often, the problems for which technical communication can most effectively contribute solutions are organizational (Francis, 2018; Lawrence et al., 2017). Engaging students with problem-solving and organizational decision-making enables them to consider the roles of distributed cognition and employee agency in the post-postmodern workplace (Wilson & Wolford, 2017); the rhetorical term I use here to encompass these two activities is phronesis. As students solve problems and begin to shape organizations' decision-making, students require expertise in ethical reasoning and inclusivity. The TPC service course has great potential to shape students' responses to workplace quandaries through issues of plain language (Willerton, 2015), accessibility (Browning & Cagle, 2017; Huntsman et al., 2019), and racial biases (Shelton, 2020) through critical engagement with technologies and systems.

Teaching students to enact inclusive practices in their decision-making is the future of the TPC service course. Sometimes, though, instructors do not have the requisite content knowledge or the time to plan their pedagogies around these evolving content areas and best practices. Instructors, particularly those without workplace experience or advanced coursework in TPC (Doan, 2019), sometimes teach TPC's genres without as much attention to its goals: communicating expertise to create accessible communications that guide decision-making and problem-solving. To support these goals, teacher-scholars must be able to both foreground content areas and prepare students for organizational decision-making. As I show below, not all theoretical orientations are appropriate to address these goals, and some approaches to rhetoric even undermine these goals. Focusing on writing practices outside of the context of organizational decision-making may not be an appropriate approach to TPC pedagogy.

To illustrate this contrast between workplace phronesis and compositionist ways of thinking, I present results from a study of ten instructors' pedagogical goals within the TPC service course, based on a content analysis of instructor interviews, their syllabi, assignment sheets for the resume and cover letter, and one section of feedback on their students' de-identified resumes and cover letters. Instructors' goals during their interviews often differed from the course outcomes on their syllabi, as less experienced instructors typically relied on terminology from composition or classical rhetoric. In contrast, instructors with both advanced graduate work in TPC pedagogy and professional experience relied on workplace phronesis to drive their pedagogy. Findings from this study illustrate the problems that arise when rhetorical concepts are not used effectively for teaching TPC, and offer a potential model grounded rhetorically by phronesis that addresses the limitations of genre theory for engaging students with workplace writing practices and with TPC's content areas: phronesis-based genre ecologies.

■ Literature Review

Over the past two decades, TPC's content areas and domains of expertise have expanded; the field has expanded and is primed to respond to a quickly evolving world in exciting and challenging ways. Those teaching TPC, especially in the service course, must now be cognizant of several domains that inform the service course's content. When students enter the TPC service course, they often expect to write instructional materials: traditionally, user manuals (Malone, 2011). User documentation now includes instructional videos (Swarts, 2012), user forums (Swarts, 2018), and chatbots (Heo & Lee, 2018). Similarly, content strategy and content management have grown more diffuse, shifting from siloed document-based strategies to more abstract and holistic approaches to information management and architecture (Getto et al., 2020). Furthermore, the TPC service course offers students insights into accessibility in writing and design that they may not be taught to consider in other parts of their education (Browning & Cagle, 2017; Huntsman et al., 2019). Within typical TPC assignments, space exists to engage students with project management (Dicks, 2010; Hackos, 2005), user experience (Chong, 2018), data visualization (Welhausen, 2017; Wolfe, 2015), and the rhetoric of health and medicine (Hannah & Arduser, 2017). Becoming an effective technical writer now means going beyond a focus on grammar or word choice and zooming out to engage larger wicked problems, such as using information literacies (Boettger et al., 2017) to create content tailored specifically to a certain audience (Doan, 2020; Spilka, 2009). TPC faculty should familiarize themselves with these content areas, even if they are "only" teaching the service course, because these areas lend exegesis to assignments and course objectives.

Embedded in field-wide issues of technology, sustainability, and ethics are issues of training TPC instructors to teach the service course. The majority of TPC service courses are taught by contingent instructors, including adjunct faculty and graduate students (Melonçon et al., 2016). The contingent status of instructors is problematic for several reasons: They may not have workplace experience, so they rely on textbooks to inform their teaching (Wolfe, 2009). Instructors might not have received pedagogical training specific to TPC (Read & Michaud, 2018), instead borrowing pedagogical practices from composition that they then apply to workplace genres (Doan, 2019). The contingent nature of many faculty and the need for more robust pedagogical training complicates service course instruction at both the instructor and programmatic levels (Knieval, 2007). In the next sections, I provide a brief overview of phronesis, which may help TPC better direct rhetorical training for instructors new to TPC.

■ Centering Workplace Phronesis

While rhetorical terms are often used to teach those new to teaching writing, rhetorical concepts hold different meanings and ways of knowing in TPC when contrasted with composition and other writing approaches (such as writing in the disciplines). Rhetoric's "function [*ergon*] is not to persuade but to see the available means of persuasion in each case" (Aristotle, 322 BCE/2007, p. 36). While other rhetorical theories, such as Aristotle's, Cicero's, and Quintilian's respective works on stasis theory (Bizzell & Herzberg, 2001), have been useful to technical communicators and those in the rhetoric of science or the rhetoric of health and medicine (Prelli, 2005; Teston, 2012; Walsh & Walker, 2016), they generally do not guide composition and—by extension—TPC pedagogy to the same extent as Neo-Aristotelian rhetoric. I am not saying that theories from Neo-Aristotelian rhetoric, such as purpose, audience, and context, are not useful. Rather, I am arguing that the separate missions and epistemologies of these fields are often overlooked when we use the same terms for different definitions.

Phronesis is a rhetorical concept that can be used to frame the goals of the TPC service course: introducing content areas and orienting students toward effective organization decision-making. Further, phronesis addresses the limitations of relying on genre theory, as over-focusing on genre obscures the high-order concerns of the TPC classroom: purposeful content, ethical reasoning, and audience awareness. In a phronesis-based genre ecology, students use multiple genres to communicate expertise and to guide decision-making. One common phronesis-based genre ecology can be seen in the employment application assignment: students write cover letters and resumes that coordinate a central argument that they would be a good fit for a specific job and company. Another is teaching students to coordinate a group project across a group charter, meeting minutes (or chat transcripts), a proposal, a presentation, a report, and group participation evaluations. These documents work together to both create and frame the project as a phronesis-based genre ecology that teaches writing through the "unofficial" genres of notes (Lawrence et al., 2017), meeting minutes, and group charters (Wolfe, 2010).

To further explain what I mean by the phronesis-based genre ecologies necessary for robust instruction within the TPC service course, I contrast the goals of

TPC and composition. While both disciplines ask students to consider real problems and solutions, they champion different skills and epistemologies through the ways that they ask students to engage with their writing processes, draw on secondary sources, and use the affordances of their modalities. Composition pedagogy foregrounds critique, persuasion, and argument-making (Booth, 2007). Discussing writing in this way is not how workplace writers operate. While these skills, or techne, are essential for both academic and technical writing, students need more experiences in phronesis, or problem-solving, that are unique to TPC. Pedagogies influence practices and vice versa when creating opportunities for TPC to respond to a rapidly changing professional world.

In composition, genres represent individual learning or public persuasion. In TPC, genre represents communicating expertise through genre ecologies. Both composition and TPC aim to foster students' "deeper understanding of how to use writing to improve students' domain-based learning, to engage them in co-constructing professional knowledge and know-how, and to socialize them into professions in ways that improve those professions and the world they serve" (Russell, 2007, p. 249). However, these fields use rhetorical theory and written genres in different ways; largely, the theories and methods for teaching composition do not transfer to teaching TPC, particularly in the service course. I say this not to disparage the work of composition instructors and scholars, but rather to point out how these fields use the same terminology and basic principles to different ends: Composition students use genre to explore a topic or persuade the public. TPC students use genres to apply for jobs, or solve a customer service problem, or ask for grant money.

■ The Limitations of Genre Theory

While genre theory affords instructors the means to explain the parts and functions of common workplace documents, deeper learning benefits from a phronesis-based approach, such as workplace genre ecologies (Doan, 2021). Genre theory has two limitations: overreliance on form and oversimplifying larger, more nebulous issues like context, kairos, and ethics. Overreliance on templates and formatting limits how much students learn about writing content across workplace genre ecologies (Lawrence, et al., 2017). The TPC service course engages with genre and genre theory differently than in many composition classrooms.

Throughout its relatively short history (McLeod, 2007), composition has been defined by its genres: the argumentative essay, the expository essay, the literacy narrative (Brodkey, 1994), the five-paragraph essay (White, 2008), the traditional research paper (Adler-Kassner & Wardle, 2015), and the multimodal project (Duffelmeyer, 2002; Yancey, 2004). Much of composition pedagogy has focused on teaching students what these genres are and how they shape an academic argument (Barnett & Kastley, 2002; Lynch et al., 1999), or how students can manage their writing processes. These composition genres are mostly genres of form, rather than genres that function in the world in the ways that TPC genres such as resumes and cover letters do. Instead of writing to learn or writing to display knowledge, when writing workplace genres, TPC practitioners write in order to solve problems or perform social actions (Miller, 1984). While instructors may use genre theory as they conceptualize, scaffold, and outline their assignment parameters, most composition students focus on the form and external expectations when writing. The assumption of transfer has been emphasized when discussing multimodal composition: to meet assignment parameters, students must use written, aural, visual, and digital elements as they craft their projects. Although the individual technes of these multimodal affordances can transfer to other places within TPC curricula (i.e., facility with Adobe InDesign), the phronesis of aligning their visuals with the task that their users want to accomplish largely does not.

As a field, TPC views genre as both form and function. Much of TPC still theorizes and enacts genres through Carolyn Miller's (1984) assertion that genres are social actions, therefore genre must reflect the writer's purpose for communicating. Over the past three-and-a-half decades, the explosive growth of digital genres (Miller, 2015; Tillery & Nagelhout, 2015) and contextualized genre theory (Devitt, 2009, 2010) has given TPC scholars lenses for viewing genres in post-postmodern situations. This theorizing, however, has not always translated into specific classroom practices, particularly within the TPC service course. For example, TPC textbooks tend to focus on genre as a series of formats or rules (Wolfe, 2009), or give heuristics for abstract problem-solving, rather than as opportunities for students to learn how professional genres can be used to solve workplace problems. Despite the field's robust theorizing about genre as social action, activity theory, and actor-network theory, these terministic screens have not always translated into actionable pedagogical methods that enable students to learn how to solve problems using workplace documents (Melonçon, 2018b; Morrison, 2017). TPC pedagogy needs to be "moving away from form-based discussions toward more productive rhetorical ones" (Lawrence et al., 2017, p. 2) by building a functional vocabulary for instructors to use when building students' information literacies (Boettger et al., 2017) and discussing content-centric writing issues (Doan, 2020; Spilka, 2009). These productive rhetorical conversations are the phroneses largely absent from a form-based approach to teaching the TPC service course. For example, how genre ecologies like post-it notes, emails, and outlines about a software project become "genre ecologies," or "sets of tools to 'transform data'" into actionable workplace genres (Spinuzzi, 2003, p. 100). Or teaching students to create usable project charters and task schedules in order to use the genres of project management to actively guide their collaborations (Wolfe, 2010). Teaching form-based or theoretical views of genre without a full consideration of the rhetorical context, then, does students a disservice. Students should be learning how genre is "driven by exigency" (Malone & Wright, 2018, p. 124) within larger communicative and social networks. TPC has its own phronesis-based genre ecologies that communicate expertise to solve problems. In the following study, I look at how instructors' approaches to theory foster their abilities to articulate and explain their pedagogical goals within the TPC service course.

■ Methods

To understand how TPC instructors articulated and enacted their pedagogical goals for the service course, I conducted interviews and content analysis of course documents. The results featured here are one part of a deep qualitative examination of instructors' feedback practices in the TPC service course, after testing these interview questions and the triangulation of the data collection in a pilot study (Doan, 2019). These results feature answers from the first ten instructors of a 20-instructor study (Doan, 2020) focusing on instructors' pedagogical goals. I use the results from the first ten instructors to make an argument about workplace phronesis in TPC teacher training with attention to genre ecologies.

With Institutional Review Board (IRB) approval #18.200 from the University of Wisconsin–Milwaukee, I recruited ten instructors through social media and professional listservs. After completing a short demographic survey, each instructor submitted their service course syllabus, the assignment sheet for their resume and/or cover letter assignment, and one section of their students' de-identified resume and cover letters with instructor feedback. Instructor interviews comprised three parts, as tested and described in my pilot study (Doan, 2019): First, instructors discussed their goals for students' learning in the service course. Second, instructors talked about their feedback workflows. Third, instructors conducted retrospective recall (Still & Koerber, 2010) to explain their rationale for writing each feedback comment on two of their students' resumes and cover letters.

During the first round of iterative coding (Tracy, 2013), I open-coded the course objectives from the first five service course syllabi, then compared these results to a single question from each instructor's interview: "What do you think your students most need to know or do when they leave your class? Why?" During the second round of coding, I coded the course objectives from Instructor 6-10's syllabi and compared those results with the interview question about what they wanted students to know after their TPC service course. The third round of coding shifted from primary-cycle coding to secondary-round coding; I used the now-established coding scheme on instructors' interviews to understand instructors' spoken beliefs about their teaching. At this point, tensions between instructors' spoken goals and their syllabi's course objectives began to emerge, as presented in the results. The differences between Instructor 6's and 10's approaches to workplace phronesis became clear during the third round of coding.

■ Limitations

While this chapter represents research with a small number of participants, collecting the survey data, pedagogical materials, and feedback on de-identified resumes and cover letters allowed for triangulation between data sources. As the first stage of a two-stage study, these results included ten instructors; the TPC articles usually include an average of 12 participants (Melonçon & St.Amant, 2018). To overcome this limitation, I conducted a four-instructor pilot study (Doan, 2019). I have triangulated my data collection and collected substantial amounts of verbal and textual data to create "thick description" of instructors' goals and feedback practices (Tracy, 2013, p. 2). These results with ten instructors come from the first stage of a two-stage study of 20 total instructors and more fully coded data (Doan, 2020). This project has two secondary limitations: race and information about workplace writing. I did not formally collect data about instructors' race or their workplace experience. During the second phase of the study, my participants included instructors of color and instructors at minority-serving institutions. To make more concrete claims about instructors' workplace experiences, I wish that I had collected more information about instructors' professional experiences and the extent to which these experiences influenced their teaching. Although Instructors 1, 5, 6, 9, and 10 volunteered this information during their interviews, having a formal interview question about instructors' workplace experiences would have given clearer viewpoints of their pedagogical goals.

■ Understanding Phronesis as a Framework for Teaching

Phronesis centers around decision-making skills and practical wisdom. Phronesis is "Aristotle's word for the mental ability to select the best course of action in situations fraught with uncertain knowledge and competing claims of morality and practicality" (Bizzell & Herzberg, 2001, p. 1633). Phronesis acts as an essential component of knowledge work—as Ancient Greeks used phronesis in warfare and rhetoric, and techne in leatherworking and pottery making. Phronesis describes how effective communicators operate in today's unstable workplaces (Wilson & Wolford, 2017), with de-contextualized texts (Swarts, 2018). Particularly when working with writing, phronesis is intuition-based, for example, when instructors judge how many comments to give on students' assignments (McMartin-Miller, 2014). Phronesis takes the norms and habits of giving feedback and translates them into an enactable strategy. When students leave the service course, we want them to use phronesis when dealing with thorny interpersonal or ethical situations. While composition classrooms engage students with phronesis, the TPC classroom uses phronesis to teach professional decision-making with subject matter experts, genre ecologies, and challenges of writing in organizations. To make an argument for the phronesis-based pedagogy that instructors should be teaching in the TPC service course, I present this study's results accompanied by examples from two associate professors who participated in the study. Both Instructors 6 and 10 grounded their pedagogical goals and feedback-giving practices within their own workplace experiences and explicitly talked about teaching students the "practical wisdom" (Instructor 10) that students would need for their future lives and careers. Other instructors in the study based their pedagogical approaches in rhetoric, but these results could use more connection to the skills and practices needed to help students become successful knowledge workers and citizens. I use the study's main results to show the limitations of these instructors' reliance on genre theory and to contrast this with the workplace phronesis that Instructors 6 and 10 used when describing their pedagogical goals and commenting on students' resumes and cover letters. These experienced instructors' focus on practical wisdom and connecting rhetorical theory to their workplace experiences lays a foundation for the types of phronesis-based genre ecologies that enhance the TPC service course, and refocuses students' and instructors' attention on each genre's content, instead of its form.

▪ Results

The results from the demographics survey show that instructors' levels of experience and pedagogical training were mixed (see Table 1.1). Working in business departments, Instructors 1 and 9 had no graduate-level pedagogy training, instead relying on their extensive business and consulting training. All other instructors had taken a course in composition pedagogy; five had taken a course in TPC pedagogy. Four instructors had additional pedagogical training: three in online teaching, one in cultural studies teaching, and one in the developmental course for students at her state university. Instructor 8 had a graduate-level certificate in pedagogy. Instructors in this study had between 3.5 and 17 years of experience teaching TPC courses.

The results of this study give a snapshot into how instructors are teaching TPC service courses as the field rapidly grows and the professional world continues to transform. In this section, I discuss results from this study that indicate that these instructors frame their course goals as rhetorical through audience, context, and purpose. Instructors' spoken pedagogical goals, however, differed from their syllabi's learning objectives: when speaking about their goals for students' learning, instructors often spoke about genre theory. When writing about their course goals in their syllabi, information literacy and content were the most common course goals except for rhetoric. Finally, these results suggest connections to explore between workplace experience; teaching experience; and a graduate degree in rhetoric, composition, or TPC.

Table 1.1. Instructor demographics

Instructor	Years Teaching TPC	Institution's Carnegie Designation	Status	Home Department
1	7	Private, 4-year, very high research activity	Clinical assistant professor	Business
2	5	Public, 4-year, master's university	Tenure track	Technical communication
3	6	Public, 4-year, master's university	Tenure track	English
4	8	Private, 4-year, high research activity	Tenure track	English
5	17	Public, 4-year, high research activity	Lecturer	English
6	16	Public, 4-year, master's university	Tenured	English
7	6	Public, 4-year, master's university	Tenure track	English
8	3.5	Public, 4-year, master's university	Tenure track	English and foreign languages
9	8	Private, 4-year, master's university	Assistant professor, non-tenure-track	Business
10	15	Public, 4-year, high research activity	Tenured	English

Note. Instructors in stage one of this study had 3.5–17 years of experience teaching TPC, came from nine different institutions, and had varying employment statuses and home departments.

During each interview, instructors' values were student-centered; they clearly cared about their students' learning and experiences in their service course. However, instructors were not always consistent with the pedagogical goals that they spoke of most frequently. For example, Instructor 1 mentioned teaching teamwork the most often, even though her main goal was to teach students to understand then apply "business communication theory." Instructors 2 and 9 mentioned audience most often, even though Instructor 9 wanted her students to understand and apply theory. Instructors 1 and 9, who taught business communication in business departments at a top-ranked business school and a small liberal arts school, respectively, both said that their students needed to understand theory, then apply that theory to business communication genres and research. Instructors 3, 5, and 10 mentioned genre most often during their interviews, even though each instructor most wanted their students to write rhetorically with attention to audience and context. Teaching engineers, Instructor 6 mentioned information and content most often, consistent with what she most wanted her service course students to know. Finally, Instructor 8 mentioned issues of tone and style most often during her interview, even though she wanted students to learn how to "communicate simply." Although Instructor 10 grounded his teaching practices in rhetorical theory like Instructor 8, he also used his workplace experience to undergird his teaching practices like Instructors 1 and 9. However, unlike Instructors 1, 8, and 9, Instructor 10 connected his workplace expertise with "practical wisdom" or phronesis. Instructors' individual pedagogical goals reflect their unique backgrounds, education, and workplace experience, along with what they want their students to take from their service courses.

When asked what their students most needed to know or do by the service course's end, each instructor had slightly different answers. Over half of the instructors in the study (Instructors 2, 3, 5, 7, 8, and 10) stated that their students needed to understand how to communicate to different audiences through the service course; for example, Instructor 5 said, "think my students need to be able to determine, depending on the circumstances, who their audience is, what their audience needs are, and what type of writing is going to communicate that best." Instructor 10 linked audience with purpose because "documents lead to actions." Instructor 4 wanted her students to know that their professional communication skills would transfer to other situations, but that students could "be effective and ethical communicators in any real context." Instructor 6 discussed writing in terms of information, framing her service course to help her engineering students "express [technical ideas] in words." These results paint a picture of how these instructors approach their service courses: introducing students to rhetorical terminology such as audience and framing information and genres that students could transfer to other contexts.

▪ Workplace Phronesis in the TPC Classroom

Experienced instructors with degrees in rhetoric, composition, or TPC (Instructors 6 and 10) used language during their interviews that was situated more firmly in workplace contexts, while still employing theoretical concepts like phronesis and transfer. Instructors 6 and 10 were best able to integrate their pedagogical philosophies across their interviews, syllabi, and feedback on students' writing because they both framed the TPC service course as an entity that has different goals and approaches than composition. Both directly credited their own professional experiences with their abilities to teach students a workplace phronesis, or "practical wisdom" (Instructor 10), instead of writing from a series of strict rules or checklists.

Instructors 6 and 10 had profound insights about the differences between composition and the TPC service course, particularly about the role of the writing instructor. Along with using his 15 years of experience teaching TPC courses and his graduate coursework in TPC, Instructor 10 deploys a rhetorical approach, but one that is specifically grounded in TPC as a field of experience and study. When asked what his students most needed to know or do at the end of the semester, Instructor 10's philosophy was inherently rhetorical:

> Probably that [students] need to approach writing texts rhetorically. So, by that, I mean that they have a sense of the audience and the purpose. That they craft the document—whatever that document is—to fit the specific audience and the specific purpose.

Purposes and audience mattered to Instructor 10's pedagogical goals. On the surface, this quote does not differ much from Instructor 8's emphasis on teaching students to "communicate things simply with co-workers." Both Instructors 8 and 10 want students to understand and communicate to their purposes and audiences. However, when Instructor 10 explains his approach to theory in his service course, a marked difference appears between his answers and those of less experienced instructors who relied on their composition training.

▪ Separating TPC from the Service Course

For Instructor 10, the service course was an opportunity for students to learn that writing had purpose and that writing could guide decision-making to produce action. During his interview, Instructor 10 spoke at length about how the rhetorical situation of his classroom differed from that of composition or literature courses:

> With a technical writing course, students are able to move away from writing a document in an attempt to please an instructor, as we have to try to do when we're in first-year writing. Or even in a literature class, where you are writing to display your knowledge or your understanding to the instructor. So yes, in a tech[nical] writing class students write to me. But I hope they try to understand that I'm not merely grading... but I'm trying to approximate what would happen to this document in a workplace.

In this quote, Instructor 10 addresses both his approach to giving feedback on students' writing and how the TPC service course differs not just from first-year composition, but from almost all other courses that students take during their undergraduate careers. To Instructor 10, the service course was not just a display of a student's knowledge, but a way to develop specific skills, or phronesis, in workplace writing.

Instructor 6 also relied on her workplace experience to inform the ways that she articulated her course goals and gave feedback on student writing. For Instructor 6, the service course provided ways for students to improve their abilities as workplace communicators and project managers. Instead of discussing the writing teacher's role like Instructor 10 did, though, Instructor 6 frames her service course in response to her engineering audience's needs. Instructor 6 was able to address these needs because of her workplace experience:

> I taught in a law school as my grad assistantship for four years. . . . So, I had some experience with writing that wasn't freshman comp essentially . . . a lot of the same principles as freshman comp certainly apply. But what I found is that it's such a different audience. That a lot of the techniques that I use in my freshman composition class—it's just not the same. . . . There are skeptics, more so than freshmen in freshman comp. I mean freshmen [in] comp are like, "Oh it's a class everyone has to take" and you know they just got out of high school and you know they just kind of get through it. This one is "I hate writing and I've already taken freshman comp. Why am I here? I'm never going to have to write. I want to be an engineer. I like math" or whatever. And so, you get an extra level of skepticism. One of the things I love is surprising them. You know like, "This is really relevant and you're really going to use this."

Teaching her students, especially the skeptical ones, that TPC skills would be relevant and useful to their education drove Instructor 6's pedagogy. She enjoyed working with her engineering students and often spoke about writing in engineering terms, such as persuading subject matter experts or tailoring information to a non-engineering audience. To help overcome her students' skepticism, Instructor 6 was very clear about telling her students how their skills would transfer to the workplace and giving students "blunt" feedback about their work. Instructors 6 and 10 asked their students to write workplace genres situated in real contexts, giving their students experiential learning opportunities that fostered workplace phronesis.

Instructors' Goals for Students' Learning: Rhetoric, Genre, and Information Literacy

Across these interviews, instructors' pedagogical goals stayed remarkably consistent: these TPC instructors relied on overtly rhetorical framing for teaching the TPC service course, both in their interviews and their syllabi's learning objectives. Rhetoric, including purpose, audience, context, and persuasion, was the most often-coded terminology in both instructors' interviews and their written learning objectives. Rhetorical theory and terminology informed instructors' approaches; instructors with fewer than six years of experience (Instructors 2, 3, 7, and 8) tended to directly apply composition or classical rhetorical theories to their TPC classrooms without considering how these theories might function differently in TPC. For example, Instructor 8 had three and a half years of experience teaching TPC courses and used rhetorical terminology as a placeholder for workplace experience in her TPC service course. When asked what students needed to be able to know or do when leaving her course, Instructor 8 answered that students "really just need to know how to communicate things simply with co-workers." The Neo-Aristotelian definitions of purpose, audience, and context often acted as a placeholder for terms specific to workplace experience or TPC theory and research. This is not to say that rhetorical terminology can never be useful, but rather to point out that overtly relying on rhetorical theory instead of workplace experience or an understanding of TPC genres and work styles diffuses the emphasis of TPC's pedagogical goals.

Implications

From this study's results, I observe the following themes: a focus on TPC content areas, workplace phronesis, and teaching students to privilege content over form. From those themes, I outline takeaways for TPC around rhetorical definitions' influences over terminology and about phronesis through experiential problem-based learning. In this section, I outline challenges and opportunities for future research on TPC pedagogy.

Theme 1: Integrating Content Areas into the TPC Service Course

The first implication of this study presents an opportunity for TPC to integrate content areas (outlined in the introduction) into the service course. Several instructors from this study used composition-based rhetorical terminology to frame their course goals and pedagogical approaches in ways that did not align with the learning outcomes and course objectives in their syllabi. While rhetoric was used consistently and remains important to TPC, this gap between the audience-, context-, and purpose-based rhetoric that these instructors are teaching creates a gap between instructors' ways of talking about TPC and their ways of writing about TPC. Tensions between rhetorically situated genre theory and teaching critical thinking or information literacy also deserve more attention. The second stage of this study revealed that instructors rarely consider teaching students to focus on writing's content as a major goal for the TPC service course, yet disproportionately often comment on students' content (Doan, 2020). More research about instructors' training could help answer these questions more effectively (Read & Michaud, 2018). How might instructors balance rhetorical terminology with teaching students to apply this terminology across TPC's content areas? Integrating these content areas more readily into the service course could also enable instructors trained outside of TPC to better enable students' preparation as professional writers.

Theme 2: Teaching a Workplace Phronesis

This study's second theme is a theoretical orientation that instructors can use to move to a workplace phronesis (Doan, 2021). Instructors 6 and 10 used their experiences in professional organizations to guide their students' attention to decision-making and organizing content that oriented readers to their purposes for writing through experiential, problem-based learning and through giving feedback that attuned students to workplace activities. However, workplace experience is not enough to produce a workplace phronesis with a theoretical component: Instructors 1 and 9, both with extensive industry experience but without graduate coursework specific to TPC, tended to rely on transmission theory, long debunked elsewhere (Slack et al., 1993). Instructors 6 and 10 present arguments for effective TPC pedagogy as the intersection between rhetorical thought, academic training, and workplace experience. Engaging students' practical wisdom with using writing to solve problems, make organizational decisions, and challenge established thinking around race, class, and gender should be primary aims of the PTC service course. The gap between rhetorical theory and phronesis in TPC should be further explored.

In connecting his experiences with rhetorical theory, Instructor 10 uses his pedagogical goals to merge theory and practice: epistemic or theoretical knowledge of rhetoric here is combined with phronesis or knowing how. "*Knowing how* is a technical sort of knowledge that falls on the wrong side of the theory-practice binary" (Sullivan & Porter, 1993, p. 409). Instructor 10's reliance on phronesis in his teaching is significant because he describes his pedagogical underpinnings of theory as technical communication theory. Of all the instructors in this study, Instructor 10 makes the most intentional effort of using theory specific to TPC both in his own interview and in his syllabus' learning objectives. In his service course syllabus, Instructor 10 wanted his students to "understand principles that inform professional communication." Instructor 10 included the rhetorical concept of "audience analysis" in his learning outcomes; he further sketched theory more broadly for his students, also wanting them to understand TPC concepts of "ethics, collaboration, graphics, and design." There is room within TPC pedagogy for pedagogical approaches that champion both rhetorical theory and the phronesis of workplace practice.

▌ Theme 3: Workplace Phronesis is Content-Centric

Teaching their students to write in workplace genres was instructors' second-most important goal during these interviews. However, the analysis of instructors' learning outcomes in their service course syllabi revealed that while rhetorical understanding and ability was still most important, critical thinking—including information literacy and teaching students to write about content—was second-most important. Despite the fact that information literacy, critical thinking, and considering content appeared as course goals in each of the ten syllabi, instructors rarely mentioned them when discussing their goals for students' learning. When discussing their comments on students' writing, however, instructors often asked students to engage with, rearrange, or revise their content (Doan, 2020). This divide between genre and information literacy points to a critical issue within TPC pedagogy: instructors often used rhetorical terminology and genre theory as placeholders for workplace phronesis that they may not have developed. Particularly for less experienced instructors, issues of purpose or genre took precedence over issues of content or detail in the service course; this result contrasts with more experienced instructors' attitudes toward phronesis (Doan, 2020). Instead, less experienced instructors relegated detail and content to lower-order issues and discussed higher-order issues such as purpose or context, when content should be considered a higher-order and high-stakes issue that could strengthen TPC's connections to industry (Boettger et al., 2017; Spilka, 2009).

Conclusion

Future research has ample ground to examine how borrowing pedagogical methods from composition leads instructors to treat phronesis as techne, instead of meeting students' higher-order needs through experiential problem-based learning (Lawrence et al., 2017; Melonçon, 2018). Within its own research, TPC should re-examine its theoretical relationship to techne and phronesis. Thus, TPC should differentiate phronesis from techne. Instead, how might instructors design experiential learning opportunities for students that ask them to demonstrate practical wisdom while balancing competing contextual demands? While technes are still important to TPC instruction, such as teaching students to use InDesign or to copyedit their written instructions, teaching phronesis should be the focus of the service course. Reducing rhetorical terminology to understanding audience or audience analysis diminishes students' opportunities to gain experience with how genres work in situations with competing moral or ethical exigencies. To enact these values, TPC must strengthen its training for new instructors, particularly through conducting empirical research about service course classrooms.

TPC has reached a critical juncture: to meet the ever-evolving needs of present and future students, TPC must continue its own rigorous tradition of pedagogical training, particularly for novice instructors. TPC should continue to rely on its own pedagogical epistemologies, rather than relying on composition pedagogy to inform its pedagogical research and new instructor training. From their separate histories, TPC and composition continue to develop different exigencies for critiquing existing problems and writing to attempt solutions. This research has raised questions about what the service course has the potential to be if instructor training in TPC focused on teaching students and instructors a workplace phronesis centered around genre ecologies.

References

Adler-Kassner, Linda & Wardle, Elizabeth A. (Eds.). (2015). *Naming what we know: Threshold concepts of writing studies.* Utah State University Press.

Aristotle. (2007). *On rhetoric: A theory of civic discourse* (G. A. Kennedy, Trans.; 2nd ed.). Oxford University Press. 322 BCE.

Barnett, Timothy & Kastley, James. (2002). From formalism to inquiry: A model of argument in Antigone. In Timothy Barnett (Ed.), *Teaching argument in the composition course: Background readings* (pp. 73–93). Bedford/St. Martin's.

Bartholomae, David. (1986). Inventing the university. *Journal of Basic Writing*, *5*(1), 4–23.

Bizzell, Patricia & Herzberg, Bruce. (Eds.). (2001). *The rhetorical tradition: Readings from classical times to the present* (2nd ed.). Bedford/St. Martin's.

Boettger, Ryan K., Lam, Chris & Palmer, Laura. (2017). Improving the data information literacies of technical communication undergraduates. In *2017 IEEE International Professional Communication Conference (ProComm)* (pp. 1–5). IEEE. https://doi.org/10.1109/IPCC.2017.8013934.

Booth, Wayne. (2007). The rhetorical stance. In T. R. Johnson (Ed.), *Teaching composition: Background readings* (3rd ed., pp. 163–171). Bedford/St. Martin's.

Brodkey, Linda. (1994). Writing on the bias. *College English*, *56*(5), 527–547. https://doi.org/10.2307/378605.

Browning, Ella R. & Cagle, Lauren E. (2017). Teaching a "critical accessibility case study": Developing disability studies curricula for the technical communication classroom. *Journal of Technical Writing and Communication*, *47*(4), 440–463. https://doi.org/10.1177/0047281616646750.

Chong, Felicia. (2018). Implementing usability testing in introductory technical communication service courses: Results and lessons from a local study. *IEEE Transactions on Professional Communication*, *61*(2), 196–205. https://doi.org/10.1109/TPC.2017.2771698.

Devitt, Amy J. (2009). Teaching critical genre awareness. In C. Bazerman, A. Bonini & D. de C. Figueiredo (Eds.), *Genre in a changing world* (pp. 337–351). The WAC Clearinghouse; Parlor Press. https://doi.org/10.37514/PER-B.2009.2324.

Devitt, Amy J. (2010). *Writing genres*. Southern Illinois University Press.

Dicks, R. Stanley. (2010). The effects of digital literacy on the nature of technical communication work. In R. Spilka (Ed.), *Digital literacy for technical communication: 21st century theory and practice* (pp. 51–81). Routledge.

Doan, Sara. (2019). Contradictory comments: Feedback in professional communication service courses. *IEEE Transactions on Professional Communication*, *62*(2), 115–129. https://doi.org/10.1109/TPC.2019.2900899.

Doan, Sara. (2020). Digging, displaying, and translating: Content-centric feedback, powered by metaphors. In *Proceedings of the IEEE Professional Communication Conference* (pp. 92–95). IEEE. https://doi.org/10.1109/ProComm48883.2020.00020.

Doan, Sara. (2021). Teaching Workplace Genre Ecologies and Pedagogical Goals Through Résumés and Cover Letters. *Business and Professional Communication Quarterly*. Advance online publication. https://doi.org/10.1177/23294906211031810.

Duffelmeyer, Barbara. (2002). Critical work in first-year composition: Computers, pedagogy, and research. *Pedagogy*, *2*(3), 357–374.

Francis, Anne Marie. (2018). A survey of assignment requirements in service technical and professional communication classes. *Programmatic Perspectives*, *10*(1), 44–76.

Getto, Guiseppe, Labriola, Jack & Ruszkiewicz, Sheryl. (Eds.). (2020). *Content strategy in technical communication*. Routledge.

Hackos, Joanne. (2005). The future of the technical communication profession: The perspective of a management consultant. *Technical Communication*, *50*, 273–276.

Hannah, M. A. & Arduser, L. (2017). Mapping the terrain: Examining the conditions for alignment between the rhetoric of health and medicine and the medical humanities. *Technical Communication Quarterly 27*(1), 1–17. https://doi.org/10.1080/10572252.2017.1402561.

Heo, Miri & Lee, Kyoung J. (2018). Chatbot as a new business communication tool: The case of Naver TalkTalk. *Business Communication Research and Practice, 1*(1), 41–45. https://doi.org/10.22682/bcrp.2018.1.1.41.

Huntsman, Sherena, Colton, Jared S. & Phillips, Christopher. (2019). Cultivating virtuous course designers: Using technical communication to reimagine accessibility in higher education. *Communication Design Quarterly, 6*(4), 12–23. https://doi.org/10.1145/3309589.3309591.

Johnson-Eilola, Johndan & Selber, Stuar A. (Eds.). (2013). *Solving problems in technical communication*. University of Chicago Press.

Knieval, Michael. (2007). Growing the service course: Anticipating problems, promise in the technical communication "mini-program." In *Proceedings of the Council for Programs on Technical and Scientific Communication*. (pp. 89–90).

Lawrence, Heidi Y., Lussos, Rachel G. & Clark, Jessica A. (2017). Rhetorics of proposal writing: Lessons for pedagogy in research and real-world practice. *Journal of Technical Writing and Communication, 49*(1), 33–50. https://doi.org/10.1177/0047281617743016.

Lynch, Dennis, George, Diana & Cooper, Marilyn. (1999). Moments of argument: Agnostic inquiry and confrontational cooperation. In L. Ede (Ed.), *On writing research: The Braddock essays 1975–1998* (pp. 390–412). Bedford/St. Martin's.

Malone, Edward. (2011). The first wave (1953–1961) of the professionalization movement in technical communication. *Technical Communication, 58*(4), 285–306.

Malone, Edward A. & Wright, David. (2018). "To promote that demand": Toward a history of the marketing white paper as a genre. *Journal of Business and Technical Communication, 32*(1), 113–147. https://doi.org/10.1177/1050651917729861.

McLeod, Susan H. (Ed.). (2007). *Writing program administration*. Parlor Press; The WAC Clearinghouse. https://wac.colostate.edu/books/referenceguides/mcleod-wpa/.

McMartin-Miller, Cristine. (2014). How much feedback is enough?: Instructor practices and student attitudes toward error treatment in second language writing. *Assessing Writing, 19*, 24–35. https://doi.org/10.1016/j.asw.2013.11.003.

Melonçon, Lisa. (2018a). Critical postscript: On the future of the service course in technical and professional communication. *Programmatic Perspectives, 10*(1), 202–230.

Melonçon, Lisa. (2018b). Programmatic locations in TPC. *Tek-Ritr*. http://tek-ritr.com/locations/.

Melonçon, Lisa, England, Peter & Ilyasova, K. Alex. (2016). A portrait of non-tenure-track faculty in technical and professional communication: Results of a pilot study. *Journal of Technical Writing and Communication, 46*(2), 206–235. https://doi.org/10.1177/0047281616633601.

Melonçon, Lisa & St.Amant, Kirk. (2018). Empirical research in technical and professional communication: A 5-year examination of research methods and a call for research sustainability. *Journal of Technical Writing and Communication, 49*(2), 1–28. https://doi.org/10.1177/0047281618764611.

Miller, Carolyn R. (1984). Genre as social action. *Quarterly Journal of Speech, 70*(2), 151–167. https://doi.org/10.1080/00335638409383686.

Miller, Carolyn R. (2015). Genre as social action (1984), revisited 30 years later (2014). *Letras & Letras, 31*(3), 56–72. https://doi.org/10.14393/LL63-v31n3a2015-5.

Morrison, Rebecca. (2017). Teaching toward the telos of critical thinking: Genre in business communication. *Business and Professional Communication Quarterly*, 80(4), 460–472. https://doi.org/10.1177/2329490617691967.

Prelli, Lawrence J. (2005). Stasis and the problem of incommensurate communication: The case of spousal violence research. In R. A. Harris (Ed.), *Rhetoric and incommensurability* (pp. 294–333). Parlor Press.

Read, Sara & Michaud, Michael. (2018). Who teaches technical and professional communication service courses?: Survey results and case studies from a national study of instructors from all Carnegie institutional types. *Programmatic Perspectives*, 10(1), 77–109.

Russell, David R. (1993). The Ethics of Teaching Ethics in Professional Communication: The Case of Engineering Publicity at MIT in the 1920s. *Journal of Business and Technical Communication*, 7(1), 84–111. https://doi.org/10.1177/1050651993007001005.

Russell, David R. (2007). Rethinking the articulation between business and technical communication and writing in the disciplines: Useful avenues for teaching and research. *Journal of Business and Technical Communication*, 21(3), 248–277. https://doi.org/10.1177/1050651907300452.

Schreiber, Joanna, Carrion, Melissa & Lauer, Jessica. (2018). Guest editors' introduction: Revisiting the service course to map out the future of the field [Special issue]. *Programmatic Perspectives*, 10(1), 1–11.

Shelton, Cecilia. (2020). Shifting out of neutral: Centering difference, bias, and social justice in a business writing course. *Technical Communication Quarterly*, 29(1), 18–32. https://doi.org/10.1080/10572252.2019.1640287.

Singleton, Meredith & Melonçon, Lisa. (2019). Introducing the feedback file for online course design in technical and professional communication. *Write Professionally*. https://writeprofessionally.org/resources/innovative-feedback-strategies/.

Slack, Jennifer D., Miller, David J. & Doak, Jeffrey. (1993). The technical communicator as author: Meaning, power, authority. *Journal of Business and Technical Communication*, 7(1), 12–36. https://doi.org/10.1177/1050651993007001002.

Spilka, Rachel. (2009). Practitioner research instruction: A neglected curricular area in technical communication undergraduate programs. *Journal of Business and Technical Communication*, 23(2), 216–237. https://doi.org/10.1177/1050651908328882.

Spinuzzi, Clay. (2003). Compound mediation in software development: Using genre ecologies to study textual artifacts. In C. Bazerman & D. R. Russell (Eds.), *Writing selves, writing societies: Research from activity perspectives*. The WAC Clearinghouse; Mind, Culture, and Activity. https://doi.org/10.37514/PER-B.2003.2317.

Still, Brian & Koerber, Amy. (2010). Listening to students: A usability evaluation of instructor commentary. *Journal of Business and Technical Communication*, 24(2), 206–233. https://doi.org/10.1177/1050651909353304.

Sullivan, Patrcia A. & Porter, Jim E. (1993). Remapping curricular geography: Professional writing in/and English. *Journal of Business and Technical Communication*, 7(4), 389–422. https://doi.org/10.1177/1050651993007004001.

Swarts, Jason. (2012). New modes of help: Best practices for instructional video. *Technical Communication*, 59(3), 195–206.

Swarts, Jason. (2018). *Wicked, incomplete, and uncertain: User support in the wild and the role of technical communication*. Utah State University Press.

Taylor, Summer S. (2011). "I really don't know what he meant by that": How well do engineering students understand teachers' comments on their writing? *Technical

Communication Quarterly, 20(2), 139–166. https://doi.org/10.1080/10572252.2011.548762.

Teston, Christa. (2012). Moving from artifact to action: A grounded investigation of visual displays of evidence during medical deliberations. *Technical Communication Quarterly*, 21(3), 187–209. https://doi.org/10.1080/10572252.2012.650621.

Tillery, Denise & Nagelhout, Ed. (Eds.). (2015). *The new normal: Pressures on technical communication programs in the age of austerity*. Baywood Publishing.

Tracy, Sarah J. (2013). *Qualitative research methods: Collecting evidence, crafting analysis, communicating impact*. Wiley-Blackwell.

Walsh, Lynda & Walker, Kenneth C. (2016). Perspectives on uncertainty for technical communication scholars. *Technical Communication Quarterly*, 25(2), 71–86. https://doi.org/10.1080/10572252.2016.1150517.

Welhausen, Candice A. (2017). Visualizing science: Using grounded theory to critically evaluate data visualizations. In H. Yu & K. M. Northcut (Eds.), *Scientific communication: Practices, theories, and pedagogies*. Routledge.

White, Edward. (2008). My five-paragraph-theme theme. *College Composition and Communication*, 59(3), 524–525.

Willerton, Russell. (2015). *Plain language and ethical action: A dialogic approach to technical content in the twenty-first century*. Routledge; Taylor & Francis Group.

Wilson, Greg & Wolford, Rachel. (2017). The technical communicator as (post-post-modern) discourse worker. *Journal of Business and Technical Communication*, 31(1), 3–29. https://doi.org/10.1177/1050651916667531.

Wolfe, Joanna. (2009). How technical communication textbooks fail engineering students. *Technical Communication Quarterly*, 18(4), 351–375. https://doi.org/10.1080/10572250903149662.

Wolfe, Joanna. (2010). *Team writing: A guide to working in groups*. Bedford/St. Martin's.

Wolfe, Joanna. (2015). Teaching students to focus on the data in data visualization. *Journal of Business and Technical Communication*, 29(3), 344–359. https://doi.org/10.1177/1050651915573944.

Yancey, Kathleen B. (2004). Made not only in words: Composition in a new key. *College Composition and Communication*, 56(2), 297–328. https://doi.org/10.2307/4140651.

Chapter 2: The Ship of Theseus: Change Over Time in Topics of Technical Communication Research Abstracts

Stephen Carradini
ARIZONA STATE UNIVERSITY

Abstract: Meta-research on technical communication's published research can contribute empirical evidence to debates about what technical communication is and what it does. In this article, I conduct a corpus analysis of 1,593 abstracts from five technical communication journals covering the years 2000–2017 to determine the topics of research article abstracts. I analyze changes over time in word usage, as measured by numbers of abstracts mentioning individual words. Increases and decreases in word frequency over time indicate three trends in the topics of technical communication research abstracts: technical communication is moving from print communication to digital communication, expanding its boundaries via the term technical and professional communication (TPC), and increasing research on core concerns of technical communicators. The digital work that featured prominently in research abstracts reflected diversified types of online work in technical communication, such as content management, user experience (UX), and social media. Words describing areas of social justice, entrepreneurship, and community-oriented work grew in usage, but these areas are still small in comparison to the number of abstracts reaffirming core concerns such as practitioners, practices, and value. Yet the rapid digital diversification of technical communication work ensures that we should always be updating what "core concerns" means in our field.

Keywords: disciplinarity, research topics, meta-research, technical communication, TPC

Debates about what technical communication is and what it does seem endless (St.Amant & Melonçon, 2016a). Are we not focused enough on practitioner issues, as some have suggested (Boettger & Friess, 2016; St.Amant & Melonçon, 2016b)? Has a fascination with the novel resulted in a decrease in work on core research questions of technical communication (Rude, 2009)? How do topics like social justice fit into the field of technical communication (Jones, 2017)? What topics are increasing or decreasing in prominence in technical communication? What do those increases or decreases in topic frequency say about the direction of the field?

One way that these questions of direction can be answered is through meta-research. "Research on research is needed," state Lisa Melonçon and Kirk St.Amant (2018), because "without a fuller understanding of what we have come to value, implicitly deduced by what has been published in the field's journals, it becomes difficult to train the next generation of students and more importantly, it becomes difficult to show what it is that we do that is unique to the field of TPC" (pp. 2, 4). Seeking what is unique to the field of technical communication is not the only reason to conduct meta-research. Researchers have conducted meta-research on published research in technical communication journals to a variety of ends. Some have investigated the type of research methods used (Boettger & Lam, 2013; Melonçon & St.Amant, 2018), the treatment of gender and feminism in technical communication (Smith & Thompson, 2002; Thompson, 1999; White et al., 2015), authorship characteristics (Lam, 2014), and citation analysis (Smith, 2000), among others. These efforts allow technical communication as a field to assess the body of work that the field has created around a certain topic and then assess the way forward to reach certain goals or initiatives related to the topic under consideration. This chapter contributes to the meta-research in technical communication by investigating the topics in technical communication journal article abstracts over time. The goal of this topical meta-research is to determine what topics are increasing and decreasing in usage, and what those changes mean for the direction of the field's research overall.

The work builds on previous meta-research on topics in technical communication. In a seminal article, Carolyn Rude (2009) conducted a content analysis of topics in technical communication books to determine the open research questions in technical communication. This oft-cited piece suggests that disciplinarity, pedagogy, practice, and social change are open questions which the field's research should continue to address. More recent articles look at the fit of research to the audiences that the research is purportedly intended for. Saul Carliner et al. (2011) analyzed the topics of five years of articles in five journals against a survey of readers' interests to find that "some alignment exists between the topics published in [*IEEE Transactions on Professional Communication*] and the preferences of participants in the survey," but that the alignment could be improved. Ryan Boettger and Erin Friess (2016) used a content analysis of topics from 1,048 articles in four technical and professional communication (TPC) journals and one practitioner magazine to determine that academic research and practitioner publications could use more alignment in topics to better help the stability of the field. Both Carliner et al. and Boettger and Friess posit that the field's research and the work of practitioners are going in different directions. All three of these articles draw conclusions and offer suggestions for the future of technical communication research based on analysis of topics.

This chapter will also focus on topics to make suggestions about the future of technical communication research, but with a chronological focus. I seek to discover what topics are increasing and decreasing over time in technical communication research journals, then assess how these changes may affect the future of technical communication research. To do this, I analyze how research topics in five technical communication research journals have changed over the years 2000–2017 by gathering a comprehensive corpus of research article abstracts published in *IEEE Transactions on Professional Communication, Journal of Business and Technical Communication, Journal of Technical Writing and Communication, Technical Communication,* and *Technical Communication Quarterly*. After dividing the abstracts into three eras (2000–2005, 2006–2011, 2012–2017), I analyzed the frequency of specific words in each era. This allowed a comparison of words increasing and decreasing in usage across the corpus; these words were descriptive of or associated with topics.

This method of topical analysis resulted in three areas of results. Words mentioned in fewer articles over time included *paper, articles, writing, rhetoric, ethical, electronic, web, engineering, information, document, write, policy, scientific, computer,* and *ethics*. Words mentioned in more articles over time included *communication, social, content, experience, online, technical, professional, user, field, projects, media, practice, practices, value,* and *community*. Words that did not appear in abstracts from the years 2000–2005 but appeared prominently in 2012–2017 abstracts included *multimodal, TPC, justice, mediated, entrepreneurs, content-management,* and *UX*.

From these findings, I argue that these changes in word frequency over time indicate three ongoing trends in the topics of technical communication research. Technical communication is

- moving from print communication to digital communication,
- increasing research on core concerns of technical communicators, and
- expanding its boundaries via the term technical and professional communication (TPC).

These three trends connect with open questions about the nature of technical communication research. Topics regarding the shift to digital reflect changes in the practice of technical communication. Changes in the practice of technical communication lead to questions regarding what the core concerns of technical communication are and should be; there is space enough for work on print and digital at the moment, but print practices are fading while digital practices are rising. These questions of core practices connect to ongoing conversations about disciplinarity brought up by the expanding boundaries of the field: the emergence of the term *technical and professional communication* shows that some researchers prefer wider boundaries in defining their field, while the term *technical communication* is still used in much larger numbers. Emerging work on how technical communication can affect social change through social justice and community action also contributes to these conversations about the boundaries of the discipline. Each of these three shifts entails its own attendant shift in pedagogy for the field. Faculty must re-skill or multi-skill to offer courses that meet emergent needs while working with practitioners to determine what the needed skills are in emerging topical areas of technical communication practice. Technical communication has not become entirely a set of emerging concepts, but emerging topical areas are growing in prominence and need to be addressed in research and pedagogy.

■ Method

To guide my chronological meta-research on topics in the field, I developed three research questions:

1. What words are decreasing over time in technical communication research abstracts?
2. What words are increasing over time in technical communication research abstracts?
3. What do increased or decreased usage of words mean for the direction(s) of technical communication research?

■ Approach

This chapter takes a meta-research approach to investigate the change over time in technical communication research topics by identifying words that reflect topics in technical communication research abstracts. Meta-research includes many approaches, including statistical meta-analysis (Graham & Perin, 2007), descriptive meta-analysis (Cardon, 2008), and content analysis (Boettger & Friess, 2016). Thomas Orr (2006) offered corpus analysis as a profitable method of professional communication research, but corpus analysis research has been used only sparingly for meta-analysis in technical communication (Carradini, 2020). I use corpus analysis for meta-research on abstracts in this chapter.

Originally called corpus linguistics due to the field commonly associated with the method, corpus analysis is a method of studying large amounts of texts in a variety of fields (Archer, 2009b; Orr, 2006). Corpus analysis can approach many types of questions; this analysis is a corpus analysis of topics in abstracts and is unconcerned with linguistics in a grammatical sense. While corpus analysis can be done qualitatively, it is primarily used to surface insights from large amounts of data that may not be easily approached via qualitative inquiry (De Groot et al., 2006; Kaufer & Ishizaki, 2006, p. 254). Researchers using corpus analysis apply quantitative approaches to investigate large numbers of texts and use the insights from these methods to further investigate and answer questions regarding the texts in the corpus. These insights can be at the level of the word or words, as in linguistics, or in larger patterns, as in this study. Multiple types of quantitative approaches can be used to discover information about the texts in the corpus, from raw frequency to statistical analysis to multi-methodological approaches (Brezina, 2018). The type of quantitative method used in each analysis corresponds to the type of question being asked about the texts in the corpus.

The results of corpus analysis should leave the purely quantitative level and point the researcher and the readers back to the texts of the corpus. The quantitative analysis (whether frequency, statistical analysis, or other methods) points out areas where the scholar should investigate the texts further (Archer et al., 2009, p. 157). Thus, the quantitative approach is a way of identifying large-scale themes that may have been difficult to identify qualitatively, and then researching those trends qualitatively. In writing studies, Derek Mueller (2017) uses the terms "distant reading" and "thin description" to describe the process of using data mining techniques to identify aspects of the discipline of rhetoric and composition that were not identifiable before, then engaging with the texts that reflect those aspects in a new way (p. 25). Mueller's study was of disciplinarity in rhetoric and composition studies, but his methods hold for other analyses of academic disciplinary data at scale. I intend to use corpus analysis to identify topics in technical communication abstracts quantitatively, assess the texts that reflect those trends qualitatively, and make arguments about the discipline at large.

I chose to use abstracts for this research because scholars in technical communication have previously employed abstract mining (White et al., 2015) and because abstracting practices including but not limited to writing journal abstracts can reveal elements of disciplinarity (Mueller, 2017, p. 62). Abstracts indicate what the article contains, previewing the language and concepts that will appear in the full article. Thus, I expect that the language in abstracts accurately represents terminology, concepts, and topics present in the full articles.

The language of the abstract is central to this effort, because I am using an approach that depends on frequency of words. High-frequency words are valuable because they have "aboutness"; they suggest what the overall textual object is about (Archer, 2009a, p. 4). The frequency of words is "a relatively objective means of uncovering lexical salience/(frequency) patterns that invite—and frequently repay—further qualitative investigation," as Dawn Archer (2009a, p. 15) states. Identifying what words often appear allows for further investigation of what the frequent appearances mean to the text. In this analysis, I chose to use the appearance of a word in an abstract as a marker that the abstract was, in some way, about that term.

While frequency of the word in the overall corpus would be the simplest way to approach frequency, I have approached frequency through the number of abstracts that contain the word, otherwise known as *range* (Bednarek, 2018, p. 98). Thus, frequency in this analysis is not relative to the length of abstracts (which showed a trend toward longer, more structured abstracts over time) or the number of words, but to whether a word appears in an abstract. Using range solved a potential methodological problem given my concern about the topics of the abstracts. I am concerned with aboutness of texts, instead of raw frequency of word usage. If a word appeared four times in a particular abstract, it could skew the number of times a word appeared in the corpus; a small number of abstracts including many repeated uses of a single word could make a topic associated with that word look prominent in research. Instead of true word frequency, I count which abstracts include the word under discussion as frequency. This allows me to see how many abstracts included a word instead of how many uses of a word exist across the corpus.

Given my interest in the topics of the abstracts (as reflected by the words in the abstract) instead of direct comparison of the frequency of words, I did not conduct analysis of statistical significance on the findings. Instead, the quantitative analysis helped me identify which words were increasing and decreasing. This analysis marked the abstracts that included those words for greater study and ultimately discussion. Further statistical research on this topic would be warranted.

■ Data Collection

I gathered 1,593 abstracts of research articles in five journals that publish articles on technical communication. I excluded other types of published work in the field, because other types of articles such as book reviews often lacked abstracts. Carliner et al. (2011) also excluded these types of articles. The 1,593 abstracts comprehensively covered the years 2000–2017; by focusing on recent research, I hope to understand what the fields look like after years of development in the 20th century. I gathered the abstracts from five top-ranked journals in technical and business communication in North America as identified by Paul Benjamin Lowry et al. (2007): *IEEE Transactions on Professional Communication*, *Journal of Business and Technical Communication*, *Journal of Technical Writing and Communication*, *Technical Communication*, and *Technical Communication Quarterly (TCQ)*. To collect abstracts from these journals, I primarily downloaded information from SCOPUS, then augmented this database using an open-source scraper tool to gather abstracts from several years of *TCQ* not included in SCOPUS. I also gathered some abstracts from *Technical Communication* manually. Researchers can download this corpus for further research use at the author's website, StephenCarradini.com.

■ Data Analysis

To analyze this data, I used corpus analysis methods and tools. Because I sought to research abstracts at a large scale, I chose the method of corpus analysis. Orr (2006) argues for more frequent use of corpus approaches in professional communication, because corpus approaches offer a fine-grained level of analytical detail and the ability to analyze at a larger scale than qualitative efforts. Orr's ideas have proven true. Corpus approaches have been used in technical communication to study use of grammar in student writing (Boettger & Wulff, 2014) and social media use for technical communicators (McGuire & Kampf, 2015), among other efforts.

To pursue this corpus analysis approach, I formatted abstracts to remove content signals (e.g., *Purpose:*, *Research Problem:*) and copyright notices where possible. I then loaded the 1,593 abstracts into the corpus analysis software AntConc (Anthony, 2017; Laursen et al., 2014). I used the software to create a full list of words from the abstracts. I used a stoplist—a list of 153 common words that carry minimal topical content such as *I, to, as, were,* and *hadn't*—to eliminate common words and facilitate the discovery of meaningful words to analyze. My stoplist came from Ranks.NL, a company that makes a webpage analyzer tool for use in search engine optimization (Ranks.NL, n.d.). While not included in the official stoplist, I manually removed from analysis words related to the reporting of information in journal articles, such as *conducted, analysis,* and *results.* These reporting words did not contain content that I deemed to be a topic or associated with a topic. While the changing over time of words used to report data can reflect methodological shifts over time (Boettger & Lam, 2013), this article is focused on the topics of the abstracts instead of methodology or other aspects of the research (Lam, 2014).

I then split the abstracts into three chronological categories to facilitate an analysis of frequency change over time. Splitting the abstracts into three categories allowed for meaningful comparisons of topic frequency between the three groups. The small number of abstracts per year would not have allowed productive year-over-year analysis that showed trends as clearly as dividing the data into three eras. The abstracts covered the 18 years of 2000–2017, so I created three even chronological eras of six years each: 2000–2005, 2006–2011, and 2012–2017. The number of articles in each era is listed in Table 2.1.

Table 2.1. Number of Abstracts Per Era

Years	Number of Abstracts
2000–2005	551
2006–2011	552
2012–2017	490

The table presents three eras of journal articles with corresponding numbers of journal articles contained in that era. The first and second eras contained almost exactly the same number of articles, while the output of the third era decreased by roughly 11% in total number of articles.

After creating these three eras of abstracts, I created a Microsoft Excel formula to analyze the number of abstracts that each word from the full corpus appeared in (also known as range). I used this formula on each era of the abstracts, creating three lists representing the range frequency of words in each era of abstracts. I then looked for trends across these three lists, re-organizing the lists based on different variables (greatest to least in 2000–2005 usage, largest percentage decrease overall, largest percentage increase overall, etc.) to find meaningful results.

Given this range methodology, I found an average increase of slightly more than one abstract per word (+1.22) over the three-era span of the corpus. The median of overall difference and mode of overall difference both resulted in 1, as well. Some of this overall average increase can be explained by an overall increase in number of words in the abstracts of the three eras, as seen in Table 2.2: the 2012–2017 era represented an increase of more than 36,000 words over the 2000–2005 era. This overall number of words per era corresponded to an increasing average abstract length over the three periods, as the 2012–2017 era's average length of abstract (175 words) was almost double the average of the 2000–2005 era (89 words). If an abstract includes more words overall than a similar abstract of previous eras, it is more likely to have increased instances of individual words than in previous eras. Even with an adjustment from raw frequency to range as the frequency method, some of this bias toward the larger number of words in the later eras is inevitable.

Table 2.2. Words in Each Era of Abstract

Years	Total Number of Words in Abstracts	Average number of words per abstract
2000–2005	49021	89
2006–2011	62880	114
2012–2017	85918	175

The table shows three eras of journal articles with corresponding numbers of total words from all abstracts in that era and the average number of words per abstract in that era. Despite Table 2.1 noting that 2012–2017 included 11% fewer abstracts than previous eras, 2012–2017 abstracts included significantly more words overall and on average per abstract than in the previous two eras.

In the results, I italicize words found in the analysis to distinguish them from words I am using to describe the concept of the word or words. I also use the language of "era" in the results: 2000–2005 is the first era, 2006–2011 is the second era, and 2012–2017 is the third era.

▪ Results

I report the results of the study by addressing words declining in usage, words rising in usage, and words that have risen from no mentions to multiple mentions over the three eras.

▪ Terms Decreasing in Use

I found 15 words trending downward in usage, appearing in fewer articles from the first era to the third era: *articles, writing, rhetoric, ethical, electronic, web, engineering, information, documents, write, policy, scientific, computer, read,* and *ethics*.

(See Table 2.3.) It is necessary to note that these are not words that dropped to no mentions in the third era,[1] but those that had the largest declines in number of abstracts in which the word appeared. These words are still included in technical communication abstracts—and in some cases many abstracts—but their usage decreased over time.

Table 2.3. Terms Decreasing in Use

Keyword	2000–2005	2006–2011	2012–2017	Percent Change
read	15	7	5	-66.7
policy	13	13	5	-61.5
electronic	20	12	8	-60
ethics	12	16	5	-58.3
write	15	22	7	-53.3
ethical	23	14	11	-52.2
articles	29	21	14	-51.7
computer	15	10	8	-46.7
engineering	31	32	20	-35.5
rhetoric	38	33	26	-31.6
web	42	44	31	-26.2
document	37	22	28	-24.3
scientific	32	34	25	-21.9
writing	102	119	90	-11.8
information	112	115	102	-8.9

Table 2.3. shows the overall percent change across three eras for keywords used in abstracts. While writing and information lost a small percentage, they lost quite a bit overall in real numbers.

The common technical communication words *information* and *writing* displayed some of the largest drops in range frequency across the eras (see Figure 2.1). *Information* went from being mentioned in 112 abstracts to 115 abstracts and

---

1. I did find words that dropped to zero uses in 2012–2017: *cross-functional, e-mail, ATTW, typeface, typography, mediate, memo, machine,* and *screens*. However, none of these words registered as a high-volume word in abstracts, and I discovered few clear content patterns in these usage-dropped-to-zero terms. *Cross-functional* featured in only six abstracts in the first era; *ATTW, e-mail, typeface,* and *typography* appeared in five abstracts; and *memo, machine, mediate,* and *screens* in four. In a minor way, these words reflect the shifts away from print (*memo*) and the expanding of the field (*cross-functional* teams may have been replaced by shifting networks of digital workers), but primarily they represent a change in how digital spaces are described and researched, which falls outside the scope of this article.

then down to 102 abstracts. While an overall loss of nine percent is not severe, the loss of ten abstracts overall places it at 12th in the list of words that lost the most abstracts in range frequency over the three eras. While *information* is still a core concept and a high-usage term, the number of abstracts that the word appears in decreased over the last two eras. The number of abstracts mentioning *writing* also decreased fairly dramatically. *Writing* increased from 102 abstracts to 119 abstracts before falling to 90 abstracts in 2012–2017. The overall loss of 12 abstracts represents only a 12 percent drop from beginning to end. However, uses of the word dropped precipitously from a 2006–2011 high of 119 to 90 in the subsequent era. This drop of 29 abstracts represented 24 percent of the 2006–2011 amount, or almost a quarter of *writing*'s highpoint lost in six years. *Write*, a corollary word to *writing*, also increased in number of abstracts before a precipitous drop, from 15 to 22 before falling to 7 abstracts in the last era. The words *rhetoric, articles, read, ethical, electronic,* and *computer* declined in usage consistently from the first era to the second era and the second era to the third era (see Figure 2.2). *Rhetoric* dropped from appearing in 38 abstracts to 33 to 26, a 31.5 percent overall drop. *Articles* dropped from 29 to 21 to 14, a 51 percent drop. *Read* dropped from 15 to 8 to 5, a 66.6 percent drop. Uses of *ethical* dropped from 23 to 14 to 11, a 52 percent drop. *Electronic* and *computer* both declined consistently over the last two eras as well.

Some words describing related fields rose or held steady in usage between the first and second eras before seeing a drop between the second and third eras (see Figure 2.3). *Engineering* saw an overall decrease of 35 percent (31 to 20) and *scientific* saw a decrease of 22 percent (32 to 25). *Policy* (13, 13, 5) held steady between eras one and two before falling. *Ethics* (12, 16, 5), a core concern of any discipline, appeared in more than ten abstracts in 2000–2006 but fell to fewer than 10 in 2012–2017.

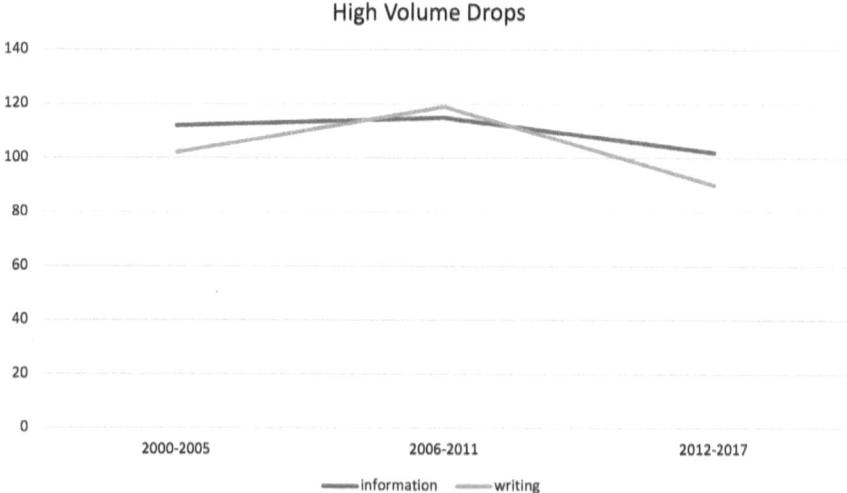

Figure 2.1. Information and writing decreased in usage overall.

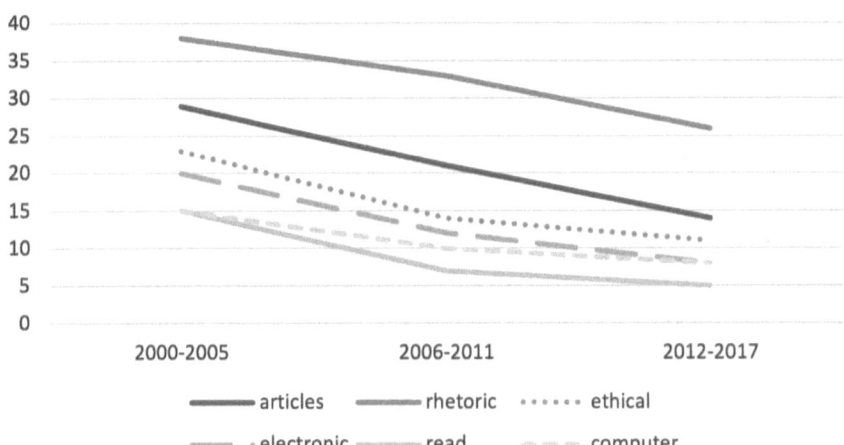

Figure 2.2. Several prominent concepts in technical communication showed two consecutive drops in number of abstracts.

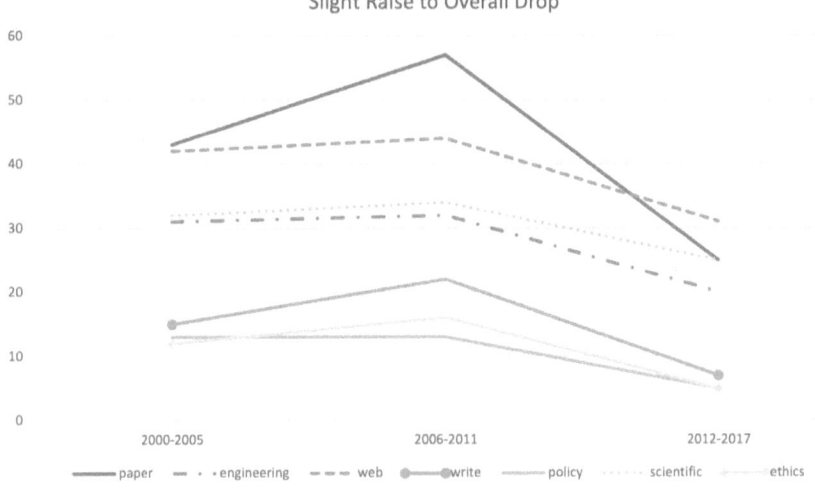

Figure 2.3. Some topics showed a rise in number of abstracts before falling in the third era.

Documents stands out as an unusual outlier in this decreasing-use section. While *documents* declined from 37 abstracts in 2000–2005 to 22 in 2006–2011, the word saw a slight resurgence to 28 abstracts in 2012–2017. The overall decline of nine abstracts masks an unusual pattern of decline and rise that no other word in this analysis displays. Overall, some previously common words lost usage share between the three eras. Words such as *read, policy, electronic, ethics*, and *write* were already lower-frequency words that saw large declines percentage-wise and by range volume.

Terms Increasing in Use

Some words increased in use over the three eras: *communication, communicators, community, content, experience, field, language, media, online, practice, practices, professional, projects, social, technical, user,* and *value.* See Table 2.4 for the number of abstracts in which each word was included.

Communication, social, and *technical* are high-volume words that increased over the three eras. (See Figure 2.4.) *Communication* increased from inclusion in 221 abstracts in 2000–2005 to 295 abstracts in 2012–2017, an increase of 74 abstracts (33.5% increase); *social* went from 41 to 88 (+47 inclusions, a 115% increase). *Technical* is used in 252 research abstracts. This number represents a 45-abstract increase over 2000–2005 (21.7% increase) despite the 2012–2017 era featuring a smaller number of articles (551 to 490). *Technical* came in second only to *communication* in the number of abstracts the word appeared in during the 2012–2017 era.

Table 2.4 shows that many of the increasing terms increase dramatically, doubling, tripling, or even quadrupling the amount of uses over the three eras.

Terms related to use of the internet grew. *Online* and *content* grew dramatically over the three eras, for an overall positive increase of 45 and 47 abstracts, respectively. The words *online* and *content* actually grew slightly faster between the first and second era than between the second and third era (see Figure 2.5). *User* and *experience* track closely together, rising modestly between the first two eras and then spiking between the second and third eras. *Media* is featured in Figure 2.6. *Media* started with a robust 31 mentions in the era of 2000–2005. It too increased slightly between eras one and two and then jumped in usage after era two.

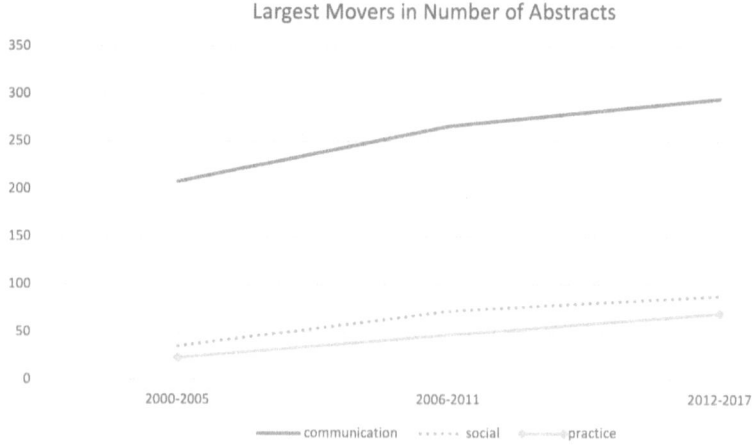

Figure 2.4. Communication, social, and technical were included in large numbers of abstracts.

Table 2.4. Terms Increasing in Use

Keywords	2000–2005	2006–2011	2012–2017	Percent Change
projects	9	17	44	388.9
experience	15	31	61	306.7
community	13	21	46	253.8
online	21	50	66	214.3
value	16	28	50	212.5
practice	23	47	70	204.4
user	20	28	58	190
media	20	31	55	175
content	27	52	74	174.1
social	35	72	88	151.4
language	24	37	55	129.2
field	36	34	71	97.2
practices	45	47	79	75.6
professional	65	103	109	67.7
communicators	61	78	92	50.8
communication	208	266	295	41.8
technical	207	244	252	21.7

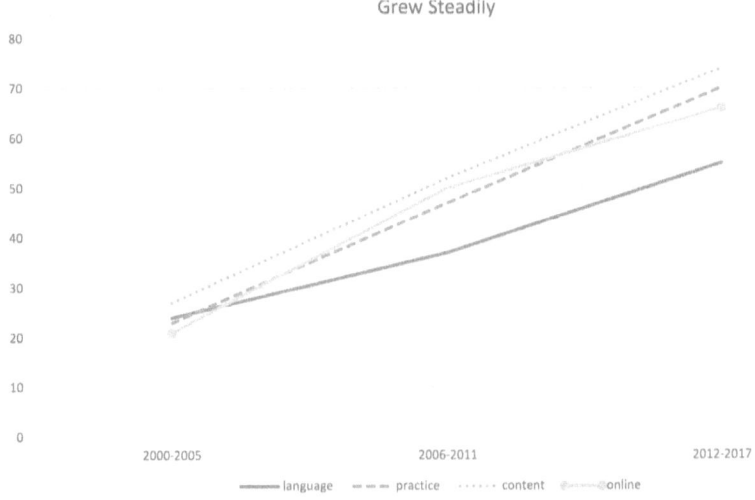

Figure 2.5. Four terms grew steadily; two reflected digital practices (online, content) while two reflect core ideas of technical communication (language, practice).

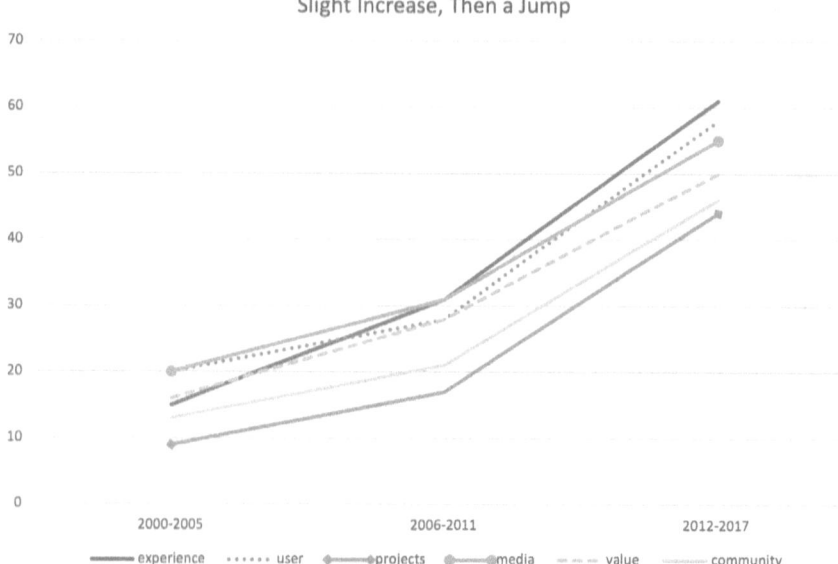

Figure 2.6. Several words experienced a slight bump between eras one and two and then a greater leap between eras two and three.

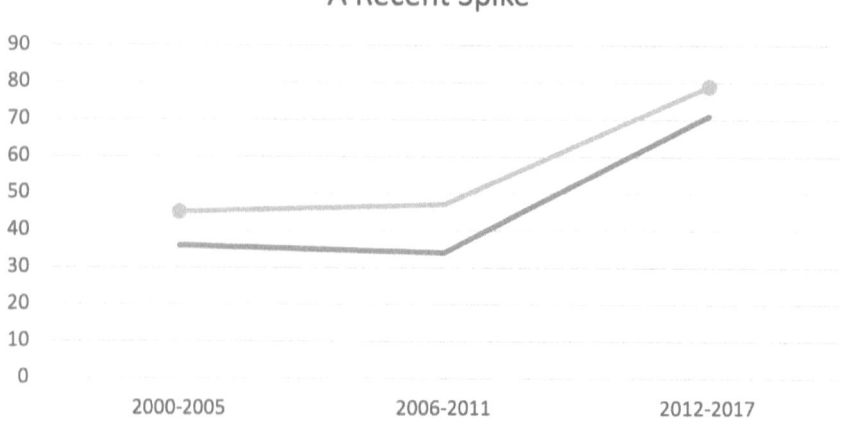

Figure 2.7. Field and practices did not increase much until between the second and third era.

Some words surrounding the traditional work of the technical communicator grew in use rapidly. Words such as *projects, community, value,* and *practice* experienced a dramatic leap in usage, with each of them more than tripling in use from the first era to the third. *Projects* almost quintupled in amount of usage. *Language* more than doubled, from 24 to 55. Use of the word *communicators* rose over the three periods, from 62 to 93 abstracts, with robust growth in use of the term: usage of *communicators* grew slightly more between the first two eras (+17) than the second two eras (+13).

Three words did not sustain rapid growth through both eras. *Professional* experienced a sharp spike between eras one and two (+38) before tapering off its rise in the next era (+6). Conversely, *field* and *practices* decreased slightly between eras one and two before eclipsing totals from eras one and two in the third era. (see Figure 2.7).

Ultimately, many words grew dramatically, either in range frequency (*communication*, +87 abstracts) or percentage (*experience*, +388.9%).

■ Terms Rising from Nothing

Words that did not appear in abstracts from the years 2000–2005 but appeared prominently in 2012–2017 abstracts included *multimodal, TPC, justice, mediated, entrepreneurs, content-management,* and *UX.* See Table 2.5 for the increase amounts.

Table 2.5. Terms Rising from Nothing*

Keywords	2000–2005	2006–2011	2012–2017
multimodal	0	4	18
TPC	0	1	13
justice	0	1	12
mediated	0	7	11
entrepreneurs	0	0	11
content-management	0	0	10
UX	0	0	10

*The table includes words that weren't used in the first era but were prominently used in the third. Because dividing by zero would create a percentage change of infinity, percent change was omitted.

Several of these words describe digital or digital-related concepts: *multimodal, mediated, content-management,* and *UX* (see Figure 2.8). *Multimodal* shows the largest overall increase in this group of words, rising from appearing in no abstracts in 2000–2005 to four in 2006–2011 to 18 abstracts in 2012–2017. This quick rise from no mentions of *multimodal* to 18 abstracts over 18 years indicates a potentially significant shift in the type of communication researched by technical communication scholars. The average word is only included in 3.2 abstracts in 2012–2017; *multimodal* is the 430th most common word in an overall list of 11,919 words. *Mediated* jumped from no abstracts to seven between the first and second eras, then tapered off its rise to only 11 in the third era. Strangely, both *content-management* and *UX* scored no hits in abstracts during the first two eras, then both appeared in ten abstracts in the third era. Because *content-management* and *UX* both appeared in zero, zero, and ten abstracts over three eras, their two lines in Figure 2.8 are the same. *Content-management's* line cannot be seen, but it is the same as *UX's.*

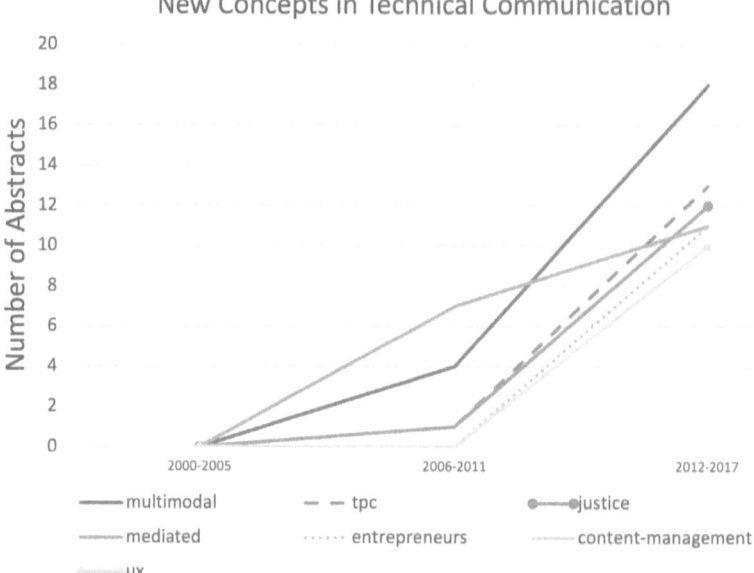

Figure 2.8. Words that increased over time from no mentions in 2000–2005 show a shift to digital, a new way of talking about the field (TPC), a new research approach (justice), and a new subject group (entrepreneurs).

Three words describe concepts that are new to the field: *TPC* and *justice* featured in only one abstract each during the second era before jumping to 13 and 12, respectively, in the third era. *Entrepreneurs* scored no hits in the first two eras before appearing in 11 abstracts in the third era, making this topic a very rapidly growing topic of research. Collectively, these seven words display a dramatic rise in amount of research in a short amount of time. These seven words appear in 84 separate abstracts (one abstract uses *TPC* and *justice* together). This number accounts for 17.14 percent of all abstracts in the 2012–2017 era—an astonishing amount considering that none of these words appeared in research during 2000–2005.

■ Analysis

I discovered three trends in word usage in abstracts from 2000–2017 as a result of this study. The first was that technical communication's research moved from a focus on print communication toward sharing that focus with digital communication. Technical communication abstracts used words describing *writing documents* and *rhetoric* less frequently over time, while using words describing *multimodal communication* and *user experience* more frequently over time. The second trend was an expansion of the field's boundaries via the term *technical and professional communication* (*TPC*). The third trend regarded increased research on core concerns of technical communicators, as reflected in the frequent and increasing use of the words *technical, communicators, value,* and *practices.* This usage pattern shows a concern with what practitioners do on a day-to-day basis and how elements of their work (and the work overall) create value. This last trend seems to be in contrast with the first two, but they occurred at the same time—the field is large enough that different groups of scholars can be focused on unique initiatives at the same time. I further explain each of the three trends in word usage below.

▪ Digital

I found that technical communication research abstracts showed increased use of words reflecting *mediated, media*-rich *multimodal* communication. This move entails a shift toward *user experience* (*UX*) while delivering reconfigurable *content* in *online* spaces via *content-management* and *social media* while turning away from *rhetoric* as the grounding concept needed to deliver *information*. While many of these words related to use of the internet (*online, content, user, experience, media*) existed in research abstracts of the first era, they grew rapidly over the next two eras.

The words *multimodal* and *mediated* both reflect the emergence of digital communication in digital spaces. *Multimodal* reflects an emphasis on communication that takes place via multiple modes. In technical communication research, this word suggests an expansion of the research area from (technical) writing to communication in digital spaces; these digital spaces consider visual, multimedia, and written modes. *Mediated*, similarly, often relates to technology or computers (as part of the term *computer-mediated communication* or *digitally mediated communication*). These words point toward digital environments as places of technical communication research.

Content-management and *UX* describe new ways of working in digital environments and technical communicators' shifting relationship to the products that they work with in those digital environments.[2] *Content-management* describes a shift away from working with *documents* and toward pieces of *content* that can be refigured into multiple environments (documents, platforms, websites, and more). Digital content-management platforms make this management possible. *UX* stands for *user experience*; user experience expands on the concept of usability by including technical communicators earlier in the design process of digital spaces and content to make sure that users can actually use the work. Both of these concepts alter the role of the technical communicator from a person writing a document as a final deliverable to creating useful knowledge and experiences in multiple modes as a final deliverable. Both *content-management* and *UX* are digital adaptations and developments of technical communication that underscore an ongoing shift to the digital. This sudden spike in research activity surrounding these two concepts reflects the speed of changes in digital spaces; concepts

---

2. User experience research can also be conducted in non-digital spaces, although it is more prominently associated with digital spaces.

emerge quickly, with research close behind. *Content-management* and *UX* join *multimodal* and *mediated* to depict shifts in technical communication research toward the study of digital communication and the study of how to work with digital communication.

The increased-usage words *online, content, user, experience, social*, and *media* further reflect the shift toward the digital. These words are often found in compound terms: *online content, user experience*, and *social media*. These phrases display new compound uses of words that have been included in all three eras, but reflect a digital turn with their new usage.

Even research on the internet is not immune to change; terminology about research on the internet seems to be changing as well. *Electronic, computer*, and *web* are words that all decreased in usage over the three eras. These words may have fallen out of use as newer words, such as digital and devices, come into play. These words may have been prominently used to describe digital spaces in Web 1.0 days. They occur less often in the Web 2.0 era that the last two eras cover (2006–2017).

▍ Changing Priorities

The change over time of words in the abstracts shows that words reflecting traditional priorities of research in the field (terms such as *writing, information*, and *documents*) are declining while words reflecting other areas of research are becoming priorities (terms such as *professional, field*, and *community*).

Any decrease in a word that is highly connected to the field's identity is important to note. For instance, the Society for Technical Communication's (2018) definition of technical communication places *information* in a central value-making role: "The value that technical communicators deliver is twofold: They make information more useable and accessible to those who need that information, and in doing so, they advance the goals of the companies or organizations that employ them." A decrease in research abstracts that mention the word *information*, then, challenges the overall paradigm that the value of technical communication lies in *information*. Potentially, that value can be created in other ways, such as developing a strong *user experience*; information is only a part of user experience. Accordingly, this decrease in use of the word *information* corresponds to an increase in *user experience, content-management*, and *content*. This shift could also be in response to the changing terminology of *content* instead of *information* to describe similar concepts. No matter the reason for the shift away from use of the word *information*, that shift is a prominent one.

Appearance of the word *writing* in abstracts decreased dramatically. The data show a large shift away from mentioning *writing* in technical communication research abstracts between 2006–2011 and 2012–2017. That *writing* lost share in abstracts (-29 abstracts) as *multimodal* communication and *user experience* gained steam in number of abstracts (+28, 18 for *multimodal* and 10 for *UX*) is a telling correlation. Technical communication has expanded its definition of what is involved in the process of communication via *multimodal* and *UX*, among others; at the same time, research abstracts mentioning *writing* have decreased. While *writing* is still often mentioned in 2012–2017 (mentioned in the 22nd most abstracts), there has been a sharp decrease percentage-wise in the number of abstracts that mention *writing*. An expanded sense of what communication is and the types of work available to the technical communicator have shifted the focus of research abstracts over the past few years. A further notable correlation is that overall uses of the word *writing* decreased in abstracts, but uses of the word *communication* continue to increase. However, *documents* showed an increase in use between eras two and three, after a steep drop between eras one and two. Perhaps the shift in use from *writing* to *communication* is a terminology shift, as documents continue to persist in research despite a shift away from writing; perhaps we *communicate* via *documents* instead of *writing documents* in contemporary technical communication research.

The overall shift away from writing and print ideas continues in the words *rhetoric, articles,* and *read*. Rhetoric has been a foundational part of technical communication since the late 1970s, if not before; this decline in use of the word in abstracts over the past two eras suggests that research interest in the topic is flagging and/or that the concept has been replaced by different grounding concepts in the work of technical communicators. As *rhetoric* emerged from work on *writing* and oral communication, it is not surprising that a decrease in abstracts mentioning the word *writing* would correspond with a shift away from using the word *rhetoric*. While digital rhetoric and the rhetoric of health and medicine are places where *rhetoric* continues to develop, the word has been used less overall in the last two eras than in the first era. The decreases in *articles* and *read* reflected a decrease in textual analysis: studies on journal articles, newspaper articles, and other types of articles declined, as did studies on how people read texts. The declines continue to indicate that theories of, genres of, and responses to *writing* are all affected by a shifting set of ideas on what communication is and what technical communicators do.

The decline in use of the words *engineering* and *scientific* is surprising, due to the central role that both of these words have played in the field historically. Engineering holds a special place in the history of technical communication as one of the founding reasons for technical communication, while scientific communication has been associated with technical communication closely enough that the Council of Programs on Technical and Scientific Communication includes the term in its name. The decline of these words in abstracts points again toward an ongoing shift in focus for technical communication research. The rise of *entrepreneurs* in technical communication research abstracts underscores the decrease in use of *scientific* and *engineering* in research. The percentage-wise decline of these two words is similar to the declines in the words *writing* and *information*. All four words represent bellwethers in thinking about how the field is shifting its attention away from previous topics and moving toward new topics—even as older words remain prominent in frequency of mentions.

The decrease in the use of the words *ethics* and *ethical* is surprising, because these seem like areas ripe for development. The number of abstracts including the words *ethics* and *ethical* decreased despite being a fundamental, grounding concept in technical communication research, pedagogy, and practice. It may be that discussions of ethics are being replaced by or subsumed by *social justice* in research—social justice mentions are increasing in technical communication research. More inquiry should investigate why the word *ethics* is declining in technical communication research abstracts; this is an unexpected and troubling finding if the concept of ethics is not being researched and foregrounded in technical communication work. Even the rise of other groundings for technical communication does not obviate the need for research on ethics. Similarly, more inquiry is needed on why the word *policy* is flagging as a research topic in technical communication abstracts; technical communication can say much about internal corporate policy as well as governmental policy. I see no clear reason from the data as to why the word *policy* is decreasing, other than (perhaps) policy's association with the also-decreasing *scientific* concerns.

These words displayed a shift away from some historically prominent words and concepts in technical communication, such as the *writing* of print *documents*. These downward trends correspond with the previously noted rise in multimodal communication in digital environments. While the digital is a rising trend, the digital is less a specific subject area than a place where subject areas happen. Other subject areas and actions are rising in prominence, particularly in ways that expand the boundaries of the field.

An expansion of technical communication's boundaries is reflected in *TPC*, *professional*, and *field*. The word *professional* is connected to the term *technical and professional communication* (*TPC*). *TPC* allows for technical communication research to include things outside the traditional scope of technical communication. This concern with expanding technical communication to include new topics and audiences is further reflected in the word *field*. Scholars in technical communication have increased their talk about the field as a whole and what can/should be included in the field. This strong interest in discussing and defining the field has grown from a constant to a phenomenon; the use of *field* held relatively steady in the first two eras, being used in 36 and 34 abstracts. However, use spiked to 71 abstracts in the third era, almost doubling its original amount from the first era. This new interest in describing/defining the *field* in the third era perhaps grew from the work of Rude (2009), as mentioned above.

The use of the words *justice* and *entrepreneurs*, another set of words that emerged in the latter two eras, shows how technical communication's research priorities continue to expand. *Justice* reflects social justice; each abstract that mentioned *justice* mentioned the word in the context of *social justice* except one abstract that mentioned it in the context of criminal justice. A social justice approach to technical communication research features different commitments and goals than other approaches to technical communication, expanding the types of research that are present in technical communication journals. One example of how new terminology and concepts are working their way into discussions of the field is shown in an abstract that offers *social justice* as an important approach for *TPC* to consider and implement (Jones, 2016).

Another expansion of the field is constituted by use of the word *community*, which spiked up 254 percent, from 13 mentions in the first era to 46 mentions in the third era. *Community* involves an expansion of the boundaries of technical communication by talking about technical and professional communication as something that is done in and for real communities, as opposed to being something in and for imagined, individualized end users. While not a new concept overall (*community* appeared in the first era), the term's use grew dramatically over the three eras. Technical communication research also recently expanded its terminological and conceptual boundaries to include *entrepreneurs* in the groups that technical communication researchers study. The word *entrepreneurs* does not appear in any abstracts for the first two eras, but appears in 11 abstracts in the third era. This professional group reflects a wider view from technical communication scholars as to who is involved in the work of technical communication. Especially as some in the field expand the name of the field to technical and professional communication, entrepreneurs represent one answer to the question of "What is professional about technical and professional communication?"

Technical communication research is expanding to include new audiences and concepts. The expansion of technical communication through the acronym *TPC* is alternately a subject of excitement and consternation, particularly in places where scholars and practitioners feel that the pursuit of the novel and interesting has crowded out other research on core issues concerning working technical communicators. Yet this research continues apace. *TPC* research pushes the boundaries to include new concepts and new constituencies into the work of the field. This work can be perceived as one outcome of the overall shift away from print toward digital. The digital space provides opportunities for many people who would not have been able to make careers on their own in the pre-digital era to make careers (Petersen, 2014, 2016). This change results in people who would otherwise work in organizations as technical communicators becoming entrepreneurs of technical communication (Lauren & Pigg, 2016a, 2016b). The acronym *TPC* suggests that professional communication of this type is something that technical communication researchers can address under the aegis of technical and professional communication.

■ Reaffirmation of Core Identity

But as much as some things change, some things stay the same. Many research abstracts in the second and third era mentioned words common in the first era, such as *technical, communication, communicators, practice, practices, projects, language,* and *value.*

The word *technical* shows that the *technical* aspects of technical communication are not going away. The use of *technical* in the names *technical communication* and *TPC* contributes to the number of uses of this word as well. While the group of people who are counted as technical communicators (or those who are eligible to be studied as technical communication research) grows, the field still uses the word *technical* in increasing amounts. Despite the expansion of the boundaries of technical communication, *technical* is still a core term.

As older words surrounding *writing* decline in use, the field has coalesced around the word *communication*. Researchers included *communication* in 208 abstracts in 2000–2005 and 295 in 2012–2017. This was an increase of 87 abstracts, but an even greater jump in percentage of abstracts: *communication* appeared in 208 of 551 abstracts (37.75%) in 2000–2005, while it appeared in 295 of 490 in 2012–2017 (60.2%). This large jump in percentage of abstracts mentioning *communication* shows that communication is becoming more central to the work described in technical communication research abstracts. Due to the previously noted rise in *user experience* and *content-management* in the field, this doubling down on the word *communication* might seem counter-intuitive. Still, this large percentage of abstracts using the word is hard to ignore as a common word that the researchers of the field can agree on.

Communicators is another particularly important word for technical communication, because one of the primary features of technical communication is the focus on a specific, definable group of people known as technical communicators. The continued use and growth of the word *communicators* indicates that research was conducted over these three periods that focused on the needs of the specific group of people that are at the core of technical communication. While the overall group of people who are counted as part of the field of technical and professional communication for research purposes is growing, the focus on the technical communicator continues to develop. *Communicators*, more than any other word, reflects that the core of technical communication research is strong and focused on practical efforts to help practitioners of technical communication, the technical communicators.

Technical, communication, and *communicators* are valuable words due to their connection to the name of the field, while *practice* and *practices* are valuable words due to the research focus that they show. The words *practice* and *practices* both increased dramatically in usage over the three eras. *Practice* more than tripled in use (23 to 70), while *practices* increased from 45 to 79 (a 75% increase). These words both point toward practical matters of work. Research on *practice* and *practices* focused on the way that technical communicators do their work. As the focus on how technical communicators do their work has been a concern of the field from the very beginning, it seems that changes in priority for the field have not significantly altered a focus on research regarding how the technical communicator's work is done.

Terms such as *projects*, *language*, and *value* are valuable to extend the idea of practices, both in what those practices are and what the goals of those practices are. *Projects* reflects two elements of technical communication research: research on pedagogy and workplace studies. The description of projects in a student context often, but not exclusively, related to service-learning projects in the community. The workplace studies usage focused on various aspects of professional projects that companies completed. *Language* also shows two aspects of technical communication: the use of language in international/intercultural contexts (both in the workplace and in English as a second language training) and *language* as a descriptor of the words used in communicating. Finally, *value* reflects technical communication's concern with developing value for employers and justifying the value that technical communicators bring to the table via communication, skills, and theories. These are areas of growth in numbers of abstracts, but also areas of field stability; technical communication research has shown a steadily growing interest in work of this type from 2000–2017.

This trend showing an increasing focus on the practical work of technical communicators seems at odds with the trend of new topics. However, these trends are both ongoing, and should be encouraged individually. The continued focus on the technical communicator allows for the core interests of the field to be continually developed and addressed.

■ Discussion

The trends in this meta-research point directly toward what technical communication did as a field in 2000–2017. Trends show technical communication research increased its use of terms that focused on the practices of technical communicators in multimodal digital spaces such as user experience, online content, content management, and social media. Researchers decreased their use of words related to topics such as information, writing, rhetoric, scientific work, and engineering. Words describing areas of social justice, entrepreneurship, and community-oriented work grew in usage, but these areas are still small in comparison to the number of abstracts including words describing more traditional concerns such as communicators, practices, and value.

This description of topics in technical communication research abstracts shows that technical communication is conducting work on at least three of the four open questions that Rude (2009) noted: practice, disciplinarity, and social change. Words describing the topic of pedagogy are less represented in this analysis due to a methodological concern that I describe below. Technical communication research is interested in the overall practice and individual practices of work, according to words whose use is rapidly growing. This finding that research on practice and practices is growing could be in response to the work of Carliner et al. (2011) and Boettger and Friess (2016), who called for technical communication researchers to focus more on the practices of technical communicators. In particular, Boettger and Friess' call for less research on rhetoric and more on practices is borne out in the research, as inclusions of the word *rhetoric* decreased from 38 to 26 abstracts over the course of the three eras, while inclusions of *practices* rose from 45 to 79 over the same span. More than three times as many abstracts mentioned *practices* than *rhetoric* in 2012–2017. This shift may be a response to the calls of both articles to align more closely with practitioner needs in research, but it may not be; the practices which researchers are conducting research on might not be the core concerns of practitioners, as stated by Boettger and Friess.

This question about "which practices?" is particularly relevant because the shift to digital changes the work that some-to-many technical communicators do. While not eliminating the need to work with documents and writing, technical communicators may be content-management professionals, user experience experts, or multimodal content creators (Brumberger & Lauer, 2015). All of these require working with language in some way, directly manipulating language, creating environments for language to be effective, or delivering language in multiple formats. So, the core concept of working with language in a technical space persists, but the actual ways of working in those spaces are shifting. Thus, the field is solidified but also shifting. Continued research efforts should be made to track how the digital affects the lives of all technical communicators, whether they are working in traditional roles with subject matter experts to create documentation and help materials for technical equipment/software or making user interfaces effective for the delivery of communication. As the type of work that technical communicators do shifts, the quest to articulate the value that technical communicators bring also must be continuously pursued (Petersen, 2017). This type of research on the practical work that technical communicators do, whether it be in traditional technical communication roles or in more far-flung digital fields, should be pursued vigorously. Research that assesses how work happens in digital spaces (Pigg, 2014) and how the digital affects traditional organizations (Spinuzzi, 2015) will require boots-on-the-ground research regarding how practitioners of technical communication do their work in a digitized and digitizing era. This sort of work takes an incredible amount of time, effort, and support from the technical communication practitioner community (Boettger & Friess, 2016). Practitioners have often given of their time and skills to research, and their sacrifices should be acknowledged as we researchers continue to ask them to be co-researchers and participants in ethnographic, interview, survey, and digital collection methods for the advancement of the shared field.

The end result of these practitioner-supported studies may be that the digital has so transformed and diversified the work of technical communication that there is no center to the field. It may be that the terms *technical communication* and *technical communicators* are the Ship of Theseus, the ship that had all its parts replaced and yet still bore the same name. The question of "Is it the same ship, even if it has had its parts interchanged?" is valid. The core concerns of technical communication (technical communicators, practices, projects, language, values, et al.) may be highly respondent to the new digital environs and thus change what it means to be a technical communicator altogether. If this shift to digital that showed in the 18 years of abstracts continues apace, technical communication may require even more multi-skilling and re-skilling in emerging skillsets than it currently requires. Thus, the practices that technical communication requires of its technical communicators should continue to be researched. The discussion as to "which practices should be researched?" is an ongoing concern, and this chapter will not conclude the discussion. While words describing traditional research areas such as *writing* have decreased in research abstracts and words describing emerging topics such as *user experience* have increased in research abstracts, use of the word *writing* has not decreased to a point where the term *user experience* is more common in research abstracts than *writing*. The balance of core, historic concerns of technical communication and emerging topics in research (and attendant pedagogy) is an open one; at the moment, the historic concerns are still more common and should be more focused on in pedagogy than the emerging concerns. This focus is not to the neglect of new concerns, which should be the continued focus of new research. At some point, there may be more user experience research than research on writing, if user experience continues to be a concept that practitioners suggest for research and/or that catches the attention of the academic field. The concerns and needs of working practitioners should be carefully considered, but the expanded boundaries of the field suggest that even "practitioners of TPC" is a category open to definition. This tension may be resolved by using the term *technical communication* to correspond to traditional concerns such as the value that practitioners bring to organizations, while using the acronym *TPC* to describe the needs of groups emerging into our research, such as entrepreneurs and social media managers. This is but one way to strike a balance between the two foci of technical communication research; others could be developed.

Research on Rude's open question of how social change can be achieved through technical communication has increased over the three eras studied. The idea of social change was not new in 2009, but the interest in various ways of implementing efforts toward social change intensified over the next eight years. Increased use of the word *community* and emergent use of the term *social justice* point toward ongoing research questions regarding how social change can be made through technical communication (Jones, 2017). Implementing social justice practices in technical communication and doing work in and for the community are ways that technical communicators can hope to affect social change; thinking equitably and communally when communicating changes the potential outcomes of communication. These two ideas stand near to and yet contrast with the concepts of ethics and users. Aspiring to a particular code of ethics and applying it to work can be a top-down approach that reduces ethics to a set of checkboxes. Social justice is an expansive concept that resists easy lists of concepts in lieu of interacting with the histories, lived experiences, and in situ practices of audiences. This approach ties into the differences between community approaches and user-focused approaches; community approaches to communication within a specific, named group of people are far different than writing for an imagined user or users. While not all communication can be done in and for specific communities, this arm of technical communication research posits a different way to make social change in the world than the traditional methods of technical communication. With use of the word *ethics* decreasing in research abstracts, one area of research is to continue to assess how technical communicators can create social change within organizations. Other ways of making social change can and should be developed in technical communication research that build on, extend, and co-exist with these ideas.

Rude's third question, regarding disciplinarity, is clearly being discussed as well. Research abstracts mentioned the words *field* and *TPC* in increased amounts, showing an interest not only in discussing the field of technical communication, but in defining it further as *technical and professional communication*. This discussion of what TPC is—and what it means to add *professional* to technical communication—is an ongoing story. The acronym *TPC*'s usage spikes from one in the first era to 13 in the last era, suggesting that it is a recent phenomenon. The emergence of the word *entrepreneurs* in the third era offers a clue as to what *TPC* might mean in practice: the expansion of the field to include other types of communicators and communication practices under the mantle of the expanded title TPC. Yet the words *technical* and *communication* have grown rapidly in use; *TPC* is still a very small percentage of the overall usage (13 uses) of the words *technical* (252 uses) and *communication* (295 uses). So while the discussion of disciplinarity has a new entrant in the acronym *TPC* and the development of the associated word *professional* (109 uses), the discussion of disciplinarity and the descriptor used for the field are both still largely focused around the term *technical communication*. While technical communication is a core identifying term, development of new topics and ideas under the mantle of technical and professional communication research should also proceed. Beyond the specific concerns of field and title, each of the changes discovered in this analysis (the shift to digital, the changing priorities, and the reaffirmation of core concerns) is related to disciplinary aspects of the field: they speak to who the research in technical communication thinks that we are.

While these findings have implications for pedagogy, Rude's fourth open question of pedagogy is less clearly covered in these findings. This is a methodological limitation. I chose to limit the analysis to words that were associated with topics in technical communication research and excluded words associated with methodology or pedagogy for purposes of scope and clarity of findings. While the specific areas of concern in technical communication pedagogy over the years 2000–2017 are not present here, the concerns of multi-skilling, re-skilling, and development of emerging skillsets to address the shift to digital and attendant shifts in technical communication priorities all fall under the realm of pedagogy. As these trends continue, research on these trends should continue to be adapted into the classes of technical communication teachers all over the world. While these skills are critical to the further development of technical communication pedagogy, the core concerns of writing, information, and documents are not gone from technical communication research abstracts or practice. The research work that expands the boundaries of the field must be set in context of a much larger amount of work focused on the core concerns of the technical communicator. Even as the words *writing, information, documents,* and *rhetoric* appeared in fewer abstracts from era one to era three, these words appeared in large numbers of abstracts—much larger numbers of abstracts than any word describing an individual emerging topic at the moment. So, the enthusiasm for what is emerging must not override the large amount of work that represents traditional concepts in technical communication.

The abstracts of 2000–2017 in technical communication research point the field toward the future: a robust path of an expanded set of practitioners working with researchers to understand and analyze the work of an increasingly-but-not-entirely digital workplace so that knowledge can make its way back to the classroom for aspiring technical and professional communication practitioners. The shift to the digital and a changing set of priorities for technical communication live in tension with a commitment to core, historical principles of technical communication. While research should continue on core concerns and emerging concepts, the rapid rise of the digital ensures that we should always be updating what "core concerns" means and what the most important practices needed in pedagogy are. The Ship of Theseus has not yet had all its parts replaced, and we may never see that occur; but we should always be checking what is on the hull.

References

Anthony, Laurence. (2017). *Antconc* (Version 3.5.0) [Computer software]. Waseda University. http://www.laurenceanthony.net/software.

Archer, Dawn. (2009a). Does frequency really matter? In Dawn Archer (Ed.), *What's in a word-list? Investigating word frequency and keyword extraction.* Ashgate Publishing.

Archer, Dawn. (2009b). Promoting the wider use of word frequency and keyword extraction techniques. In Dawn Archer (Ed.), *What's in a word-list? Investigating word frequency and keyword extraction.* Ashgate Publishing.

Archer, Dawn, Culpeper, Jonathan & Rayson, Paul. (2009). Love—'a familiar or a devil'? An exploration of key domains in Shakespeare's comedies and tragedies. In Dawn Archer (Ed.), *What's in a word-list? Investigating word frequency and keyword extraction.* Ashgate Publishing.

Bednarek, Monika. (2018). *Language and television series: A linguistic approach to TV dialogue.* Cambridge University Press.

Boettger, Ryan K. & Friess, Erin. (2016). Academics are from Mars, practitioners are from Venus: Analyzing content alignment within technical communication forums. *Technical Communication, 63*(4), 314–327.

Boettger, Ryan K. & Lam, Chris. (2013). An overview of experimental and quasi-experimental research in technical communication journals (1992–2011). *IEEE Trans-*

actions on *Professional Communication, 56*(4), 272–293. https://doi.org/10.1109/Tpc.2013.2287570.

Boettger, Ryan K. & Wulff, Stefanie. (2014). The naked truth about the naked *this*: Investigating grammatical prescriptivism in technical communication. *Technical Communication Quarterly, 23*(2), 115–140. https://doi.org/10.1080/10572252.2013.803919.

Brezina, Vaclav. (2018). *Statistics in corpus linguistics: A practical guide*. Cambridge University Press.

Brumberger, Eva & Lauer, Claire (2015). An evolution of technical communication: An analysis of industry job postings. *Technical Communication, 62*(4), 224–243.

Cardon, Peter W. (2008). A critique of Hall's contexting model: A meta-analysis of literature on intercultural business and technical communication. *Journal of Business and Technical Communication, 22*(4), 399–428. http://doi.org/10.1177/1050651908320361.

Carliner, Saul, Coppola, Nancy, Grady, Helen & Hayhoe, George (2011). What does the Transactions publish? What do Transactions' readers want to read? *IEEE Transactions on Professional Communication, 54*(4), 341–359. https://doi.org/10.1109/TPC.2011.2173228.

Carradini, Stephen. (2020). A comparison of research topics associated with technical communication, business communication, and professional communication, 1963–2017. *IEEE Transactions on Professional Communication, 63*(2), 118–138. https://doi.org/10.1109/TPC.2020.2988757.

De Groot, Elizabeth B., Korzilius, Hubert, Nickerson, C. & Gerritsen, Marinel. (2006). A corpus analysis of text themes and photographic themes in managerial forewords of Dutch-English and British annual general reports. *IEEE Transactions on Professional Communication, 49*(3), 217–235. https://doi.org/10.1109/TPC.2006.880755.

Graham, Steve & Perin, Dolores. (2007). A meta-analysis of writing instruction for adolescent students. *Journal of Educational Psychology, 99*(3), 445–476. http://doi.org/10.1037/0022-0663.99.3.445.

Jones, Natasha N. (2016). The technical communicator as advocate: Integrating a social justice approach in technical communication. *Journal of Technical Writing and Communication, 46*(3), 342–361. https://doi.org/10.1177/0047281616639472.

Jones, Natasha N. (2017). Modified immersive situated service learning: A social justice approach to professional communication pedagogy. *Business and Professional Communication Quarterly, 80*(1), 6–28. https://doi.org/10.1177/2329490616680360.

Kaufer, David & Ishizaki, Suguru. (2006). A corpus study of canned letters: Mining the latent rhetorical proficiencies marketed to writers-in-a-hurry and non-writers. *IEEE Transactions on Professional Communication, 49*(3), 254–266. https://doi.org/10.1109/TPC.2006.880743.

Lam, Chris. (2014). Where did we come from and where are we going? Examining authorship characteristics in technical communication research. *IEEE Transactions on Professional Communication, 57*(4), 266–285. https://doi.org/10.1109/TPC.2014.2363892.

Lauren, Benjamin & Pigg, Stacey. (2016a). Networking in a field of introverts: The egonets, networking practices, and networking technologies of technical communication entrepreneurs. *IEEE Transactions on Professional Communication, 59*(4), 342–362. http://doi.org/10.1109/Tpc.2016.2614744.

Lauren, Benjamin & Pigg, Stacey. (2016b). Toward multidirectional knowledge flows: Lessons from research and publication practices of technical communication entrepreneurs. *Technical Communication, 63*(4), 299–313.

Laursen, Anne Lise, Mousten, Birthe, Jensen, Vigdis & Kampf, Constance. (2014). Using an ad-hoc corpus to write about emerging technologies for technical writing and translation: The case of search engine optimization. *IEEE Transactions on Professional Communication*, *57*(1), 56–74. https://doi.org/10.1109/TPC.2014.2307011.

Lowry, Paul Benjamin, Humpherys, Sean LaMarc, Malwitz, Jason & Nix, Joshua. (2007). A scientometric study of the perceived quality of business and technical communication journals. *IEEE Transactions on Professional Communication*, *50*(4), 352–378. https://doi.org/10.1109/Tpc.2007.908733.

McGuire, Mark & Kampf, Constance. (2015). *Using social media sentiment analysis to understand audiences: A new skill for technical communicators?* [Paper presentation]. 2015 IEEE International Professional Communication Conference, Limerick, Ireland. https://doi.org/10.1145/2775441.2775472.

Melonçon, Lisa & St.Amant, Kirk. (2018). Empirical research in technical and professional communication: A 5-year examination of research methods and a call for research sustainability. *Journal of Technical Writing and Communication*, *49*(2), 128–155. http://doi.org/10.1177/0047281618764611.

Mueller, Derek N. (2017). *Network sense: Methods for visualizing a discipline*. The WAC Clearinghouse; University Press of Colorado. https://doi.org/10.37514/WRI-B.2017.0124.

Orr, Thomas. (2006). Introduction to the special issue: Insights from corpus linguistics for professional communication. *IEEE Transactions on Professional Communication*, *49*(3), 213–216. https://doi.org/10.1109/tpc.2006.880750.

Petersen, Emily January. (2014). Redefining the workplace: The professionalization of motherhood through blogging. *Journal of Technical Writing and Communication*, *44*(3), 277–296. https://doi.org/10.2190/TW.44.3.d.

Petersen, Emily January. (2016, October 2–5). *Reterritorializing workspaces: Entrepreneurial podcasting as situated networking, connected mediation, and contextualized professionalism* [Paper presentation]. 2016 IEEE International Professional Communication Conference, Austin, TX, USA. https://doi.org/10.1109/IPCC.2016.7740516.

Petersen, Emily January. (2017). Articulating value amid persistent misconceptions about technical and professional communication in the workplace. *Technical Communication*, *64*(3), 210–222.

Pigg, Stacey. (2014). Coordinating constant invention: Social media's role in distributed work. *Technical Communication Quarterly*, *23*(2), 69–87. https://doi.org/10.1080/10572252.2013.796545.

Ranks.NL. (n.d.). *Stopword lists*. https://www.ranks.nl/stopwords.

Rude, Carolyn D. (2009). Mapping the research questions in technical communication. *Journal of Business and Technical Communication*, *23*(2), 174–215. https://doi.org/10.1177/1050651908329562.

Smith, Elizabeth Overman. (2000). Strength in the technical communication journals and diversity in the serials cited. *Journal of Business and Technical Communication*, *14*(2), 131–184. https://doi.org/10.1177/105065190001400201.

Smith, Elizabeth Overman & Thompson, Isabelle. (2002). Feminist theory in technical communication: Making knowledge claims visible. *Journal of Business and Technical Communication*, *16*(4), 441–477. https://doi.org/10.1177/105065102236526.

Society for Technical Communication. (2018). *Defining technical communication*. https://www.stc.org/about-stc/defining-technical-communication/.

Spinuzzi, Clay. (2015). *All edge: Inside the new workplace networks*. University of Chicago Press.

St.Amant, Kirk & Melonçon, Lisa. (2016a). Addressing the incommensurable: A research-based perspective for considering issues of power and legitimacy in the field. *Journal of Technical Writing and Communication, 46*(3), 267–283. https://doi.org/10.1177/0047281616639476.

St.Amant, Kirk & Melonçon, Lisa. (2016b). Reflections on research: Examining practitioner perspectives on the state of research in technical communication. *Technical Communication, 63*(4), 346–364.

Thompson, Isabelle. (1999). Women and feminism in technical communication: A qualitative content analysis of journal articles published in 1989 through 1997. *Journal of Business and Technical Communication, 13*, 154–178. https://doi.org/10.1177/1050651999013002002.

White, Kate, Rumsey, Suzanne Kesler & Amidon, Stevens. (2015). Are we "there" yet? The treatment of gender and feminism in technical, business, and workplace writing studies. *Journal of Technical Writing and Communication, 46*(1), 27–58. https://doi.org/10.1177/0047281615600637.

Chapter 3: Mapping Technical Communication as a Field: A Co-Citation Network Analysis of Graduate-Level Syllabi

Michael J. Faris
TEXAS TECH UNIVERSITY

Greg Wilson

Abstract: Echoing their earlier 2001 commentary, Johndan Johnson-Eilola and Stuart A. Selber (2004) wrote in the introduction of *Central Works in Technical Communication* that technical communication must develop "a coherent body of disciplinary knowledge" in order to become a mature discipline and profession (p. xxvii). We revisit the question of the field's coherence and maturity, providing an update on Elizabeth Overman Smith's (2000a, 2004) citation analyses of the field in which she provided a set of "points of reference." We might look to such an identifiable body of core texts as an argument for coherence, as core texts are essential to defining a discipline. This chapter provides a co-citation network analysis of texts assigned in 60 graduate syllabi for courses on the foundations of technical communication. We use social network and citation analysis tools to identify 82 core texts that we argue constitute "a coherent body of disciplinary knowledge" and signal adequate maturity in our field to move past our disciplinary anxiety of inadequacy and underdevelopment.

Keywords: co-citation, social network analysis, disciplinarity, graduate education, syllabus

In the 1970s and 1980s, technical communication emerged as an academic field that studied, theorized, justified, defined, and developed pedagogy for the professional practice of technical communication. Early work like Carolyn R. Miller's (1979) "A Humanistic Rationale for Technical Writing" and David Dobrin's (1983) "What's Technical About Technical Writing?" sought to differentiate the study of technical communication from other academic English studies and to complicate the teaching of technical writing as more than the direct conveyance of facts. Workplace studies by Jack Selzer (1983), Dorothy Winsor (1990), Stephen Doheny-Farina (1986), and others explored and established methods for understanding and modeling how technical professionals used language to accomplish technical tasks on the job.

Over the decades following these foundational arguments, scholars in technical communication continued to be concerned with both the status of technical communication practitioners (e.g., Hart-Davidson, 2001; Henry, 2000; Johnson-Eilola, 1996; Kynell-Hunt & Savage, 2003–2004; Savage, 1999; Slack et al., 1993; Wilson, 2001; Wilson & Wolford, 2017) and the legitimacy and status of technical communication as an academic discipline (Grove & Zimmerman, 1997; Johnson-Eilola & Selber, 2001, 2004; Pinelli & Barclay, 1992; Rude, 2009; Smith 2000a, 2000b, 2004; Staples, 1999; Wahlstrom, 1997). Scholars expressed concern about the identity, coherence, and institutional locations of technical communication. For instance, Johndan Johnson-Eilola and Stuart A. Selber (2001) noted that the field lacked "a coherent body of knowledge in both the academy and workplace" (p. 407). To respond to this problem, they argued for a model of graduate education that "organizes the field by locating its modes of analysis in the three-dimensional space of thinking, doing, and teaching" (p. 405). In the preface to their much-used collection *Central Works in Technical Communication* (*CWTC*), Johnson-Eilola and Selber (2004) reiterated their concerns about the field's lack of coherence: they identified technical communication as an "intellectual enterprise" having the proto-elements of a discipline, but lacking a "coherent ... framework" around which these elements could coalesce (p. xv). They wrote, "Our field will not achieve the status of a mature profession until it can come to grips with a coherent body of disciplinary knowledge" (p. xxvii). The goal of *CWTC*, then, was to identify and organize a set of scholarly papers that can be a coherent discursive center for understanding technical communication as a discipline.

Nearly two decades after the publication of *CWTC*, we ask, How coherent or dispersed is technical communication as a scholarly field? Has the field matured, developing a shared body of knowledge, shared modes of thinking and methods, and shared broad research questions that help to develop the field as a discipline or "mature profession" (Johnson-Eilola & Selber, 2001, p. 408)? One way to approach this question is through methods of citation analysis. Elizabeth Overman Smith (2000a) analyzed citations in five technical communication journals over a period of ten years (1988–1997). By studying over 25,000 citations, she identified a list of 163 heavily cited texts that constituted shared "points of reference" for the field, or those texts that have been influential in shaping the field and "are representative of the knowledge base for technical communication" (p. 452). In a follow-up study, Smith (2004) narrowed this list of 163 points of reference down to 26 texts to provide "an important, magnified view" of the field (p. 53). Drawing on Stephen Toulmin (1972), Smith (2004) understood points of reference as a *transmit*: "a group of texts that record the conversations of the members of the discipline and their use of the concepts and the procedures that make up the discipline's activities" (p. 51). From her analysis, Smith (2000a) proposed that these texts show the field's shared interest in certain topics: "discussions of professional issues (defining technical communication, pedagogy, and research methods), rhetoric and the rhetorics of communities, document design and technology issues, and workplace communication" (p. 438). Further, she argued, "As a discipline, technical communication has developed depth and rigor" with a broad, interdisciplinary research and theoretical base (2000b, p. 131). While now two decades old, Smith's analyses showed that in the 1980s and 1990s, technical communication as a field was developing a strong interdisciplinary approach to research and journals in the field were increasing in relevance and prestige. Further, her analyses showed that there was a corpus of texts that seemed central to scholarship in technical communication.

This chapter provides an update on Smith's work and presents a co-citation network analysis of 60 graduate syllabi for courses on the foundations of technical communication. While Smith (2000a, 2000b) relied on raw citation counts in her studies, we turn to co-citation network analysis, which combines the approaches of citation analysis in information science with the approaches of social network analysis developed in sociology (De Bellis, 2009; de Solla Price, 1965; Healy, 2013; Otte & Rousseau, 2002; Small, 1973; Wang, 2012). We propose co-citation network analysis as a problem-solving approach that "maps" the field: the citation maps of syllabi that we develop in this article show us what scholarship we value, how coherent or diffuse the field is, and what graduate-level teachers hope to pass on to graduate students entering the field. In *The Structure of Scientific Revolutions,* Thomas Kuhn (1970) explained that a discipline develops coherence through a shared "research tradition" that is transmitted to new members of the field through an agreed-upon body of scholarship (p. 11). As Collin Gifford Brooke (2011) explained, scholars have mental maps of a discipline as a network, privileging certain texts as more central to the discipline and making connections between texts. We hope to explore how these networked maps are transmitted to graduate students and if there are shared understandings of the discipline (that is, coherence) across the field.

In this chapter, we use the concepts *coherent* and *diffuse* to discuss disciplinarity. While disciplines are described and defined in a variety of (sometimes conflicting) ways, one common identifier is the coherence of a shared body of knowledge or texts. For example, Annette Shelby (1996) wrote, "the notion of a discipline implies the existence of a coherent—though necessarily dynamic—body of knowledge organized around central theoretical propositions and paradigms that are subject to ongoing challenges and necessary revision" (p. 99). These theoretical propositions and paradigms are often conveyed through a collection of texts, which are sites of knowledge-making practices for disciplines (Hyland, 2004) and assist in the work of enculturating new members of the field into the discipline (Kuhn, 1970; Toulmin, 1972). Thus, we understand disciplinary *coherence* as marked by agreement about a set of texts foundational to the field, what Smith (2000a, 2004) called "points of reference." While we use *diffuse* somewhat in contrast to *coherence*, we also want to caution that these two concepts are not dichotomous, as many disciplines are both coherent and diffuse. A healthy discipline, we believe, has a degree of coherence around a recognizable body of shared disciplinary knowledge and a degree of diffusion. As Gwendolynne Reid and Carolyn R. Miller (2018) observed, "all disciplines can usefully be thought of as 'diffuse'" because of new avenues of research and overlap or networked relationships with other disciplines (p. 105). Thus, the question is not whether technical communication as a field is diffuse, but rather if it is *too diffuse* so that it lacks coherence.

We begin this chapter by providing a sketch of concerns about technical communication's coherence over the last few decades. We then provide a discussion of our methods and methodology; we argue for a mapping approach to understanding a scholarly field that draws on the methods of co-citation network analysis. While we are not analyzing citation networks, and are instead analyzing what texts are assigned in graduate-level courses, we find the methods of co-citation network analysis useful in mapping the landscape of a discipline. After we overview our data collection methods and present network graphs created in Gephi (open-source network analysis software), we develop co-citation maps to determine the field's coherence and to locate an updated list of points of reference that help to constitute technical communication as a discipline. We then link our findings to the questions about technical communication's coherence as a field. By using co-citation network analysis, we can better understand the maturity of the field.[1]

■ Technical Communication: Coherent or Too Diffuse?

As with any new discipline, technical communication has grappled with how coherent or diffuse its body of scholarship is: is there a shared textual tradition that provides the field with coherence, or is the discipline too diffuse and dispersed with a wide array of interdisciplinary traditions that prevent a shared research agenda? In their *Technical Communication* article, for instance, Thomas E. Pinelli and Rebecca O. Barclay (1992) questioned if technical communication was too interdisciplinary, lacking "a substantial, coherent, and esoteric body of

---

1. We became interested in this project after reading Dan Wang's (2012) co-citation analysis of sociology syllabi during Collin Gifford Brooke's "Rhetorics and Networks" workshop at the 2015 Rhetoric Society of America Summer Institute. Greg was slated to teach Foundations of Technical Communication in Fall 2015, and we thought this was an opportunity to not only study the field from another angle, but also introduce graduate students to both the complexity of the field and the challenges of data collection, entry, and coding. After we collected an initial sample of 24 syllabi, we worked with graduate students (at both the M.A. and Ph.D. level) in Greg's course to create a data entry schema and asked each student to enter data into spreadsheets for one or two syllabi. Michael then cleaned some of the data (ensuring consistency across the data) and shared some initial findings from the social network analysis with the class later in the semester. This activity was a useful one for students, as it helped reveal that the field is interpreted in different ways by different teachers, yet there are also recognizable trends in how to approach introducing graduate students to the field. Additionally, it served as an introduction for many graduate students to replicable methods and data coding and entry.

specialized knowledge," without which research is "fragmented" (p. 528). While some scholars in the late 1990s and early 2000s argued that the field had developed "disciplinary maturity" (Staples, 1999, p. 153), that technical communication journals had "become more academically rigorous" (Smith, 2000b, p. 169), and that doctoral students' research was robust and thriving (Rainey, 1999), concerns about disciplinary coherence, and thus legitimacy and identity, continued. Billie Wahlstrom (1997) noted that despite these successes, research in technical communication lacked "a unifying vision . . . [which] has hurt technical communication's development of a coherent and rigorous research agenda" (p. 307). Laurel Grove and Donald Zimmerman (1997) wrote, "technical communication needs to emerge as a legitimate and respected academic research discipline" (p. 157), suggesting that the field "must identify the body of knowledge that summarizes its most influential and scientifically sound research and practical application guidelines" (pp. 158–159). To do otherwise, they argued, would risk technical communication remaining "undisciplined" (p. 159). And in the introduction to *Reshaping Technical Communication*, Barbara Mirel and Rachel Spilka (2002) wrote about the field's "identity crisis," expressing concern that disparate research projects wouldn't cohere "toward a common objective" (p. 4). Gerald Savage (1999) added that "academics and practitioners are not clearly related by a common body of knowledge" (p. 369). While prospects were good for continued robust research, technical communication scholars were still concerned about the field's coherence, status, and value around the turn of the century.

If more recent scholarship is any indication, these concerns continue today. Respondents to Ann Blakeslee's (2009) questionnaire about technical communication research expressed that the field lacked a coherent research agenda. Carolyn Rude (2015) challenged the field to mend the growing gap between research and practice and expressed concern about the diffuse research in technical communication: "Diffusion comes with the cost of identity and impact" (p. 370) and "Isolated projects do not readily create a coherent whole that contributes to what we mean academically by technical communication" (p. 375). Elsewhere, Rude (2009) suggested that the field lacked coherence in part because it had not yet identified a set of "overriding research questions" (p. 174). Most recently, Kirk St.Amant and Lisa Melonçon (2016) observed that the field "has a problem of incommensurability" "due to a lack of common, unifying goals"; thus, there's a need for the field to develop some "common ground" (p. 270). Clearly, the coherence of technical communication as a field continues to be a concern of scholars: without this coherence, the field lacks disciplinary identity and status.

Importantly, new members of a discipline are enculturated into the field through graduate education. Learning the shared concepts, questions, methods and—significantly—textual traditions occurs in part during graduate school. Particularly, a course like Foundations of Technical Communication makes an argument to students (as well as to other stakeholders) that *this is the tradition of the field from which we build*. Courses like this introduce students to texts that serve as transmits—students are introduced to points of reference that help to enculturate them into the discipline (Toulmin, 1972). We propose that one way to study the field's coherence or diffusion—and thus status—is to attend to the arguments made by graduate-level syllabi about what constitutes the field and what kind of scholarly conversations graduate students are introduced to. In this chapter, we study these syllabi and "map" the field through methods of co-citation network analysis.

■ Mapping a Discipline and Co-Citation Network Analysis

Cartographic metaphors and mapping practices have become common methodological approaches and metaphors for understanding fields, disciplines, and curricula in both technical and professional communication and rhetorical studies (e.g., Glenn, 1997; Mueller, 2017; Peeples & Hart-Davidson, 2012; Rude, 2009; Slack, 2003; Sullivan & Porter, 1993, 1997; Tirrell, 2012; Unger & Sánchez, 2015; Yeats & Thompson, 2000). Following Patricia Sullivan and James E. Porter, we understand mapping as a postmodern methodology that doesn't seek to represent a "static reality" (1997, p. 79) but rather allows for "a dynamic pluralism" (1993, p. 392). Thus, we attempt to map technical communication as a discipline by *locating* its textual traditions rather than attempting to provide a "common meaning" of technical communication that "exclud[es] enriching diversities" (Sullivan & Porter, 1993, p. 391).

Mapping, too, has been a common approach in information sciences, where researchers map scholarship using formal methods to provide "spatial representation[s] among disciplines, fields, specialties, and individual papers (or authors)" (De Bellis, 2009, p. 142). In his overview of bibliometrics and citation analysis, Nicola De Bellis (2009) explained that mapping methods help to describe "the intellectual structure of a research area" by "tracing and evaluating the relative position and strength of the actors on a stage" and to empirically test "such abstract constructs as 'discipline,' 'specialty,' 'paradigm,' and 'scientific community'" (p. 142). In order to metaphorically map the terrain of technical communication as a discipline, we deploy the methods of co-citation network analysis, which we borrow from information sciences.

Co-citation analysis, first proposed by Henry Small (1973), explores the relationships between documents (or authors or journals) that are cited together in subsequent texts. By the time of Small's innovation in the 1970s, information science scholars had been studying bibliometric citations in order to evaluate the impact and importance of scientific literature for nearly two decades. Eugene Garfield's 1955 article "Citation Indexes for Science" had argued that an index of citations would better reflect knowledge production than subject heading indexes, which relied heavily (in the pre-digital print era) on a limited terminology for subjects developed by professional indexers. Citation indexes helped information science scholars to situate authors and texts within networks of knowledge production and "to evaluate the significance of a particular work and its impact on the literature and thinking of the period" (Garfield, 1955, p. 109). Subsequent scholars began to analyze citations in terms of *networks*. Derek J. de Solla Price's (1965) influential work analyzed citation distribution in scientific papers, showing how "each group of new[ly published] papers is 'knitted' to a small, select part of the existing scientific literature but connected rather weakly and randomly to a greater part" (p. 149). That is, de Solla Price's analyses of citation networks showed that there was a body of work within the network of scientific literature that was heavily cited—"classic" literature—and the more "ephemeral" work that composed the majority of scientific literature but was not heavily cited (p. 149).

Citations (and, in our study's case, reading lists on syllabi), we contend, are important for understanding a discipline because they help to reveal how a discipline acknowledges a tradition and builds off this tradition. Citation analysis is not a new method to rhetoric and composition or technical communication scholars. Rhetoric and composition scholars have studied citation counts in *College Composition and Communication* (*CCC*) and *Rhetoric Society Quarterly* to explore disciplinary questions about composition studies (Detweiler, 2015; Goggin, 2000; Mueller, 2012; Phillips et al., 1993). Derek Mueller's (2012) work has perhaps been most influential: by graphing the frequency of authors cited in a 25-year span of *CCC* articles, he showed that rhetoric and composition has a "long tail" of cited scholars, which suggests that the field has become diffuse with disciplinary breadth and specializations, an aspect of the discipline that must be grappled with in graduate education (pp. 207–219). Technical and professional communication scholars have also turned to citation analysis in order to explore disciplinary status and the maturation of disciplinary journals (see Reinsch & Lewis, 1993; Reinsch & Reinsch, 1996). Smith's (2000a, 2004) work has perhaps been most ambitious, mapping technical communication through citation analysis and developing the field's major points of reference in scholarship in the 1980s and 1990s. Scholars like Smith, Mueller, and others have largely focused on citation counts of journals, authors, or texts in their citation analyses, and they have mostly relied on tables and bar, line, and plot graphs to visualize their data. Smith, whose citation analysis used percentages and comparisons of how frequently journals and serials were cited in the pages of technical communication journals, encouraged scholars to turn to other analytic methods to study citations and "map" connections (2000b, p. 175).

■ Co-Citation Network Analysis as an Inventional Heuristic

In contrast to these approaches, we draw on co-citation network analysis to study texts assigned in graduate-level syllabi. Small's (1973) proposal was that studying co-citation networks might help to develop a more detailed map of a field than crude citation counting. As he wrote, "If it can be assumed that frequently cited papers represent the key concepts, methods, or experiments in a field, then co-citation patterns can be used to map out in great detail the relationships between these key ideas" (pp. 265–266). Co-citation network analysis draws on the analytic methods of social network analysis in order "to trace the map of relationships among . . . key documents/key concepts, to outline and graphically visualize the structure of a research field, its connections with other fields, and its articulation into subfields and new research fronts" (De Bellis, 2009, p. xxvi). Social network analysis is comprised of a set of analytic strategies and theoretical approaches used to study the relationships of a set of "nodes" that are connected by links or "edges" (Barabási, 2002; Frith, 2014; Kadushin, 2012; Scott, 2012). Scholars in information and library science and in the digital humanities have analyzed citations using social network analysis, understanding citations as a form of network building (De Bellis, 2009; Healy, 2013; Otte & Rousseau, 2002; Wang, 2012). While citation network analysis has historically focused on citations in scholarly journals, Dan Wang (2012) proposed that studying syllabi instead of scholarship is helpful in exploring questions of disciplinarity for three reasons: 1) Unlike published scholarship, syllabi are meant to introduce the contours of a field to newcomers, 2) "syllabi offer insight into the courser divisions of a field because they are meant to summarize major research agendas," and 3) syllabi impact the development of a field "by forming consensus about the origin of ideas within a field" (p. 2).

Wang (2012) created a co-citation network of texts assigned in 52 syllabi from sociology courses to answer the question "Is there a canon in economic sociology?" Co-citation analysis explores the frequency of how often two texts or authors are cited together in later works. Whereas a traditional citation network includes *directed* edges from an article to a text it cites, a co-citation network creates an *undirected* edge between two texts if they are cited together. The motivations for using co-citation networks to study citations are that co-citation networks move us beyond crude citation counts (though these can be useful, as the studies cited above show) and allow us to visualize conversations or important topics in a field. Wang used a co-citation network in his study to calculate texts' relevance to the network: using algorithms to measure a text's authority (how often other texts linked to it) and status as a hub (how often it linked to texts with authority), Wang identified "a rather select canon of references in economic sociology" (2012, p. 4).

Of course, co-citation network analysis, like other quantitative approaches, risks flattening complex relationships (Frith, 2014; Fuhse & Mützel, 2011; Johnson, 2015). Just as there are a variety of reasons to cite a text in an article—to situate an argument, to build on the ethos of other scholars, to mark a claim as tentative (rather than a fact), to meet the perceived expectations of a journal editor or reviewers, to engage in-depth with another's ideas, and so forth (Cozzens, 1989)—there are many reasons to include a text on a syllabus: it may be foundational to a scholarly conversation, it may provide an example of a method or approach, it may be future-oriented and lay out a research agenda, it may provide a synthesis of research or perspectives to help orient students to a field, and so on.

A quantitative approach to citation network analysis ignores these complexities and particularities.

However, while social network analysis certainly risks missing nuance and context, it also provides a heuristic for researchers to invent and generate new questions. Mueller (2017) suggested that methods that map, graph, or otherwise visualize a field can serve as an inventional heuristic to raise questions about a field or discipline, providing "inventive and generative capacity" (p. 105). Co-citation network maps of the field can help us develop what Mueller called "network sense," "incomplete but nevertheless vital glimpses of an interconnected disciplinary domain focused on relationships that define and cohere widespread scholarly activity" (2017, p. 3). As Mueller explained, such maps can help scholars to recognize patterns in a field or discipline, "foster[ing] network sense" and offering us the opportunity to see a field differently and raise new questions about the field (p. 62). Thus, as we analyze data from our corpus and use graphs of our co-citation network, we use these visualizations to raise questions about texts and the field, attending to what Mueller and digital humanities scholar Matthew G. Kirschenbaum (2007) called "provocations," those "invitations to invent" that arise from data, rather than seeing the data as a form of "proof" about the field (Mueller, 2017, p. 4).

■ Methods: Data Collection

To collect graduate-level syllabi for foundations courses in technical communication, we conducted a web search and requested syllabi through an IRB-approved process (protocol #505361 at Texas Tech). We searched the web pages or online course catalogs of 110 Ph.D., master's, and graduate certificate programs to see which programs offer graduate courses that provide students with a scholarly focused introduction to the field. We were ultimately looking for the types of courses that Johnson-Eilola and Selber (2001) identified as those that "provide new members of the field with a broad (if fluid) map that helps them develop new knowledge in the context of other knowledges" (p. 420). Thus, we were not interested in more specialized courses (e.g., rhetoric of health and medicine or publications management), courses that focused primarily on technical communication practices or genres, or practicum courses designed to cover the day-to-day teaching of technical or professional communication. Of the 110 programs we searched—a list we developed from Lisa Melançon's (2009) and Dave Yeats and Isabelle Thompson's (2010) lists of programs and by searching additional programs we felt might include such a course—a maximum of 77 programs offer this type of course. This number is likely higher than actual offerings: course descriptions are often vague and many programs do not include syllabi online, so it was not always possible to tell if a graduate course titled something like "Introduction to Technical Communication" was more likely to be practice-based or to be more "three-dimensional," introducing students to the "thinking, doing, and teaching" of technical communication (Johnson-Eilola & Selber, 2001, p. 415). To keep our dataset current, we limited syllabi that we would include to the eight-year period between Fall 2008 and Spring 2017.

We collected syllabi using three methods: 1) we searched the web for syllabi that had been posted publicly online; 2) we sent a request for syllabi out on disciplinary listservs and through our personal social media accounts (Facebook and Twitter); and 3) we emailed professors directly to request syllabi from programs that were not represented in our web search or initial public requests. Our requests explained that we were looking for graduate-level syllabi with titles such as Foundations of Technical Communication, Research and Theory in Technical/Professional Communication, and History of Technical Communication. We specified that we were looking for syllabi that included both a course description and reading list of assigned texts.

Our web search and solicitations resulted in a corpus of 60 syllabi from 45 institutions, representing 49 different courses taught by 56 different professors. Table 3.1 represents our search for syllabi and the results of that search. Our dataset represents 50 percent of programs we searched with a Ph.D. program and 33.9 percent of programs we searched that have a master's program or graduate certificate but no Ph.D. program. Eleven programs are represented twice in our dataset and two programs are represented three times because we received or found syllabi that we deemed substantially different. Five of these programs had two different courses that met our criteria (often one more theory focused and one more pedagogically focused). The other eight have one course, but we found or received two or three syllabi taught by different instructors. (One syllabus in our dataset is a course revision proposal.)

Table 3.2 shows the various foci of the courses based on the course titles. The variety of course titles reveals a lack of consensus on the name of the field: *technical* versus *professional* and *communication* versus *writing*. (While Sullivan and Porter (1993) argued for understanding professional and technical communication as different fields, with professional writing more aligned with humanism and English studies and technical writing more aligned with technical fields, Melonçon (2009) noted that "this distinction does not necessarily hold in terms of degree names," nor is it "one reflected in curriculum" (p. 138). Also, see Melonçon for a discussion of degree program names regarding "writing" versus "communication" and the inclusion of "rhetoric" in degree names.) Additionally, nine of the courses focused, at least in part, on teaching technical or professional communication/writing. Table 3.3 shows the programmatic locations of these courses: most of these courses were housed in English departments, though some courses were from engineering, humanities, interdisciplinary, or stand-alone technical communication programs. It is also worth noting that some programs do not have, or do not require, a foundations-style course. In Melonçon's (2009) study of master's programs, 62 percent required an introduction to the field of technical communication course, 7 percent offered the course as a concentration, and 1 percent offered the course as an elective.

Table 3.1. Number of programs we searched and numbers of programs, syllabi, courses, and instructors represented in our study

	Program type		Total
	Master's program or a graduate certificate in technical or professional writing (and no Ph.D. program)	Ph.D. program in English, rhetoric, technical communication, or similar field	
Programs searched	62	48	110
Programs that may have a foundations-style course	37	34	71
Programs included in our study	21	24	45
Number of syllabi from included programs	27	33	60
Courses represented in syllabi	23	26	49
Instructors represented in syllabi	24[a]	32	56

[a] One syllabus from an M.A. program was a course revision proposal, so we did not attribute it to a specific instructor.

Table 3.2. The foci of courses, based on course titles, in our study

Course focus based on the course title	Number of syllabi represented
Technical writing or communication (including prefixes like *foundations in*, *introduction to*, or *principles of*, and including terms like *theory*, *research*, *history*, or *practice*)	20
Professional writing or communication (including terms like *rhetoric*, *theory*, or *research*)	19
Professional and technical communication or writing (including terms like *theory* or *practice*)	5
Teaching technical communication or writing	3
Teaching professional writing (including terms like *theory*)	3
Teaching technical and professional writing (including terms like *theory* or *methods*)	2
Teaching business and technical writing	1
Other foci (these course titles usually affixed an additional key term to a title above, like *technology studies*, *scientific communication*, *writing studies*)	7
Total	60

Table 3.3. Program and institution types of courses represented in our study

Institution type (Carnegie Classification)	Program type	Number of institutions represented	Number of syllabi collected
Research institution (R1, R2, or R3)	Ph.D. in rhetoric and composition or rhetoric and writing (including listed as an emphasis or concentration) in an English department	14	18
	Ph.D. in English (emphasis or concentration not listed on program's website)	1	1
	Ph.D. in technical communication and rhetoric or rhetoric and professional communication (including listed as an emphasis or concentration) in an English department	5	9
	Ph.D. in engineering, interdisciplinary Ph.D. program, or other non-English field	3	4
	Ph.D. in technical communication in a technical communication department	1	1
	Master's in English (may have a technical communication graduate certificate)	4	4
	Master's in rhetoric and writing in a rhetoric and writing program	1	1
	Master's in technical or professional communication in an English department	3	4
	Master's in technical and/or professional communication in an interdisciplinary, engineering, or technical communication department	3	4
	Master's in communication in a communication department	1	1
Master's degree granting institutions	Master's in English or in writing in an English department	2	2
	Master's in technical and/or professional communication in an English department	3	6
	Master's in technical and/or professional communication in an interdisciplinary or technical communication department	3	4
Bachelor's degree granting institutions	Master's in writing in an English departmentæ	1	1
Total		45	60

Once syllabi were collected, all identifying information (like institution, instructor name, office hours, contact information, and similar information) was removed and syllabi were renamed "Syllabus A" through "Syllabus Z," and then doubling and then tripling letters (e.g., AA, BB, . . . AAA, BBB). With the help of students in Greg's 2015 graduate seminar, Foundations of Technical Communication, we entered each syllabus' assigned readings into a spreadsheet. The spreadsheet included columns for the following information:

- syllabus name (e.g., "Syllabus A")
- reading assigned
- the reading's original publication date
- the name of the anthology if the reading was a reprint or in an edited collection

We developed a scheme for entering the assigned readings into our spreadsheet so that our software (Gephi) would understand each entry of a reading as the same. When syllabi were unclear about the title of a reading, we were often able to make inferences about which text was assigned, and we occasionally contacted instructors to ask for clarification on an assigned text. We excluded texts from the spreadsheet that were listed as optional but included readings that were assigned to individual students. For example, if an instructor listed ten readings that she or he assigned to individual students to read and present on to the class, those texts were included in our data.

Readings were entered in the spreadsheet as *Author last name, First four words of the title*. In order to be consistent with these entries, we developed the following rules:

- Use sentence case for titles (only capitalizing first words and proper nouns).
- Use ampersands and Oxford commas when there were two or three authors.
- Use the first author's name and *et al.* without a comma if there were four or more authors.
- End titles before punctuation other than commas (e.g., colons and dashes).
- Remove prepositions, conjunctions, and articles from the end of excerpted titles.

So, for example, Miller's "A Humanistic Rationale for Technical Writing" was entered as "Miller, A humanistic rationale," and Slack, Miller, and Doak's "The Technical Communicator as Author: Meaning, Power, Authority" was entered as "Slack, Miller & Doak, The technical communicator." Some texts required us to deviate from this practice. For example, Thralls and Blyer's "The Social Perspective and Pedagogy in Technical Communication" and "The Social Perspective and Professional Communication" would have resulted in the same node title. In this instance, we added two words—"and pedagogy" and "and professional," respectively. In other instances where confusion might arise, we added parenthetical years to the entry. We then proofread the spreadsheet to ensure consistently entered titles.

The 60 syllabi in our dataset assigned a total of 1,956 texts, averaging 32.6 texts per syllabus. The amount of reading assigned varied considerably: the syllabus with the most readings included 81 texts, eight syllabi assigned between 50 and 75 texts, ten assigned 40–49 texts, 11 assigned 30–39 texts, 15 assigned 20–29 texts, and the remaining 15 assigned fewer than 20 texts. One syllabus included only one required reading (Michael Hughes and George Hayhoe's *A Research Primer*) and had many readings listed as "to be announced." The 1,956 readings amounted to 978 unique texts (articles, book chapters, and monographs). The vast majority—720 of them—were assigned only once each. Of the remaining 258 texts, 103 were assigned in two syllabi, 46 were assigned three times, 33 were assigned four times, 41 were assigned five to ten times, 19 were assigned 11–15 times, and six were assigned 16 or more times.

■ Methods: Creating the Co-Citation Network

To develop our co-citation network, we reorganized our data into a comma-separated values (CSV) file. Each line in this file represented a pair of readings that was assigned together on the same syllabus. For example, Syllabus A assigned 63 different texts. When this data was entered into our CSV file, data from Syllabus A resulted in 1,953 combinations of texts that were assigned together. The resulting CSV file for the whole dataset, which included 39,714 entries connecting co-cited texts, was then entered into Gephi, an open-source social networking analytic software.

Once in Gephi, we applied a variety of social network analytics to the dataset. Of particular importance to our study, we applied the following:

- Degree and weighted degree for texts in the network. A text's (or, in network terminology, a node's) degree in a co-citation network tells us how many other texts it was assigned with in the network. Its weighted degree tells us how frequently it was assigned along with those other texts (Scott, 2012). For example, Miller's "A Humanistic Rationale" was the most frequently assigned text in the dataset (assigned 35 times). In the co-citation network, it had a degree of 648, meaning it was assigned in syllabi along with 648 other texts. Its weighted degree was 1,273, meaning that it was assigned with the same texts multiple times (e.g., Miller's article was assigned with Katz's "The Ethic of Expediency" 21 times).
- Authority algorithms. In social network analysis, authority algorithms measure how important and influential a node is to a network. Authority algorithms (like Google's PageRank) measure a node's importance based on the importance of the other nodes it's connected to. These algorithms calculate authority by analyzing the link or edge structure of a network, determining authority through recursively analyzing the data (Kadushin, 2012; Wang, 2012). To determine a text's authority, we used Jon M. Kleinberg's (1999) Hypertext Induced Topic Selection (HITS) algorithm in Gephi.
- Community detection algorithms. Community detection algorithms determine "communities" or subgraphs within a network. In Gephi, we used

Vincent D. Blondel et al.'s (2008) modularity class algorithm to determine "sub-units or communities, which are sets of highly interconnected nodes" (p. 2) in the co-citation network. Using a community detection algorithm allowed us to see if groups of texts seemed to be assigned together quite frequently, and to speculate if there are "conversations" or common areas of interest or topics in the dataset.

The Co-Citation Network and Authoritative Texts in Technical Communication

The resulting co-citation network is visualized in Figure 3.1. Because this co-citation network is quite large (978 texts, or nodes, connected by 31,936 edges, or links), we have applied a filter to the visualization to make it more legible and less cluttered. Figure 3.1 displays nodes only if they have an edge weight of at least two (that is, they were assigned together at least twice) and consequently only displays 247 of the 978 texts in the network.

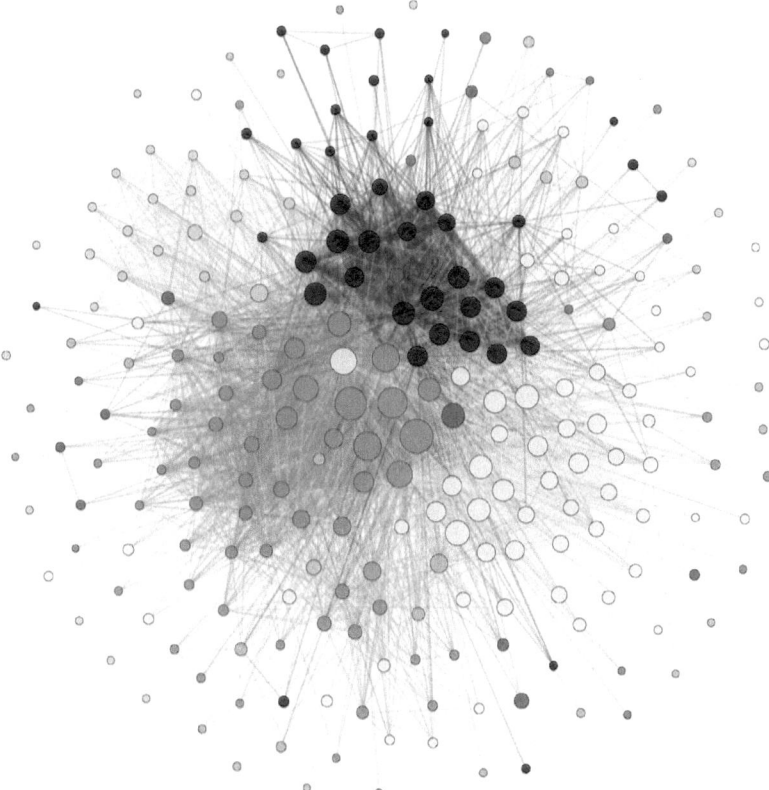

Figure 3.1. The co-citation network for our dataset, filtered to show nodes only if they have an edge weight of at least two. Different colors represent different modularity classes, and node size is larger if the text has more authority in the network.

While this graph is still visually busy, making it difficult to say too much about it without turning to analytic data, it does provide a high-level "map" of the discipline that allows us to quickly observe a few takeaways: First, many texts in the dataset are assigned together infrequently, resulting in many texts on the edges of this visualization that aren't as central or authoritative. Second, there does appear to be a group of texts that are more authoritative to the network than others. And third, some texts are assigned together quite frequently, and some communities of text seem to have emerged in this network. Figure 3.2 provides a more focused visualization of the co-citation network, showing just the 102 most authoritative texts in the network that were assigned in at least four syllabi in the dataset. In Figure 3.2 and Table 3.4, we chose to include only texts that were assigned by at least four syllabi because we wanted to mitigate the influence of some syllabi that assigned many different texts. One limitation of using HITS authority algorithms (and other algorithms as well that measure authority, centrality, or influence) is that syllabi that assign more texts have more influence on the co-citation network than syllabi that assign fewer texts. For example, Yrjö Engeström's "Activity Theory and Individual and Social Transformation" and Clay Spinuzzi's *Network* both have strong HITS authority scores but were only assigned on two syllabi each. They earned high authority scores in the algorithm because they were assigned along with many other texts that were assigned frequently in the network: a syllabus with 75 readings assigned both texts; another with 81 readings (the most in the dataset) assigned Engeström's chapter; and a third that assigned *Network* had 52 readings. Consequently, we decided to include only texts that were assigned by at least four syllabi in Figure 3.2 and Table 3.4.

Table 3.4 provides a list of these 102 texts, along with their original publication date and whether they were included in Smith's (2000a) list of 163 points of reference for the field and her subsequent (2004) list of 26 points of reference. Notably, only 21 of these 102 texts were in Smith's (2000a) list of 163 points of reference. And of the 26 texts Smith (2004) listed in her more "magnified view" (p. 53) of the field, only nine continue in our list. If our sampling of graduate-level syllabi is any indication, the field has changed in the two decades since Smith's citation analyses. (We speculate on reasons why later in this chapter.) But also, a few texts have remained quite central to the field over the years. For instance, 35 of the 60 syllabi we collected assigned Miller's "A Humanistic Rationale," and it is the most authoritative text in the co-citation network. The status of Miller's essay in this network is unsurprising: Smith's (1997) analysis of intertextual connections to "A Humanistic Rationale" showed just how influential the essay was to knowledge creation in the field. Scholars in technical communication would likely express no surprise at other texts that have also remained central to the field since the late 1990s. For example,

Robert Connors' essay, "The Rise of Technical Writing Instruction in America," has been influential in understanding the history of technical communication instruction. Articles by the likes of David Dobrin, Stephen Katz, Carolyn R. Miller, Cezar Ornatowski, Russell Rutter, and Dale Sullivan have also shaped the field's views of the rhetorical and ethical aspects of technical communication. Johnson-Eilola's and Slack, Miller, and Doak's articles have influenced how we understand the role of technical communicators as knowledge workers. And Doheny-Farina's, Selzer's, and Winsor's studies of workplace writing helped to shift the field's attention from pedagogy to the contexts of writing in professional settings.

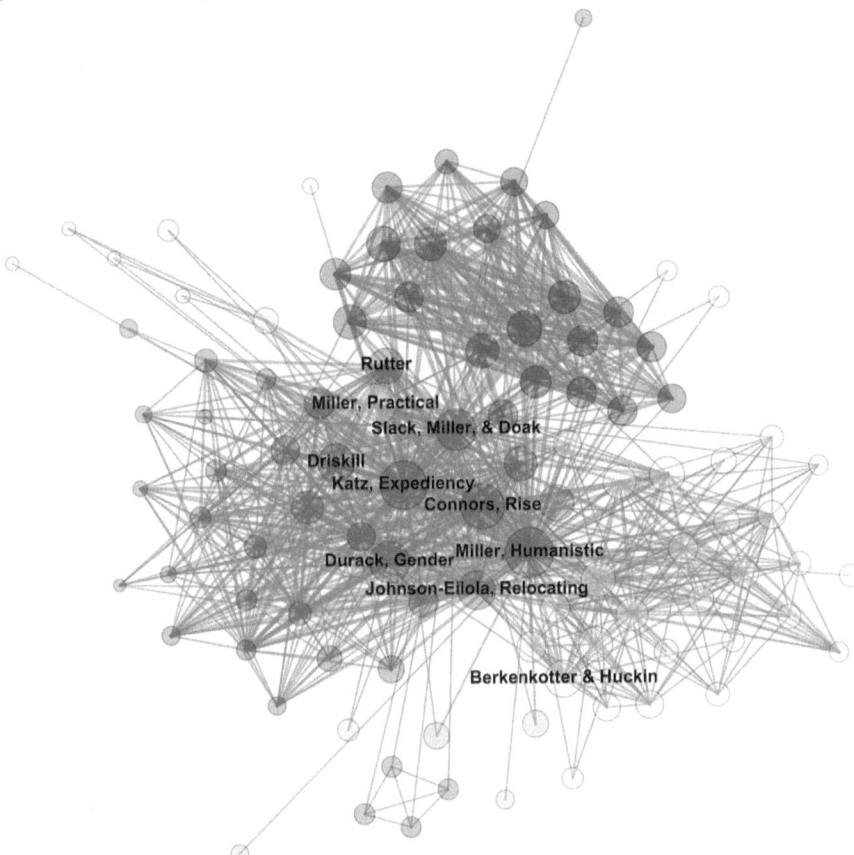

Figure 3.2. The co-citation network for our dataset, filtered to show the 102 most authoritative texts in the network that were also assigned in at least four syllabi. Different colors represent different modularity classes, and node size is larger if the text has more authority in the network. The ten most authoritative texts in the network are labeled.

Table 3.4. The 102 most authoritative texts that were assigned at least four times in our dataset, ranked by HITS authority score (Kleinberg, 1999)

Rank by HITS authority score					
In entire dataset	In subgraph without SPTC[a]	Text (original publication year)	Times assigned in dataset	In Smith's (2000a) list of 163 texts	In Smith's (2004) list of 26 texts
1	1	Miller, A humanistic rationale (1979)	35	x	x
2	2	Katz, The ethic of expediency (1992)	28	x	x
3	3	Connors, The rise of technical (1982)	23	x	
4	4	Durack, Gender, technology (1997)	16		
5	7	Slack, Miller & Doak, The technical communicator (1993)	22	x	
6	8	Miller, What's practical about technical (1989)	14	x	x
7	15	Johnson-Eilola, Relocating the value (1996)	17		
8	5	Driskill, Understanding the writing context (1989)	11		
9	10	Berkenkotter & Huckin, Rethinking genre (1993)	9		
10	11	Rutter, History, rhetoric, and humanism (1991)	15	x	
11	-	Hart-Davidson, What are the work (2013)	14		
12	18	Miller, Genre as social action (1984)	8	x	
13	9	Allen, The case against defining (1990)	13		
14	-	Mirel, How can technical communicators () 2013	13		
15	16	Breuch, Thinking critically about technological (2002)	9		
16	13	Moore, Myths about instrumental discourse (1999)	6		
17	*	Selfe & Selfe, What are the boundaries (2013)	14		
18	6	Dobrin, What's technical about technical (1983)	15	x	x

Rank by HITS authority score					
In entire dataset	In subgraph without SPTC[a]	Text (original publication year)	Times assigned in dataset	In Smith's (2000a) list of 163 texts	In Smith's (2004) list of 26 texts
19	-	Ceraso, How can technical communicators (2013)	11		
20	47	Blakeslee, Bridging the workplace (2001)	8		
21	*	Porter, How can rhetoric theory (2013)	14		
22	-	Schriver, What do technical communicators (2013)	12		
23	14	Johnson, Audience involved (1997)	12		
24	22	Redish, What is information design (2000)	7		
25	-	Spinuzzi, How can technical communicators (2013)	10		
26	-	Cargile Cook et al., How can technical communicators (2013)	9		
27	-	Scott, How can technical communicators (2013)	8		
28	-	Henry, How can technical communicators (2013)	10		
29	17	Selzer, The composing process (1983)	12	x	x
30	31	Rude, Mapping the research questions (2009)	11		
31	59	Lay, Feminist theory (1991)	9	x	
32	-	St. Amant, What do technical communicators (2013)	8		
33	-	Burnett, Cooper & Welhausen, What do technical communicators (2013)	11		
34	40	Freedman & Adam, Learning to write professionally (1996)	8		
35	*	Henze, What do technical communicators (2013)	11		
36	-	Blakeslee & Savage, What do technical communicators (2013)	9		

Rank by HITS authority score					
In entire dataset	In subgraph without SPTC[a]	Text (original publication year)	Times assigned in dataset	In Smith's (2000a) list of 163 texts	In Smith's (2004) list of 26 texts
37	-	Longo & Fountain, What can history teach (2013)	8		
38	38	Johnson, Complicating technology (1998)	6		
39	34	Thrush, Multicultural issues in technical (1997)	6		
40	48	Mirel, Advancing a vision (2002)	8		
41	50	Winsor, Engineering writing (1990)	10		
42	-	Wysocki, What do technical communicators (2013)	7		
43	20	Barton & Barton, Ideology and the map (1993)	15		
44	45	Bernhardt, Teaching for change, vision (1995)	6		
45	-	Johnson-Eilola & Selber, Introduction (2013)	10		
46	53	Kramer & Bernhardt, Teaching text design (1996)	7		
47	-	Mehlenbacher, What is the future (2013)	8		
48	23	Cargile Cook, Layered literacies (2002)	9		
49	55	Johnson, Johnson responds (1999)	4		
50	60	Johnson, User-centered technology (1998)	10[b]		
51	-	Swarts, How can work tools (2013)	8		
52	35	Selber, Beyond skill building (1994)	7		
53	43	Carliner, Computers and technical communication (2009)	5		
54	67	Jackson, The rhetoric of design (2000)	4		
54	67	Fukuoka, Kojima & Spyridakis, Illustrations in user manuals (1999)	4		

Rank by HITS authority score					
In entire data-set	In sub-graph without SPTC[a]	Text (original publication year)	Times assigned in dataset	In Smith's (2000a) list of 163 texts	In Smith's (2004) list of 26 texts
56	51	Ornatowski, Between efficiency and politics (1992)	8	x	
57	74	Gurak & Bayer, Making gender visible (1994)	4		
58	61	Herndl, Teaching discourse and reproducing (1993)	5		
59	32	Wilson & Herndl, Boundary objects as rhetorical (2007)	4		
60	82	Spilka, Communicating across organizational boundaries (1995)	4		
61	83	Russell, The ethics of teaching (1993)	4	x	
61	83	Porter, The exercise of critical (1998)	4		
63	-	Dicks, How can technical communicators (2013)	7		
64	12	Sullivan, Political-ethical implications (1990)	10	x	x
65	*	Spinuzzi, Pseudotransactionality, activity theory (1998)	5		
66	29	Bernhardt, The shape of text (1993)	9		
67	*	Selber, Johnson-Eilola & Selfe, Contexts for faculty professional (1995)	4		
68	36	Sullivan & Porter, On theory, practice (1998)	11		
69	52	Zoetewey & Staggers, Teaching the Air Midwest (2004)	5		
70	42	Bitzer, The rhetorical situation (1968)	4		
71	65	Spilka, Orality and literacy (1990)	4		
72	28	Dubinsky, Becoming user-centered, reflective practitioners (2004)	4		
73	*	Moses & Katz, The phantom machine (2006)	4		

Rank by HITS authority score		Text (original publication year)	Times assigned in dataset	In Smith's (2000a) list of 163 texts	In Smith's (2004) list of 26 texts
In entire dataset	In subgraph without SPTC[a]				
74	39	Selfe & Selfe, The politics (1994)	9		
75	24	Blakeslee, Addressing audiences (2009)	4		
75	24	Salvo & Rosinsky, Information design (2009)	4		
75	24	Thatcher, Understanding digital literacy (2009)	4		
78	21	Thralls & Blyler, The social perspective and professional (1993)	9		
79	27	Clark, Shaped and shaping tools (2009)	4		
80	58	Rude, The report for decision (1995)	8		
81	*	Paradis, Text and action (1991)	7	x	
82	*	Henry, Writing workplace cultures (2001)	4		
83	*	Selfe & Hawisher, A historical look (2002)	4		
84	88	Thralls & Blyler, The social perspective and pedagogy (1993)	7		
85	30	Charney, Empiricism is not (1996)	11		
86	62	Doheny-Farina, Writing in an emerging (1986)	8	x	x
87	44	Longo, *Spurious coin* (2000)	4		
88	63	MacKinnon, Becoming a rhetor (1993)	4	x	
89	41	Brasseur, Contesting the objectivist paradigm (1993)	6		
90	85	Wolfe, How technical communication textbooks (2009)	6		
91	*	Hallenbeck, User agency, technical communication (2012)	5		
92	*	Howard, Who "owns" electronic texts (1996)	6		
93	75	Dragga & Voss, Cruel pies (2001)	4		

Rank by HITS authority score		Text (original publication year)	Times assigned in dataset	In Smith's (2000a) list of 163 texts	In Smith's (2004) list of 26 texts
In entire dataset	In subgraph without SPTC[a]				
94	37	Allen et al., What experienced collaborators say (1987)	5	x	x
95	*	Grabill & Simmons, Toward a critical rhetoric (1998)	4		
96	72	Bosley, Cross-cultural collaboration (1993)	7	x	
97	*	Harrison, Frameworks for the study (1987)	4	x	
98	*	Anson & Forsberg, Moving beyond the academic (1990)	4	x	
99	*	Katz, Writing review (1998)	4		
100	*	Foss, Foss & Trapp, Perspectives on the study (1985)	4		
101	*	Faigley, Nonacademic writing (1985)	5	x	x
102	*	Mirel, Writing and database technology (1996)	4		

[a] *Texts no longer in the network when syllabi that assigned* Solving Problems in Technical Communication (SPTC) *were removed are marked with a (-) and texts ranking below 100th are marked with (*).*

[b] *Johnson's* User-Centered Technology *(or chapters from it) was assigned in seven syllabi, and his chapter reprinted in Peeples's (2003)* PWR *was assigned in three syllabi.*

Also notable in Table 3.4 is the presence of every chapter and the introduction from Johnson-Eilola and Selber's (2013) *Solving Problems in Technical Communication (SPTC)*. Because of the dominance in the co-citation network of this relatively new collection, which "is for students who are learning about the field" (p. 1) and synthesizes scholarship in the field for new practitioners, we included a column in Table 3.4 that lists texts' authority ranking if syllabi that assigned *SPTC* were excluded from the network. Since *SPTC* was published recently, it may have been assigned frequently because teachers are testing out the book; it has not yet passed the test of time, and a reproduction of this study in a few years might find that the book has fallen off of syllabi. Another possibility is that professors are using this collection because of the chapters' strong synthesis of prior scholarship. Not only do the authors provide useful overviews of research and helpful heuristics, but they also model how scholarship can deploy literature reviews to do intellectual work. Fifteen of the 60 syllabi assigned *SPTC*, so the co-citation network is different if these syllabi are excluded: chapters from *SPTC* become less authoritative or aren't in the network at all (a few syllabi didn't assign *SPTC* but did assign photocopies of scans of a few chapters). Indeed, anthologies like *SPTC* and *CWTC* have quite a strong influence on this co-citation network—and consequently on graduate education in the field, as we address in the next section.

Communities in the Co-Citation Network and the Influence of Anthologies

As we mentioned in the previous section, we applied community detection algorithms (Blondel et al., 2008) to the co-citation network, hoping to learn if there were subsets of the co-citation network that might reveal "communities" within the field or perhaps even differing views of the field. For example, if many communities were detected that revealed complete separate sets of texts not connected to the rest of the graph, this would tell us that the field is rather disperse with little shared understanding of what shared texts constitute the field's "transmit" (Toulmin, 1972; Smith, 2004). Or, if communities were detected that seemed to be heavily connected to each other (usually through texts that served as hubs) but with quite a few texts not linked to each other, this would tell us that there was a core set of texts that the field largely shares but quite disperse ways of approaching the field outside of those texts. And potentially, these communities could tell us, based on the texts in the community, something about ways teachers of graduate courses understand the network of the field.

The latter of these two potential findings proved true: we identified 16 communities in the co-citation network, most of which were highly connected to each other, and interestingly, the community detection algorithm highlighted the influence of anthologies on this co-citation network (and thus, on graduate education in the field). Five anthologies or collections proved to be particularly influential: Johnson-Eilola and Selber's (2004) *CWTC* was assigned in 16 syllabi; their (2013) *SPTC* was assigned in 15 syllabi; James Dubinsky's (2004) *Teaching Technical Communication* (*TTC*) was assigned 11 times; J. Blake Scott, Bernadette Longo, and Katherine V. Wills' (2006) *Critical Power Tools: Technical Communication and Cultural Studies* (*CPT*) was assigned 5 times; and Tim Peeples' (2003) *Professional Writing and Rhetoric* (*PWR*) was assigned by 4 syllabi. A sixth collection, Rachel Spilka's (2009) *Digital Literacy for Technical Communication,* was also somewhat influential, as it was required in three syllabi, and some other syllabi assigned scans or photocopies from particular chapters. Some syllabi required more than one of these books; some listed them as suggested readings or books to own; and some listed assigned readings in ways that made it clear that texts were scanned or photocopied from these books (rather than provide the original when they were reprinted in these anthologies or collections).

The community detection algorithm reveals just how powerful anthologies and collections are in shaping how graduate courses transmit the field to graduate students. The algorithm detected 16 different communities in the network (based on the defaults of the algorithm in Gephi; admittedly, changing these defaults would detect smaller, and thus more, or larger, and thus fewer, communities). Some of these communities were small—a collection of texts assigned in only one syllabus. But others were quite large and centered around either these anthologies and collections (and texts assigned along with them) or around some central approaches in the field. Table 3.5 lists the major communities in the network along with the central topics and representative texts in the field.

Table 3.5. Communities in the co-citation network[a]

Color in Figures 3.1 & 3.2 (size)	Topics & description	Representative or notable texts
Blue (172 texts)	Johnson-Eilola & Selber's (2013) *SPTC*	Chapters from *SPTC*
Orange (164 texts)	Dubinsky's (2004) *TTC* and portions of Scott et al.'s (2006) *CPT*	Most chapters from *TTC* and *CPT* Cargile Cook, "Layered Literacies" Wilson & Herndl, "Boundary Objects as Rhetorical" Zoetewey & Staggers, "Teaching the Air Midwest"
Green (129 texts)	Johnson-Eilola & Selber's (2004) *CWTC* and Spilka's (2009) *Digital Literacy*	All chapters from *CWTC* and some chapters from Spilka's *Digital Literacy* Note: Chapters that were reprinted in both *CWTC* and either *TTC* or *PWR* were in this community.
Purple (73 texts)	Peeples' (2003) *PWR*	Most chapters from *PWR*
Brownish-Green (71 texts)	Rhetorical theory and method	Miller, "Genre as Social Action" Engeström, "Activity Theory and Individual" Ong, "The Writer's Audience" MacKinnon, "Becoming a Rhetor" Dragga & Voss, "Cruel Pies" Diehl et al., "Grassroots" Dicks, "Cultural Impediments to Understanding" Locker, "Will Professional Communication be" Bazerman, *Shaping Written Knowledge* Faber, "Professional Identities" Graham & Whalen, "Mode, Medium, and Message"

Color in Figures 3.1 & 3.2 (size)	Topics & description	Representative or notable texts
Sky Blue (62 texts)	Technology, methodology, and curriculum	Johnson, *User-Centered Technology*
		Longo, *Spurious Coin*
		Wolfe, "How Technical Communication Textbooks"
		Melonçon & Henschel, "Current State of U.S."
		Buchanan, "Declaration by Design"
		Potts, *Social Media in Disaster*
		Simmons & Zoetewey, "Productive Usability"
		Blakeslee, "The Technical Communication Research"
		Sullivan & Porter, *Opening Spaces*
		Porter & Sullivan, "Remapping Curricular Geography"
		Johnson et al., "User-Centered Technology in Participatory"
		Sullivan & Porter, "Remapping Curricular Geography"
		Palmeri, "Disability Studies, Cultural Analysis"
		Teston, "Moving from Artifact"
Red (53 texts)	Rhetoric: Methods, audience, and authorship	Ede & Lunsford, "Audience Addressed"
		Hughes & Hayhoe, *A Research Primer*
		Ornatowski, "Technical Communication and Rhetoric"
		Coney, "The Implied Author"
		Redish, "Understanding People"
		Coney, "Technical Communication Theory"
Dark green (51 texts)	Miscellaneous texts	Carliner, "Computers and Technical Communication"
Gray (34 texts)	Rhetoric and cultural studies, including chapters from *CPT*	Bitzer, "The Rhetorical Situation"
		Hellenbeck, "User Agency, Technical Communication"
		Chapters from *CPT*

[a] *Modularity class determined by Blondel et al.'s (2008) algorithm. Communities that represent texts assigned only once or twice are excluded from this list.*

One reason, then, that the list of the 102 most authoritative texts differs from Smith's (2000a) 163 points of reference in the field is because of the influence of these anthologies and collections. The publication of collections like *CWTC* and *TTC* made accessing and assigning texts easier, and the editors of these collections have likely had a strong hand in shaping what the field considers foundational texts. In this way, editors serve as what Maureen Daly Goggin (2000) called "discipliniographers," or those who help to write the discipline through their authority and role in publishing processes (p. 148). For instance, Jo

Allen's "The Case Against Defining Technical Communication," first published in 1990, is authoritative in our network but is not in Smith's (2000a) list, likely because Dubinsky (2004) chose to anthologize the essay in *TTC*. Ben Barton and Marthalee Barton's 1993 "Ideology and the Map" serves as a similar case: not in Smith's (2000a) list of points of reference but frequently assigned in these graduate courses because of its inclusion in *CWTC*.

These discipliniographers have been so influential on graduate courses in our dataset that, of the 102 most authoritative texts in the network listed in Table 3.4, all but ten of the texts were reprinted or originally published in one of the six collections listed previously. Two of these texts have been central to rhetorical theory more broadly: Miller's (1984) "Genre as Social Action" and Lloyd Bitzer's (1968) "The Rhetorical Situation." Four of these texts are from the first few years of the 21st century, published just before or as most of these anthologies were published: Kelli Cargile Cook's (2002) "Layered Literacies," Meredith Zoetewey and Julie Staggers' (2004) "Teaching the Air Midwest," Longo's (2000) *Spurious Coin*, and Sam Dragga and Dan Voss' (2001) "Cruel Pies." (We count Robert Johnson's 1998 *User-Centered Technology* as reprinted in these anthologies because one of the chapters was reprinted in *PWR*.) Only four articles in Table 3.4 were published after 2006 (except those from Spilka, 2009): Carolyn Rude's (2009) "Mapping the Research Questions," Greg Wilson and Carl Herndl's (2007) "Boundary Objects," Joanna Wolfe's (2009) "How Technical Communication Textbooks," and Sarah Hallenbeck's (2012) "User Agency." Likely in part because of the influence of anthologies in this dataset, the foundational texts in the field skew toward the 1980s, 1990s, and early 2000s.

Table 3.6 in the next section provides our updated list of points of reference for the field based on our dataset. We don't intend this list to be definitive, but rather a tentative list of texts that are influential on the field and are seen by teachers and publishers as important to transmit to graduate students entering the field. Table 3.6 also includes the publication locations of these texts based on the source of the text in teachers' syllabi, showing just how influential collections—especially *CWTC* and *TTC*—are on the field. For example, while the original printing of Miller's "A Humanistic Rationale" was in *College English*, it was most frequently assigned from *CWTC* or *TTC*. (Sometimes it was unclear what the source of a text was when it was assigned, so we included that as "unknown" in the final column of Table 3.6.)

The Field's Coherence and Diffusion: Toward Points of Reference in Technical Communication

While a vast number of texts in our dataset were assigned only once—720 of them, in fact—we want to propose that the core set of authoritative texts suggests that the field of technical communication has developed some coherence. As social network scholars have argued, most social networks are structured with a "long tail," meaning that a network is often structured around a few highly connected nodes and many nodes that are less connected (Anderson, 2004; Barabási, 2002; Kadushin, 2012; Mueller, 2012). In his study of citations in *CCC*, Mueller (2012) expressed concern that a long tail of cited authors might represent a diffuse field in rhetoric and composition with many different specializations and conversations. As he noted, one can see the discipline as coherent by focusing on the heavily cited authors in rhetoric and composition, or see a diffuse field by focusing on the long tail. It is only by looking at the relationship between heavily cited texts and the long tail that one can begin to describe the field. While Mueller took a diachronic approach to see how citation patterns in *CCC* shifted over time, we turn to social network analytics—particularly authority and community detection—to help us explore the relationships between texts in the network.

We don't automatically share anxieties that technical communication is diffuse based solely on the presence of a long tail in this network. For one, this distribution is typical of networks. But more specifically, a long tail makes sense: instructors might include a text on a syllabus for a variety of reasons, leading to numerous texts not shared across the dataset. An article could be assigned because it provides an example of a theoretical approach, not because it is foundational to the field. Additionally, instructors will have different theoretical or methodological approaches. For example, one syllabus assigned historical textbooks from the 19th and 20th centuries—ones not assigned by anyone else. Another syllabus is heavier on critical theory, assigning Althusser, Foucault, and Derrida, theorists not included in other syllabi. Further, some instructors might be exemplifying approaches through working on shared projects that require readings they wouldn't assign in another iteration of the course. Thus, we are not so concerned about the presence of a long tail in this network.

Most citation analyses in technical communication and in rhetoric and composition have focused on citation counts, asking which authors, texts, or journals are most frequently cited in a corpus of scholarship (Detweiler, 2015; Goggin, 2000; Mueller, 2012; Phillips et al., 1993; Reinsch & Lewis, 1993; Reinsch & Reinsch, 1996; Smith, 2000a, 2000b, 2004). Citation counts are useful for determining which texts are cited most frequently (or assigned most frequently in our dataset), but do not tell us much about how central or authoritative these texts are to the network. That is, citation counts do little to tell us about how a text is situated *in relation* to the rest of the network. Social network analytics—like authority algorithms—can provide us data that tells us how central or authoritative a text is in a network.

Table 3.6 lists our points of reference for technical communication in graduate education. This list is based on Table 3.4, removing chapters from *SPTC* and listing the 82 most authoritative texts in the co-citation network that were assigned at least four times. We have provided the full citation for each text (citations for other texts we mention in the dataset but that aren't listed in Table 3.6 are provided in the appendix). As we observed previously, collections and anthologies are highly influential on this list.

Mapping Technical Communication as a Field 97

Table 3.6. An updated list of 82 points of reference in technical communication and their publication locations in the dataset

Text	CWTC 2004	TTC 2004	CPT 2006	PWR 2003	Original, elsewhere, or unknown
Allen, Jo. (1990). The case against defining technical communication. *Journal of Business and Technical Communication, 4*(2), 68–77. https://doi.org/10.1177/105065199000400204	-	7	-	-	6
Allen, Nancy, Atkinson, Dianne, Morgan, Meg, Moore, Teresa & Snow, Craig. (1987). What experienced collaborators say about collaborative writing. *Journal of Business and Technical Communication, 1*(2), 70–90. https://doi.org/10.1177/105065198700100206	5	-	-	-	-
Anson, Chris M. & Forsberg, L. Lee. (1990). Moving beyond the academic community: Transitional stages in professional writing. *Written Communication, 7*(2), 200–231. https://doi.org/10.1177/0741088390007002002	-	-	-	3	1
Barton, Ben F. & Barton, Marthalee S. (1993). Ideology and the map: Toward a postmodern visual design practice. In Nancy Roundy Blyler & Charlotte Thralls (Eds.), *Professional communication: The social perspective* (pp. 49–78). Sage.	10	-	-	-	5
Berkenkotter, Carol & Huckin, Thomas N. (1995). Rethinking genre from a sociocognitive perspective. In *Genre knowledge in disciplinary communication: Cognition/culture/power* (pp. 1–25). Lawrence Earlbaum.	-	7	-	-	2
Bernhardt, Stephen A. (1993). The shape of text to come: The texture of print on screens. *College Composition and Communication, 44*(2), 151–175. https://doi.org/10.2307/358836	7	-	-	2	-
Bernhardt, Stephen A. (1995). Teaching for change, vision, and responsibility. *Technical Communication, 42*(4), 600–602.	-	6	-	-	-
Bitzer, Lloyd F. (1968). The rhetorical situation. *Philosophy & Rhetoric, 1*(1), 1–14.	-	-	-	-	4
Blakeslee, Ann M. (2001). Bridging the workplace and the academy: Teaching professional genres through classroom–workplace collaborations. *Technical Communication Quarterly, 10*(2), 169–192. https://doi.org/10.1207/s15427625tcq1002_4	-	7	-	-	1

Text	CWTC 2004	TTC 2004	CPT 2006	PWR 2003	Original, elsewhere, or unknown
Blakeslee, Ann M. (2009). Addressing audiences in a digital age. In Rachel Spilka (Ed.), *Digital literacy for technical communication: 21st century theory and practice* (pp. 199–229). Routledge.	-	-	-	-	4
Bosley, Deborah S. (1993). Cross-cultural collaboration: Whose culture is it anyway? *Technical Communication Quarterly, 2*(1), 51–62. https://doi.org/10.1080/10572259309364523	6	-	-	-	1
Brasseur, Lee E. (1993). Contesting the objectivist paradigm: Gender issues in the technical and professional communication curriculum. *IEEE Transactions on Professional Communication, 36*(3), 114–123. https://doi.org/10.1109/47.238051	5	-	-	-	1
Breuch, Lee-Ann Kastman. (2002). Thinking critically about technological literacy: Developing a framework to guide computer pedagogy in technical communication. *Technical Communication Quarterly, 11*(3), 267–288. https://doi.org/10.1207/s15427625tcq1103_3	9	-	-	-	-
Cargile Cook, Kelli. (2002). Layered literacies: A theoretical frame for technical communication pedagogy. *Technical Communication Quarterly, 11*(1), 5–29. https://doi.org/10.1207/s15427625tcq1101_1	-	-	-	-	9
Carliner, Saul. (2009). Computers and technical communication in the 21st century. In Rachel Spilka (Ed.), *Digital literacy for technical communication: 21st century theory and practice* (pp. 21–50). Routledge.	-	-	-	-	5
Charney, Davida. (1996). Empiricism is not a four-letter word. *College Composition and Communication, 47*(4), 567–593. https://doi.org/10.2307/358602	10	-	-	-	1
Clark, Dave. (2009). Shaped and shaping tools: The rhetorical nature of technical communication technologies. In Rachel Spilka (Ed.), *Digital literacy for technical communication: 21st century theory and practice* (pp. 85–102). Routledge.	-	-	-	-	4
Connors, Robert J. (1982). The rise of technical writing instruction in America. *Journal of Technical Writing and Communication, 12*(4), 329–352.	13	7	-	-	4[a]

Text	CWTC 2004	TTC 2004	CPT 2006	PWR 2003	Original, elsewhere, or unknown
Dobrin, David N. (1983). What's technical about technical writing? In Paul V. Anderson, R. John Brockman & Carolyn R. Miller (Eds.), *New essays in technical and scientific communication: Research, theory, practice* (pp. 227–250). Baywood.	10	2	-	-	3
Doheny-Farina, Stephen. (1986). Writing in an emerging organization: An ethnographic study. *Written Communication, 3*(2), 158–185. https://doi.org/10.1177/0741088386003002002	6	-	-	-	2
Dragga, Sam & Voss, Dan. (2001). Cruel pies: The inhumanity of technical illustrations. *Technical Communication, 48*(3), 265–274.	-	-	-	-	4
Driskill, Linda. (1989). Understanding the writing context in organizations. In Myra Kogen (Ed.), *Writing in the business professions* (pp. 125–145). National Council of Teachers of English.	6	-	-	4[b]	2
Dubinsky, James M. (2004). Becoming user-centered, reflective practitioners. In James M. Dubinsky (Ed.), *Teaching technical communication: Critical issues for the classroom* (pp. 1–10). Bedford/St. Martin's.	-	4	-	-	-
Durack, Katherine T. (1997). Gender, technology, and the history of technical communication. *Technical Communication Quarterly, 6*(3), 249–260. https://doi.org/10.1207/s15427625tcq0603_2	11	3	-	-	2
Faigley, Lester. (1985). Nonacademic writing: The social perspective. In Lee Odell & Dixie Goswami (Eds.), *Writing in nonacademic settings* (pp. 231–248). The Guilford Press.	-	-	-	3	2
Foss, Sonja K., Foss, Karen A. & Trapp, Robert. (1985). Perspectives on the study of rhetoric. In *Contemporary perspectives on rhetoric* (pp. 1–10). Waveland Press.	-	-	-	4	-
Freedman, Aviva & Adam, Christine. (1996). Learning to write professionally: "Situated learning" and the transition from university to professional discourse. *Journal of Business and Technical Communication, 10*(4), 395–427. https://doi.org/10.1177/1050651996010004001	-	8	-	-	-

Text	CWTC 2004	TTC 2004	CPT 2006	PWR 2003	Original, elsewhere, or unknown
Fukuoka, Waka, Kojima, Yukiko & Spyridakis, Jan H. (1999). Illustrations in user manuals: Preference and effectiveness with Japanese and American readers. *Technical Communication, 46*(2), 167–176.	-	4	-	-	-
Grabill, Jeffrey T. & Simmons, W. Michelle. (1998). Toward a critical rhetoric of risk communication: Producing citizens and the role of technical communicators. *Technical Communication Quarterly, 7*(4), 415–441. https://doi.org/10.1080/10572259809364640	-	-	-	2	2
Gurak, Laura J. & Bayer, Nancy L. (1994). Making gender visible: Extending feminist critiques of technology to technical communication. *Technical Communication Quarterly, 3*(3), 257–270. https://doi.org/10.1080/10572259409364571	-	4	-	-	-
Hallenbeck, Sarah. (2012). User agency, technical communication, and the 19th-century woman bicyclist. *Technical Communication Quarterly, 21*(4), 290–306. https://doi.org/10.1080/10572252.2012.686846	-	-	-	-	5
Harrison, Teresa M. (1987). Frameworks for the study of writing in organizational contexts. *Written Communication, 4*(1), 3–23. https://doi.org/10.1177/0741088387004001001	4	-	-	-	-
Henry, Jim. (2001). Writing workplace cultures. *College Composition and Communication, 53*(2). https://library.ncte.org/journals/CCC/issues/v53-2	-	-	2	-	2
Herndl, Carl G. (1993). Teaching discourse and reproducing culture: A critique of research and pedagogy in professional and non-academic writing. *College Composition and Communication, 44*(3), 349–363. https://doi.org/10.2307/358988	4	-	-	-	1
Howard, Tharon W. (1996). Who "owns" electronic texts? In Patricia Sullivan & Jennie Dautermann (Eds.), *Electronic literacies in the workplace: Technologies of writing* (pp. 177–198). National Council of Teachers of English.	4	-	-	2	-
Jackson, Lisa Ann. (2000). The rhetoric of design: Implications for corporate intranets. *Technical Communication, 47*(2), 212–219.	-	4	-	-	-

Text	CWTC 2004	TTC 2004	CPT 2006	PWR 2003	Original, elsewhere, or unknown
Johnson, Robert R. (1997). Audience involved: Toward a participatory model of writing. *Computers and Composition, 14*(3), 361–376. https://doi.org/10.1016/S8755-4615(97)90006-2	10	-	-	-	2
Johnson, Robert R. (1998). Complicating technology: Interdisciplinary method, the burden of comprehension, and the ethical space of the technical communicator. *Technical Communication Quarterly, 7*(1), 75–98. https://doi.org/10.1080/10572259809364618	-	5	-	-	1
Johnson, Robert R. (1998). *User-centered technology: A rhetorical theory for computers and other mundane artifacts*. State University of New York Press.	-	-	-	3[c]	7
Johnson, Robert R. (1999). Johnson responds. *Technical Communication Quarterly, 8*(2), 224–226. https://doi.org/10.1080/10572259909364662	-	4	-	-	-
Johnson-Eilola, Johndan. (1996). Relocating the value of work: Technical communication in a post-industrial age. *Technical Communication Quarterly, 5*(3), 245–270. https://doi.org/10.1207/s15427625tcq0503_1	10	6	-	-	1
Katz, Steven B. (1992). The ethic of expediency: Classical rhetoric, technology, and the Holocaust. *College English, 54*(3), 255–275. https://doi.org/10.2307/378062	12	-	-	4[b]	13
Katz, Susan M. (1998). Writing review as an opportunity for individuation. In *The dynamics of writing review: Opportunities for growth and change in the workplace* (pp. 73–98). Ablex.	-	-	-	3	1
Kramer, Robert & Bernhardt, Stephen A. (1996). Teaching text design. *Technical Communication Quarterly, 5*(1), 35–60. https://doi.org/10.1207/s15427625tcq0501_3	-	7	-	-	-
Lay, Mary M. (1991). Feminist theory and the redefinition of technical communication. *Journal of Business and Technical Communication, 5*(4), 348–370. https://doi.org/10.1177/1050651991005004002	7	3[d]	-	-	-
Longo, Bernadette. (2000). *Spurious coin: A history of science, management, and technical writing*. State University of New York Press.	-	-	-	-	4

Text	CWTC 2004	TTC 2004	CPT 2006	PWR 2003	Original, elsewhere, or unknown
MacKinnon, Jamie. (1993). Becoming a rhetor: Developing writing ability in a mature, writing-intensive organization. In Rachel Spilka (Ed.), *Writing in the workplace: New research perspectives* (pp. 41–55). Southern Illinois University Press.	-	-	-	3	1
Miller, Carolyn R. (1979). A humanistic rationale for technical writing. *College English, 40*(6), 610–617.	13	9[d]	-	-	14
Miller, Carolyn R. (1984). Genre as social action. *Quarterly Journal of Speech, 70*(2), 151–167. https://doi.org/10.1080/00335638409383686	-	-	-	-	8
Miller, Carolyn R. (1989). What's practical about technical writing? In Bertie E. Fearing & W. Keats Sparrow (Eds.), *Technical writing: Theory and practice* (pp. 14–24). Modern Language Association.	-	6	-	3	5
Mirel, Barbara. (1996). Writing and database technology: Extending the definition of writing in the workplace. In Patricia Sullivan & Jennie Dautermann (Eds.), *Electronic literacies in the workplace: Technologies of writing* (pp. 91–114). National Council of Teachers of English.	4	-	-	-	-
Mirel, Barbara. (2002). Advancing a vision of usability. In Barbara Mirel & Rachel Spilka (Eds.), *Reshaping technical communication* (pp. 165–188). Lawrence Earlbaum.	-	8	-	-	-
Moore, Patrick. (1999). Myths about instrumental discourse: A response to Robert R. Johnson. *Technical Communication Quarterly, 8*(2), 210–223. https://doi.org/10.1080/10572259909364661	-	5	-	-	1
Moses, Myra G. & Katz, Steven B. (2006). The phantom machine: The invisible ideology of email (a cultural critique). In J. Blake Scott, Bernadette Longo & Katherine V. Wills (Eds.), *Critical power tools: Technical communication and cultural studies* (pp. 71–105). State University of New York Press.	-	-	4	-	-

Text	CWTC 2004	TTC 2004	CPT 2006	PWR 2003	Original, elsewhere, or unknown
Ornatowski, Cezar M. (1992). Between efficiency and politics: Rhetoric and ethics in technical writing. *Technical Communication Quarterly, 1*(1), 91–103. https://doi.org/10.1080/10572259209359493	-	-	-	5	3
Paradis, James. (1991). Text and action: The operator's manual in context and in court. In Charles Bazerman & James Paradis (Eds.), *Textual dynamics of the professions: Historical and contemporary studies in writing in professional communities* (pp. 256–278). University of Wisconsin Press.	7	-	-	-	3
Porter, James E. (1998). The exercise of critical rhetorical ethics. In *Rhetorical ethics and internetworked writing* (pp. 133–147). Ablex.	-	3	-	-	1
Redish, Janice C. (2000). What is information design? *Technical Communication, 47*(2), 163–166.	-	7	-	-	-
Rude, Carolyn D. (1995). The report for decision making: Genre and inquiry. *Journal of Business and Technical Communication, 9*(2), 170–205. https://doi.org/10.1177/1050651995009002002	5	-	-	-	3
Rude, Carolyn D. (2009). Mapping the research questions in technical communication. *Journal of Business and Technical Communication, 23*(2), 174–215. https://doi.org/10.1177/1050651908329562	-	-	-	-	11
Russell, David R. (1993). The ethics of teaching ethics in professional communication: The case of engineering publicity at MIT in the 1920s. *Journal of Business and Technical Communication, 7*(1), 84–111. https://doi.org/10.1177/1050651993007001005	-	4	-	-	-
Rutter, Russell. (1991). History, rhetoric, and humanism: Toward a more comprehensive definition of technical communication. *Journal of Technical Writing and Communication, 21*(2), 133–153. https://doi.org/10.2190/7BBK-BJYK-AQGB-28GP	11	-	-	-	4
Salvo, Michael J. & Rosinski, Paula. (2009). Information design: From authoring text to architecting virtual space. In Rachel Spilka (Ed.), *Digital literacy for technical communication: 21st century theory and practice* (pp. 103–127). Routledge.	-	-	-	-	4

Text	CWTC 2004	TTC 2004	CPT 2006	PWR 2003	Original, elsewhere, or unknown
Selber, Stuart A. (1994). Beyond skill building: Challenges facing technical communication teachers in the computer age. *Technical Communication Quarterly, 3*(4), 365–390. https://doi.org/10.1080/10572259409364578	6	-	-	-	1
Selber, Stuart A., Johnson-Eilola, Johndan & Selfe, Cynthia L. (1995). Contexts for faculty professional development in the age of electronic writing and communication. *Technical Communication, 42*(4), 581–584.	-	4	-	-	-
Selfe, Cynthia L. & Hawisher, Gail E. (2002). A historical look at electronic literacy: Implications for the education of technical communicators. *Journal of Business and Technical Communication, 16*(3), 231–276. https://doi.org/10.1177/1050651902016003001	-	4	-	-	-
Selfe, Cynthia L. & Selfe, Richard J., Jr. (1994). The politics of the interface: Power and its exercise in electronic contact zones. *College Composition and Communication, 45*(4), 480–504. https://doi.org/10.2307/358761	6	-	-	-	3
Selzer, Jack. (1983). The composing process of an engineer. *College Composition and Communication, 34*(2), 178–187.	10	-	-	-	3[a]
Slack, Jennifer Daryl, Miller, David James & Doak, Jeffrey. (1993). The technical communicator as author: Meaning, power, authority. *Journal of Business and Technical Communication, 7*(1), 12–36. https://doi.org/10.1177/1050651993007001002	11	7	3	5[b]	4
Spilka, Rachel. (1990). Orality and literacy in the workplace: Process- and text-based strategies for multiple audience adaptation. *Journal of Business and Technical Communication, 4*(1), 44–67. https://doi.org/10.1177/1050651990004001003	-	-	-	3	1
Spilka, Rachel. (1995). Communicating across organizational boundaries: A challenge for workplace professionals. *Technical Communication, 42*(3), 436–450.	-	4	-	-	-

Text	CWTC 2004	TTC 2004	CPT 2006	PWR 2003	Original, elsewhere, or unknown
Spinuzzi, Clay. (1996). Pseudotransactionality, activity theory, and professional writing instruction. *Technical Communication Quarterly*, 5(3), 295–308. https://doi.org/10.1207/s15427625tcq0503_3	-	4	-	-	1
Sullivan, Dale L. (1990). Political–ethical implications of defining technical communication as a practice. *Journal of Advanced Composition*, 10(2), 375–386.	10	-	-	-	-
Sullivan, Patricia & Porter, James E. (1993). On theory, practice, and method: Toward a heuristic research methodology for professional writing. In Rachel Spilka (Ed.), *Writing in the workplace: New research perspectives* (pp. 220–237). Southern Illinois University Press.	10	-	-	-	2[a]
Thatcher, Barry. (2009). Understanding digital literacy across cultures. In Rachel Spilka (Ed.), *Digital literacy for technical communication: 21st century theory and practice* (pp. 169–198). Routledge.	-	-	-	-	4
Thralls, Charlotte & Blyler, Nancy Roundy. (1993). The social perspective and pedagogy in technical communication. *Technical Communication Quarterly*, 2(3), 249–269. https://doi.org/10.1080/10572259309364540	-	7	-	-	-
Thralls, Charlotte & Blyler, Nancy Roundy. (1993). The social perspective and professional communication: Diversity and directions in research. In Nancy Roundy Blyler & Charlotte Thralls (Eds.), *Professional communication: The social perspective* (pp. 3–34). Sage.	9	-	-	-	-
Thrush, Emily A. (1997). Multicultural issues in technical communication. In Katherine Staples & Cezar Ornatowski (Eds.), *Foundations for teaching technical communication: Theory, practice, and program design* (pp. 161–178). Ablex.	-	6	-	-	-

Text	CWTC 2004	TTC 2004	CPT 2006	PWR 2003	Original, elsewhere, or unknown
Wilson, Greg & Herndl, Carl G. (2007). Boundary objects as rhetorical exigence: Knowledge mapping and interdisciplinary cooperation at the Los Alamos National Laboratory. *Journal of Business and Technical Communication, 21*(2), 129–154. https://doi.org/10.1177/1050651906297164	-	-	-	-	4
Winsor, Dorothy A. (1990). Engineer writing/writing engineering. *College Composition and Communication, 41*(1), 58–70.	7	-	-	-	3
Wolfe, Joanna. (2009). How technical communication textbooks fail engineering students. *Technical Communication Quarterly, 18*(4), 351–375. https://doi.org/10.1080/10572250903149662	-	-	-	-	6
Zoetewey, Meredith W. & Staggers, Julie. (2004). Teaching the Air Midwest case: A stakeholder approach to deliberative technical rhetoric. *IEEE Transactions on Professional Communication, 47*(4), 233–243. https://doi.org/10.1109/TPC.2004.837969	-	-	-	-	5

[a] One syllabus provided both the original of these texts and its reprint in CWTC.
[b] One syllabus listed these texts in both CWTC and PWR.
[c] One chapter of Johnson's book is reprinted in PWR.
[d] One syllabus listed these texts in both CWTC and TTC.

Of course, while Smith (2000a, 2004) studied citations in journals, the points of reference in Table 3.6 are drawn from graduate-level syllabi. We make no claims about shifts in citation practices in scholarly articles (though we speculate that some of these shifts would be paralleled in scholarly citations). One likely explanation of differences in our list is that while scholars are likely to cite monographs frequently, teachers are less likely to assign monographs in a course that is meant to introduce students to an entire field: teachers can cover a much broader ground with articles and edited collections than with monographs. While many syllabi included a monograph, monograph choices were diverse. Michael Hughes and George Hayhoe's *A Research Primer* was required by six syllabi; Bernadette Longo's *Spurious Coin* was required four times; Robert Johnson's *User-Centered Technology* was required five times (and other syllabi assigned chapters from it); and Clay Spinuzzi's *Network* was assigned twice. Other monographs were assigned only once in the dataset.

∎ Conclusions

What, we ask, has changed over the last decade or two that might help to explain changes from Smith's lists to ours? We have three speculations about the reasons for these shifts. First, some scholarly conversations that have become touchstones in the field occurred toward the end of Smith's study, such as the exchange between Robert Johnson and Patrick Moore in the 1998 and 1999 volumes of *Technical Communication Quarterly*. Thirty of the 82 texts in Table 3.6 were published after 1997. Another 15 were published from 1995–1997, which means they were unlikely to be cited much by the end of Smith's (2000a) original study (texts printed from 1988–1997). Of course, the field has changed over the last two decades, and new publications replace older publications as touchstones for scholars and teachers.

Second, we point out that as the discipline has matured, it has relied far less on composition scholarship as touchstones for concepts and approaches to the field. As early as 2000, Smith (2000a) was observing a shift away from composition theory in citations in the field. Many of the publications that we would identify as composition scholarship on Smith's list of points of reference (e.g., Lisa Ede and Andrea Lunsford's "Audience Addressed/Audience Invoked") do not continue on our list of points of reference (though some are still in our dataset, including Ede and Lunsford's article). Of the texts in Table 3.6, only seven were published in *CCC*. Importantly, six of these are reprinted in *CWTC* and one (Henry, 2001) was reprinted with revisions in *CPT*, and most of them were assigned from these collections. Technical communication appears to have (at least in part) moved away from composition studies as a touchstone for our research and methods. One might even wonder if these texts would be so heavily assigned if they weren't reprinted in *CWTC*, and if other texts about workplace studies, research methods, and digital media might take their place in syllabi.

Which leads to our third speculation, which we have discussed above: the publication of anthologies like *CWTC* and *TTC* made accessing and assigning texts easier, and the editors of these collections seem to have had a strong hand in shaping what the field considers points of reference. It seems that *CWTC* and *TTC* have been particularly influential in shaping the field as it is presented to graduate students. These two collections might be one reason the majority of the points of reference in Table 3.6 are from the 1990s and early 2000s. The continued use of these edited collections has contributed to a list of points of reference that seem to cohere around a body of work published between 1989 and 2002.

Certainly, *CWTC* and *TTC* have done and continue to do much useful disciplinary work, as is evidenced by their prevalent use in the syllabi in our dataset. As Johnson-Eilola and Selber (2004) perhaps intended, *CWTC*, along with *TTC*, helped to provide scholars and teachers "with a coherent body of disciplinary knowledge" (p. xxvii). Now, nearly two decades after the publication of these two collections, we might ask about the sort of work they do now. Do they, we ask, provide the sorts of points of reference that help the field move forward in research and scholarship when they are used in graduate education? Or do they introduce new scholars to conversations that, now two or three decades old, might prevent (or make more difficult) asking new research and teaching questions that more recent scholarship might provoke? We don't believe we have the answers to these questions, but we believe that the answer might be a little bit of yes and a little bit of no to each one. We certainly can't deny that it's useful for graduate students to read now canonical texts anthologized in these two collections. But we also wonder if an effect of these two collections isn't to flatten the historicity of the anthologized articles. We are certain that teachers likely provide context to students as they read (situating the article in historical context, discussing how the field has responded to questions and problems raised by older works). And most teachers placed these works in conversation with more recent scholarship. For example, one syllabus paired Miller's "A Humanistic Rationale" with Byron Hawk's *Technical Communication Quarterly* article "Toward a Post-Technê"; another paired Emily Thrush's "Multicultural Issues in Technical Communication" with more recent work on race, like Angela Haas' 2012 article "Race, Rhetoric, and Technology." But, we wonder, does the material space of a printed anthology do at least some flattening of the dynamics of scholarly conversations over time? Perhaps it is this flattening that ultimately gets recognized as disciplinarity.

Our examination of this dataset of 60 syllabi updates and expands understanding of the coherence of technical communication as a field. We used the dataset to characterize where and how the foundations of technical communication are taught in graduate curricula. We employed citation and social network analysis to demonstrate the presence of a core set of texts and updated Smith's (2000a, 2004) points of reference for the field by providing a list of the 82 most authoritative texts in our co-citation network. Overall, our characterization implies that our field has "come to grips with a coherent body of disciplinary knowledge" as Johnson-Eilola and Selber (2004, p. xxvii) indicated was necessary for the field to achieve maturity and coherence. We argue that technical communication has achieved adequate maturity to move past our disciplinary anxiety of inadequacy and underdevelopment and to begin to ask new questions.

With this disciplinary maturity comes opportunities for growth and diversity. In closing, we'd like to call attention to a so-far unremarked-upon aspect of our dataset: texts in our updated list of points of reference are authored predominantly by White scholars. In fact, readers might notice that the citation network of this chapter (that is, who we as authors have entered conversation with) is also predominantly White. Rebecca Walton et al. (2019) have observed "the lack of scholarly work by minority scholars" in technical communication, asking the field to consider "how and whose knowledge we legitimize in the field" (pp. 2–3). Their discussion in *Technical Communication After the Social Justice Turn* prompts us to echo their call "to diversifying our field in its foundational theories, its professoriate, its programs, and its citation practices" (p. 3). One avenue forward (among many) is more diverse representation of theories and sites of study in our graduate courses. Certainly, courses like the ones in our study—courses designed to introduce graduate students to the field—must cover some of the foundational work that the field has come to recognize as transmits or points of reference. But there is room for including materials in those courses that enter into conversation with our field's foundational history. And further, we might follow Brooke's (2011) suggestion of teaching how "to read the citation network of the discipline" explicitly in graduate courses (p. 98). For example, a course might pair reading and discussing Godwin Agboka's (2012) "Liberating Intercultural Technical Communication from 'Large Culture' Ideologies" alongside earlier, more foundational work on cross-cultural technical communication in order to not only understand Agboka's critique of "large culture" ideologies but also to explore how Agboka enters into (and constructs through his writing) the conversation about "culture" in the field. Consequently, also following Brooke's (2011) suggestion for rhetoric and composition courses, we might suggest that our graduate courses do not have to be driven by a coverage model, and that many could instead have students "study a topic or issue as it unfolds in the discipline" and attend to the "epistemic practices" of texts as they join in conversation with each other (p. 102). These are just a few suggestions for graduate education in the field—increasing the diversity of the texts we assign and explicitly teaching technical communication (at least as a scholarly discipline) as a network.

In this chapter, we have used co-citation network analysis to map the field of technical communication. As we have shown, this method of mapping can be useful to identify points of reference central to a field, but as we admit, such a quantitative approach can be limited because it risks reducing the complexity of relationships within a network. As we hope we have shown, these quantitative methods help to abstract the field and allow us to ask questions about the nature of the field. In closing, we suggest that these methods can be combined with other qualitative and quantitative social network analyses, like those advocated by Jordan Frith (2014) in technical communication and Nathan Johnson (2015) in rhetoric and composition, to develop thicker and richer maps of these disciplines. For instance, we might ask how the location of a scholar's graduate training and who they trained under affects their views of the field and how they transmit those views onto their graduate students. Social network analysis can also be used to trace how new ideas or projects develop and spread within a field. Mapping technical communication and rhetoric and composition through social network analysis can help us to see the field differently and thus confirm or question our assumptions about the field.

■ Acknowledgments

We wish to acknowledge that this chapter was written and revised within the historical territories of the Teya, Jumano, Apache, and Comanche peoples. We would also like to thank students in Greg's graduate seminar for helping to code and enter data into Excel spreadsheets and Craig Snoeyink for his assistance automating some of the re-organizing of data into CSV files for the co-citation network. We are also grateful to our colleagues who provided feedback on earlier drafts of this chapter: Kelli Cargile Cook, Amy Koerber, Kristen Moore, Abigail Selzer King, and Rachel Wolford.

■ References

Agboka, Godwin. (2012). Liberating intercultural technical communication from "large culture" ideologies: Constructing culture discursively. *Journal of Technical Writing and Communication*, *42*(2), 159–181. https://doi.org/10.2190/TW.42.2.e.

Anderson, Chris. (2004, October). The long tail. *Wired*. https://www.wired.com/2004/10/tail/.

Barabási, Albert-László. (2002). *Linked: The new science of networks*. Perseus.

Blakeslee, Ann M. (2009). The technical communication research landscape. *Journal of Business and Technical Communication*, *23*(2), 129–173. https://doi.org/10.1177/1050 651908328880.

Blondel, Vincent D., Guillaume, Jean-Loup, Lambiotte, Renaud & Lefebvre, Etienne. (2008). Fast unfolding of communities in large networks. *Journal of Statistical Mechanics: Theory and Experiment*, P10008. https://doi.org/10.1088/1742-5468/2008/10/P10008.

Brooke, Collin Gifford. (2011). Discipline and punish: Reading and writing the scholarly network. In Sidney I. Dobrin (Ed.), *Ecology, writing theory, and new media: Writing ecology* (pp. 92–105). Routledge.

Cozzens, Susan E. (1989). What do citations count? The rhetoric-first model. *Scientometrics*, *15*(5–6), 437–447. https://doi.org/10.1007/BF02017064.

De Bellis, Nicola. (2009). *Bibliometrics and citation analysis: From the* Science Citation Index *to cybermetrics*. The Scarecrow Press.

de Solla Price, Derek J. (1965). Networks of scientific papers. *Science*, *149*(3683), 510–515. https://doi.org/10.1126/science.149.3683.510.

Detweiler, Eric. (2015). "/" "and" "-"?: An empirical consideration of the relationship between "rhetoric" and "composition." *Enculturation: A Journal of Rhetoric, Writing, and Culture*, *20*. http://enculturation.net/an-empirical-consideration.

Dobrin, David N. (1983). What's technical about technical writing? In Paul V. Anderson, R. John Brockman & Carolyn R. Miller (Eds.), *New essays in technical and scientific communication: Research, theory, practice* (pp. 227–250). Baywood.

Doheny-Farina, Stephen. (1986). Writing in an emerging organization: An ethnographic study. *Written Communication*, *3*(2), 158–185. https://doi.org/10.1177/0741088386003002002.

Dubinsky, James M. (Ed.). (2004). *Teaching technical communication: Critical issues for the classroom*. Bedford/St. Martin's.

Frith, Jordan. (2014). Social network analysis and professional practice: Exploring new methods for researching technical communication. *Technical Communication Quarterly*, *23*(4), 288–302. https://doi.org/10.1080/10572252.2014.942467.

Fuhse, Jan & Mützel, Sophie. (2011). Tackling connections, structure, and meaning in networks: Quantitative and qualitative methods in sociological network research. *Quality & Quantity*, *45*(5), 1067–1089. https://doi.org/10.1007/s11135-011-9492-3.

Garfield, Eugene. (1955). Citation indexes for science: A new dimension in documentation through association of ideas. *Science, 122*(3159), 108–111. https://doi.org/10.1126/science.122.3159.108.

Glenn, Cheryl. (1997). *Rhetoric retold: Regendering the tradition from antiquity through the Renaissance.* Southern Illinois University Press.

Goggin, Maureen Daly. (2000). *Authoring a discipline: Scholarly journals and the post-World War II emergence of rhetoric and composition.* Lawrence Earlbaum.

Grove, Laurel K. & Zimmerman, Donald E. (1997). Introduction: Bridging communication science to technical communication—Advancing the profession. *IEEE Transactions on Professional Communication, 40*(3), 157–167. https://doi.org/10.1109/TPC.1997.649552.

Hart-Davidson, William. (2001). On writing, technical communication, and information technology: The core competencies of technical communication. *Technical Communication, 48*(2), 145–155.

Healy, Kieran. (2013, June 18). A co-citation network for philosophy. *Kieran Healy.* https://kieranhealy.org/blog/archives/2013/06/18/a-co-citation-network-for-philosophy/.

Henry, Jim. (2000). *Writing workplace cultures: An archaeology of professional writing.* Southern Illinois University Press.

Hyland, Ken. (2004). *Disciplinary discourses: Social interactions in academic writing* (Michigan Classics ed.). The University of Michigan Press.

Johnson, Nathan. (2015). Modeling rhetorical disciplinarity: Mapping the digital network. In Jim Ridolfo & William Hart-Davidson (Eds.), *Rhetoric and the digital humanities* (pp. 96–107). University of Chicago Press.

Johnson-Eilola, Johndan. (1996). Relocating the value of work: Technical communication in a post-industrial age. *Technical Communication Quarterly, 5*(3), 245–270. https://doi.org/10.1207/s15427625tcq0503_1.

Johnson-Eilola, Johndan & Selber, Stuart A. (2001). Sketching a framework for graduate education in technical communication. *Technical Communication Quarterly, 10*(4), 403–437. https://doi.org/10.1207/s15427625tcq1004_3.

Johnson-Eilola, Johndan & Selber, Stuart A. (Eds.). (2004). *Central works in technical communication.* Oxford University Press.

Johnson-Eilola, Johndan & Selber, Stuart A. (Eds.). (2013). *Solving problems in technical communication.* University of Chicago Press.

Kadushin, Charles. (2012). *Understanding social networks: Theories, concepts, and findings.* Oxford University Press.

Kirschenbaum, Matthew G. (2007). The remaking of reading: Data mining and the digital humanities. *NGDM 07: National Science Foundation Symposium on the Next Generation of Data Mining and Cyber-Enabled Discovery for Innovation.* https://pdfs.semanticscholar.org/9b33/4177e179ba9783a74533169bdc8d3d07a7aa.pdf.

Kleinberg, Jon M. (1999). Authoritative sources in a hyperlinked environment. *Journal of the ACM, 46*(5), 604–632. https://doi.org/10.1145/324133.324140.

Kuhn, Thomas S. (1970). *The structure of scientific revolutions* (2nd ed.). University of Chicago Press.

Kynell-Hunt, Teresa & Savage, Gerald J. (Eds.). (2003–2004). *Power and legitimacy in technical communication* (2 vols.). Baywood.

Melonçon, L. (2009). Master's programs in technical communication: A current overview. *Technical Communication, 56*(2), 137–148.

Miller, Carolyn R. (1979). A humanistic rationale for technical writing. *College English, 40*(6), 610–617.

Mirel, Barbara & Spilka, Rachel. (2002). Introduction. In Barbara Mirel & Rachel Spilka (Eds.), *Reshaping technical communication: New directions and challenges for the 21st century* (pp. 1–6). Lawrence Erlbaum.

Mueller, Derek. (2012). Grasping rhetoric and composition by its long tail: What graphs can tell us about the field's changing shape. *College Composition and Communication, 64*(1), 195–223.

Mueller, D. N. (2017). *Network sense: Methods for visualizing a discipline*. The WAC Clearinghouse; University Press of Colorado. https://doi.org/10.37514/WRI-B.2017.0124.

Otte, Evelien & Rousseau, Ronald. (2002). Social network analysis: A powerful strategy, also for the information sciences. *Journal of Information Science, 28*(6), 441–453. https://doi.org/10.1177/016555150202800601.

Peeples, Tim. (Ed.). (2003). *Professional writing and rhetoric: Readings from the field*. Longman.

Peeples, Tim & Hart-Davidson, Bill. (2012). Remapping professional writing: Articulating the state of the art and composition studies. In Kelly Ritter & Paul Kei Matsuda (Eds.), *Exploring composition studies: Sites, issues, and perspectives* (pp. 52–72). Utah State University Press.

Phillips, Donna Burns, Greenberg, Ruth & Gibson, Sharon. (1993). *College Composition and Communication*: Chronicling a discipline's genesis. *College Composition and Communication, 44*(4), 443–465.

Pinelli, Thomas E. & Barclay, Rebecca O. (1992). Research in technical communication: Perspectives and thoughts on the process. *Technical Communication, 39*(4), 526–532.

Rainey, Kenneth T. (1999). Doctoral research in technical, scientific, and business communication, 1989–1998. *Technical Communication, 46*(4), 501–531.

Reid, Gwendolynne & Miller, Carolyn R. (2018). Classification and its discontents: Making peace with blurred boundaries, open categories, and diffuse disciplines. In Rita Malenczyk, Susan Miller-Cochran, Elizabeth Wardle & Kathleen Blake Yancey (Eds.), *Composition, rhetoric, and disciplinarity* (pp. 87–110). Utah State University Press.

Reinsch, N. L., Jr. & Lewis, Phillip V. (1993). Author and citation patterns for *The Journal of Business Communication*, 1978–1992. *Journal of Business Communication, 30*(4), 435–462. https://doi.org/10.1177/002194369303000404.

Reinsch, N. Lamar, Jr. & Reinsch, Janet W. (1996). Some assessments of business communication scholarship from social science citations. *Journal of Business and Technical Communication, 10*(1), 28–47. https://doi.org/10.1177/1050651996010001002.

Rude, Carolyn D. (2009). Mapping the research questions in technical communication. *Journal of Business and Technical Communication, 23*(2), 174–215. https://doi.org/10.1177/1050651908329562.

Rude, Carolyn D. (2015). Building identity and community through research. *Journal of Technical Writing and Communication, 45*(4), 366–380. https://doi.org/10.1177/0047281615585753.

Savage, Gerald J. (1999). The process and prospects for professionalizing technical communication. *Journal of Technical Writing and Communication, 29*(4), 355–381. https://doi.org/10.2190/7GFX-A5PC-5P7R-9LHX.

Scott, John. (2012). *Social network analysis* (3rd ed.). SAGE.

Scott, J. Blake, Longo, Bernadette & Wills, Katherine V. (Eds.). (2006). *Critical power tools: Technical communication and cultural studies*. State University of New York Press.

Selzer, Jack. (1983). The composing process of an engineer. *College Composition and Communication, 34*(2), 178–187.

Shelby, Annette. (1996). A discipline orientation: Analysis and critique. *Management Communication Quarterly, 10*(1), 98–105. https://doi.org/10.1177/0893318996010001006.

Slack, Jennifer Daryl. (2003). The technical communicator as author? A critical postscript. In Teresa Kynell-Hunt & Gerald J. Savage (Eds.), *Power and legitimacy in technical communication, volume I: The historical and contemporary struggle for professional status* (pp. 193–207). Baywood.

Slack, Jennifer Daryl, Miller, David James & Doak, Jeffrey. (1993). The technical communicator as author: Meaning, power, authority. *Journal of Business and Technical Communication, 7*(1), 12–36. https://doi.org/10.1177/1050651993007001002.

Small, Henry. (1973). Co-citation in the scientific literature: A new measure of the relationship between two documents. *Journal of the American Society for Information Science, 24*(4), 265–269. https://doi.org/10.1002/asi.4630240406.

Smith, Elizabeth Overman. (1997). Intertextual connections to "A humanistic rationale for technical writing." *Journal of Business and Technical Communication, 11*(2), 192–222. https://doi.org/10.1177/1050651997011002003.

Smith, Elizabeth Overman. (2000a). Points of reference in technical communication scholarship. *Technical Communication Quarterly, 9*(4), 427–453. https://doi.org/10.1080/10572250009364708.

Smith, Elizabeth Overman. (2000b). Strength in the technical communication journals and diversity in the serials cited. *Journal of Business and Technical Communication, 14*(2), 131–184. https://doi.org/10.1177/105065190001400201.

Smith, Elizabeth Overman. (2004). Points of reference contributing to the professionalization of technical communication. In Teresa Kynell-Hunt & Gerald J. Savage (Eds.), *Power and legitimacy in technical communication, volume II: Strategies for professional status* (pp. 51–72). Baywood.

Spilka, Rachel (Ed.). (2009). *Digital literacy for technical communication: 21st century theory and practice*. Routledge.

St.Amant, Kirk & Melonçon, Lisa. (2016). Addressing the incommensurable: A research-based perspective for considering issues of power and legitimacy in the field. *Journal of Technical Writing and Communication, 46*(3), 267–283. https://doi.org/10.1177/0047281616639476.

Staples, Katherine. (1999). Technical communication from 1950–1998: Where are we now? *Technical Communication Quarterly, 8*(2), 163–164. https://doi.org/10.1080/10572259909364656.

Sullivan, Patricia A. & Porter, James E. (1993). Remapping curricular geography: Professional writing in/and English. *Journal of Business and Technical Communication, 7*(4), 389–422. https://doi.org/10.1177/1050651993007004001.

Sullivan, Patricia & Porter, James E. (1997). *Opening spaces: Writing technologies and critical research practices*. Ablex.

Tirrell, Jeremy. (2012). A geographical history of online rhetoric and composition journals. *Kairos: A Journal of Rhetoric, Technology, and Pedagogy, 16*(3). http://kairos.technorhetoric.net/16.3/topoi/tirrell/.

Toulmin, Stephen. (1972). *Human understanding* (Vol. 1). Clarendon.

Unger, Don & Sánchez, Fernando. (2015). Locating queer rhetorics: Mapping as an inventional method. *Computers and Composition, 38*, 96–112. https://doi.org/10.1016/j.compcom.2015.09.011.

Wahlstrom, Billie. (1997). Designing a research program in scientific and technical communication: Setting standards and defining the agenda. In Katherine Staples & Cezar Ornatowski (Eds.), *Foundations for teaching technical communication: Theory, practice, and program design* (pp. 299–315). Ablex.

Walton, Rebecca, Moore, Kristen R. & Jones, Natasha N. (2019). *Technical communication after the social justice turn: Building coalitions for action.* Routledge.

Wang, Dan. (2012, May). Is there a canon in economic sociology? *Accounts: ASA Economic Sociology Newsletter, 11*(2), 1–8.

Wilson, Greg. (2001). Technical communication and late capitalism: Considering a postmodern technical communication pedagogy. *Journal of Business and Technical Communication, 15*(1), 72–99. https://doi.org/10.1177/1050651901015000104.

Wilson, Greg & Wolford, Rachel. (2017). The technical communicator as (post-postmodern) discourse worker. *Journal of Business and Technical Communication, 31*(1), 3–29. https://doi.org/10.1177/1050651916667531.

Winsor, Dorothy A. (1990). Engineer writing/writing engineering. *College Composition and Communication, 41*(1), 58–70.

Yeats, Dave & Thompson, Isabelle. (2000). Mapping technical and professional communication: A summary and survey of academic locations for programs. *Technical Communication Quarterly, 19*(3), 225–261. https://doi.org/10.1080/10572252.2010.481538.

■ Appendix: Additional Texts from Our Dataset

Below are the full APA citations for the texts from our dataset that we mention in the chapter but are not listed in in Table 3.6. We do not include the 19 chapters and introduction from Johnson-Eilola and Selber's (2013) *SPTC* (listed in Table 3.4) in this list.

Althusser, Louis. (1971). Ideology and ideological state apparatuses (notes towards an investigation). In *Lenin and philosophy and other essays* (Ben Brewster, Trans., pp. 127–186). Monthly Review Press.

Bazerman, Charles. (1988). *Shaping written knowledge: The genre and activity of the experimental article in science.* University of Wisconsin Press.

Buchanan, Richard. (1985). Declaration by design: Rhetoric, argument, and demonstration in design practice. *Design Issues, 2*(1), 4–22. https://doi.org/10.2307/1511524.

Coney, Mary B. (1984). The implied author in technical discourse. *JAC, 5*, 163–172.

Coney, Mary B. (1997). Technical communication theory: An overview. In Katherine Staples & Cezar Ornatowski (Eds.), *Foundations for teaching technical communication: Theory, practice, and program design* (pp. 1–15). Ablex.

Derrida, Jacques. (1978). Structure, sign, and play in the discourse of the human sciences. In *Writing and difference* (Alan Bass, Trans., pp. 278–293). University of Chicago Press.

Dicks, R. Stanley. (2002). Cultural impediments to understanding: Are they surmountable? In Barbara Mirel & Rachel Spilka (Eds.), *Reshaping technical communication* (pp. 13–25). Lawrence Erlbaum.

Diehl, Amy, Grabill, Jeffrey T., Hart-Davidson, William & Iyer, Vishal. (2008). Grassroots: Supporting the knowledge work of everyday life. *Technical Communication Quarterly, 17*(4), 413–434. https://doi.org/10.1080/10572250802324937.

Ede, Lisa & Lunsford, Andrea. (1984). Audience addressed/audience invoked: The role of audience in composition theory and pedagogy. *College Composition and Communication, 35*(2), 155–171.

Engeström, Yrjö. (1999). Activity theory and individual and social transformation. In Yrjö. Engeström, Reijo Miettinen & Raija-Leena Punamäki (Eds.), *Perspectives on activity theory* (pp. 19–39). Cambridge University Press.

Faber, Brenton. (2002). Professional identities: What is professional about professional communication? *Journal of Business and Technical Communication, 16*(3), 306–337. https://doi.org/10.1177/10506519020160003.

Foucault, Michel. (1972). The discourse on language. In *The archaeology of knowledge and the discourse on language* (A. M. Sheridan Smith, Trans., pp. 215–237). Pantheon.

Graham, S. Scott & Whalen, Brandon. (2008). Mode, medium, and genre: A case study of decisions in new-media design. *Journal of Business and Technical Communication, 22*(1), 65–91. https://doi.org/10.1177/1050651907307709.

Haas, Angela M. (2012). Race, rhetoric, and technology: A case study of decolonial technical communication theory, methodology, and pedagogy. *Journal of Business and Technical Communication, 26*(3), 277–310. https://doi.org/10.1177/1050651912439539.

Hawk, Byron. (2004). Toward a post-technê—or, inventing pedagogies for professional writing. *Technical Communication Quarterly, 13*(4), 371–392. https://doi.org/10.1207/s15427625tcq1304_2.

Hughes, Michael A. & Hayhoe, George F. (2008). *A research primer for technical communication: Methods, exemplars, and analyses*. Lawrence Erlbaum.

Johnson, Robert R., Salvo, Michael J. & Zoetewey, Meredith W. (2007). User-centered technology in participatory culture: Two decades "beyond a narrow concept of usability testing." *IEEE Transactions on Professional Communication, 50*(4), 320–332. https://doi.org/10.1109/TPC.2007.908730.

Melonçon, Lisa & Henschel, Sally. (2013). Current state of U.S. undergraduate degree programs in technical and professional communication. *Technical Communication, 60*(1), 45–64.

Ong, Walter. (1975). The writer's audience is always a fiction. *PMLA, 90*(1), 9–21.

Ornatowski, Cezar M. (1997). Technical communication and rhetoric. In Katherine Staples & Cezar Ornatowski (Eds.), *Foundations for teaching technical communication: Theory, practice, and program design* (pp. 31–52). Ablex.

Palmeri, Jason. (2009). Disability studies, cultural analysis, and the critical practice of technical communication pedagogy. *Technical Communication Quarterly, 15*(1), 49–65. https://doi.org/10.1207/s15427625tcq1501_5.

Porter, James E. & Sullivan, Patricia A. (2007). "Remapping curricular geography": A retrospection. *Journal of Business and Technical Communication, 21*(1), 15–20. https://doi.org/10.1177/1050651906293507.

Potts, Lisa. (2014). *Social media in disaster response: How experience architects can build for participation*. Routledge.

Redish, Janice C. (1997). Understanding people: The relevance of cognitive psychology to technical communication. In Katherine Staples & Cezar Ornatwoski (Eds.),

Foundations for teaching technical communication: Theory, practice, and program design (pp. 67–84). Ablex.

Spinuzzi, Clay. (2008). *Network: Theorizing knowledge work in telecommunications*. Cambridge University Press.

Sullivan, Patricia A. & Porter, James E. (1993). Remapping curricular geography: Professional writing in/and English. *Journal of Business and Technical Communication, 7*(4), 389–422. https://doi.org/10.1177/1050651993007004001.

Sullivan, Patricia & Porter, James E. (1997). *Opening spaces: Writing technologies and critical research practices*. Ablex.

Teston, Christa. (2012). Moving from artifact to action: A grounded investigation of visual displays of evidence during medical deliberations. *Technical Communication Quarterly, 21*(3), 187–209. https://doi.org/10.1080/10572252.2012.650621.

Part Two: Reflection and Maintenance of Major Concepts

Chapter 4: "Visualize a Triangle." What's Professional About Professional Communication?

Brenton Faber
WORCESTER POLYTECHNIC INSTITUTE

Abstract: Research into occupational rhetoric has promoted professional communication as an aspirational discourse by conflating occupational and professional forms and activities. As such, professional communication has become a general term that encircles most forms of workplace, business, technical, or organizational communication. Yet, historically, the professions have played an important role in mediating the regulatory and capitalist forces of government and business. Here, professional discourse is not an aggregate or aspirational form of workplace communication but a separate field motivated to promote cognitive concepts associated with health, justice, science, and knowledge and to constrain the excesses of capitalist and regulatory discourses. Conflating professional discourse with business, regulatory, or other forms of workplace communication obscures the conditions, ethics, and intentions that motivate each sector and the real and important tensions between these sectors. Examining professional discourse as a function rather than an occupational status opens up situational research that could investigate specific professional activities within competing discourses. Such moments and spaces could show where and how discourses are deployed as a correction to capitalist or regulatory over-reach. Such a project could investigate how rhetorical agents modulate discourses while retaining and deploying legitimacy, credibility, and the ability to enact social and economic power.

Keywords: professionalism, discourse, intention, ethics, curation, modulation

It is a particularly good time to revisit the art and science of professional communication. On December 21, 2018, the editorial board of *The New York Times* reported that Judge Emmet G. Sullivan of the Federal District Court in Washington reprimanded the former Attorney General, Jeff Sessions, for not adhering to professional legal standards with regards to federal asylum activities. As the editorial board put it, Judge Sullivan told Sessions to "follow the law." The case being considered involved actions Sessions took in June 2018 to reject potential immigrants' claims of domestic and gang-related violence as criteria for seeking asylum. As the editorial board wrote, "In his ruling on Wednesday, the judge . . . all but accused Mr. Sessions of taking the law into his own hands. By creating a system that categorically denied these claims, the judge wrote, 'the attorney general has failed to stay within the bounds of statutory authority.'" In other words, Sessions had acted unprofessionally.

America's 2016 overlaying of government and capitalism and the ensuing challenges by the professions are clarifying and prescient spaces for professional communication research. The events of June 2018 show a legal profession pushing back against a government official whose actions broke the law and violated his purpose, scope, and privilege as a legal professional acting within government. A hopeful account of Judge Sullivan's actions, as well as other judicial and medical challenges since 2016, would posit that a much-weakened professional sector appears to be reasserting itself in the face of a similarly weakened government sector. Judge Sullivan's decision thus reasserted the professions' role within the necessary and dynamic tensions among government, capitalism, and professions. As Eliot Krause (1996) would have it, physicians challenging the treatment of immigrant children at the southern border and lawyers challenging multiple federal government environmental, immigration, and ethical actions demonstrate the professions asserting themselves to "influence and confront the power of both capitalism and the state" (p. 2) —or in this case, a state that has been overrun by self-interested and self-styled capitalists. After decades during which the American economy has deprofessionalized specialized knowledge-based work, distributed the occupational authority typically associated with the professions to other semi- and non-specialized groups, and diluted the professions' social power, it could be that the professions' authority, knowledge, and system of societal checks and balances may again be finding social purpose and resolve.

This chapter revisits the findings from my 2002 study, "Professional Identities: What's Professional About Professional Communication?" and the reception and influence the study has had on professional communication teaching and research. In short, the study was not able to hold back what has appeared to be an ongoing desire to enfold a good deal of non-fiction and occupational writing within the realm of professional communication. The critique in this initial section of the chapter is that while writers have desired and claimed professional status, what has been missing in these claims has been the reciprocal necessity of professional accountability. In other words, what has been missing is an articulation of the specifically professional purpose enacted by a particular discursive form and the social responsibilities that are aligned with that purpose.

As route to better understanding the purpose of professional communication, this chapter then returns to Elliot Krause's (1996) distinctions among professional, regulatory, and capitalist domains. A robust democracy, according to Krause, requires a productive tension among the three domains as each sector holds the other two within productive boundaries. Krause uses a triangle metaphor here, with each sector sustaining, restricting, and defining the other two. To demonstrate how Krause's model applies to communication scholarship, the chapter next offers two short case studies showing professional discourse operating as a check and balance against capitalist impulses first within an institution and next within the free market.

The chapter concludes by overtly switching frames from occupational status in the 2002 study to the purpose or, otherwise put, the intent of specific communicative forms. Professional communication is professional when it influences and confronts either the unfettered power of capitalism, the regulatory power of the state, or both. At the same time, professional communication is more than protest and advocacy: the initial findings from "Professional Identities" continue to hold. The professions still rely on individual clients, they have a social responsibility to and are accountable for a specific and exclusive knowledge base, and they have an ethical obligation to work on behalf of and be subject to that same knowledge base.

What "Professional Identities" did not sufficiently articulate is the layering and integration of these characteristics with communicative intent. When a professional works with an individual audience (i.e., patient, client, student), the activity is intended to adjudicate professional knowledge as it relates to the audience's particular circumstance. By articulating professional communication through the frame of professional intent, professional communication may not be restricted to particular occupations, guilds, or settings. Instead, professional communication could be seen as the enactment of crucial checks and balances at particular, necessary, and strategic moments.

Professional Identities: What's Professional About Professional Communication?

"Professional Identities" (Faber, 2002) was written to mark and respond to a growth in writing and communication programs that aligned themselves with the art and science of the professions. The concern that led to the project was a perception that this growth and alliance was occurring without a concomitant attention to the concepts of *professional* or *professionalism*. While researchers had articulated specific functional and categorical definitions that were consistent with the sociological literature on the professions (e.g., Couture, 1992; Geisler, 1994; Savage, 1999; Sullivan & Porter, 1993), these portraits had little influence on pedagogy, program development, or studies and articulations of workplace communication. As the article showed, rhetorical studies of workplaces largely conflated all forms of occupational writing as "professional."

At the time, Sullivan and Porter (1993) had articulated an alternative frame for understanding the unique roles associated with professional communication as something different from other forms of workplace writing. Working from Eliot Freidson's (1970, 1986) studies of medicine, Sullivan and Porter emphasized that the professions apply knowledge gained from esoteric education to serve the essential needs of the public (p. 417). Thus, professional communication within a corporate or institutional context would be oriented not to promote a specific company, product, or service, but towards "helping the company better understand the needs and interests of the public" (Sullivan & Porter, 1993, p. 414). In this characterization, the professional communicator was presented as an independent advocate for the public within or outside of the institutional confines of a corporation or government entity.

Working from Sullivan and Porter's (1993) study, I researched "Professional Identities" by examining 34 articles published in *The Journal of Business and Technical Communication* (*JBTC*), *The Journal of Technical Writing and Communication* (*JTWC*), and *Technical Communication Quarterly* (*TCQ*) between 1990 and 1999 in an effort to characterize how writers in the field advanced the concept of *professional* in professional communication. The 34 articles were comprehensive of all articles that included the phrase *professional communication* in the title or abstract and provided conclusions that spoke to curricular or research implications for professional communication, professional writing, or professional communicators. As I wrote at the time, the goal of the study was to "examine what the authors seemed to imply through their use of the term *professional* and, thus, how scholars in the field have conceptualized this term" (Faber, 2002, p. 310).

The study offered three findings that articulated what rhetorical scholars presented when they used the term *professional* to discuss professional communication.

■ 1. Audience Relationship

Professionals were viewed as workers who have an integral relationship with a specific and known audience. Professionals rarely communicated with anonymous audiences, larger (mass) groups of people, or people with whom they did not have a known and deliberate relationship. For example, a lawyer's professional responsibility is to represent a specific client. The lawyer may provide free legal advice on a website or blog but in that capacity will note that such communication is not professional advice, but it is educational or informative writing. Similarly, responsible medical blogs or websites do not claim to be diagnostic but are informational, and their writers advise readers to seek *professional* (individualized) medical assistance from a physician.

■ 2. Social Responsibility

Professionals were portrayed as people who work in occupations that have specific social and community obligations and responsibilities. These obligations and responsibilities are knowledge-based and serve larger conceptual categories such as "justice," "health," "knowledge," or "learning" rather than practical, immediate, materialist, or rule-bound objectives. The professional's social responsibility also informs client relationships in that professionals provide advice and direction clients are not always obligated to follow. However, in situations where a particular client explicitly violates or endangers the obligate arenas protected by professional powers, the professional has a duty to act. Thus, academics are required to enforce penalties for plagiarism offenses since intellectual dishonesty is an explicit breach of the cognitive realm for which academics have assumed responsibility. Physicians have a duty to act if a patient's health is endangered by institutional or even other practitioner actions.

3. Ethical Awareness

Professionals were viewed as members of an occupational group who have unique and specific ethical obligations to their specialty knowledge. A professional is ultimately responsible and accountable to professional knowledge as established and certified by other members of the profession. The professional is evaluated by the codes of conduct, duties, and performance expectations established by other professionals rather than by institutional authority, clients, or customers, even if payment is rendered by these other groups. While professionals may be paid by a customer or may work within a large institution (hospital, university), professionals break the traditional capitalist contract in that they do not see themselves as ultimately accountable to the people who pay them. Similarly, professionals break traditional bureaucratic hierarchies in that the rules of the profession supersede the rules of the workplace.

The Professionalization of Everyone

In the time since "Professional Identities" was published, writers have offered alternatively careful and creative propositions and defenses for situating as *professional* the rhetorical activities of occupational, hobbyistic, and personal pursuits. Not comprehensively, these have included accounts detailing the activities of writers of online product reviews (Mackiewicz, 2010); women providing online advice about motherhood (Petersen, 2014; Rogers & Green, 2015); women podcasters (Petersen, 2016); craft beer artisans and people who write about, advertise, and promote the craft beer industry (Rice, 2016); Pre-hospital care providers (Angeli, 2018); lesbian, gay, bisexual, and transgender (LGBT) individuals working in corporate (retail) management (Cox, 2019); and physicians who reasserted autonomy and power as they simultaneously adapted to potentially deprofessionalizing workplace changes (Del Canale, 2012).[1] While not comprehensive and with some

---

1. In full disclosure, while I served as a reviewer for a number of these projects, my review did not address whether or not the particular occupation or activity chosen by the writer might or should qualify as a profession, a semi-profession, the professional-managerial class, or another occupational arrangement. In part, I did not envision that as my role. Instead, I have been more interested in how the fields that study occupational rhetorics have articulated an understanding of the professions and whether or not that articulation is accompanied by rhetorical forms or activities that delineate purposes that are unique to a *professional* disposition.

exceptions, this discussion has largely focused on whether or not particular writers' forms and actions could be considered (or should be considered) professional.

As the list above suggests, regardless of (or despite) efforts towards a more restrictive accounting of a specifically *professional* communication, researchers aligned with the occupational practices of technical and other workplace-specific communication have continued to aggregate nearly all workplace rhetorics with professional activities (Bridgeford et al., 2014; Coppola, 2012; Rosén, 2013; Spigelman & Grobman, 2006). Several writers, following the distinction made by Barbara Couture and Jone Rymer (1993) have continued to promote a distinction between "professionals who write" and "career writers" (Couture & Rymer, 1993, p. 5; see for example, Artemeva & Fox, 2014; Bhatia & Bremner, 2014; Henry, 2000). Yet, as Cindy Sing-Bik Ngai (2018) has demonstrated in her recent review of the research literature in professional communication, it remains common to conflate "occupational" and "professional" without drawing distinctions in the rhetorical purpose, form, intent, audience, or action different occupations or actors may enact or promote in their communication. Ngai's (2018) review is insightful as she shows that *professional communication* has emerged as a generalizing term that encircles any form of workplace, business, technical, or organizational talk, writing, and communication. At the same time, Ngai also documents specific context-specific studies in business, education, engineering, engineering management (marketing, collaboration), and medicine that also self-identify as professional communication.

■ Professional Communication as Aspirational Discourse

Over the two decades since "Professional Identities," when the research literature has differentiated *professional* communication from *occupational*, the distinction has appeared to be aspirational rather than conceptual, functional, or categorical. Advocates of particular discourses have made distinct cases to argue that a specific practice be considered *professional*. The form here has been to claim an aspiration to achieve professional status and then subsequently detail particular shortcomings that need to be (or have been) overcome before the practice could reach full professionalization.

Terry Skelton and Shirley Andersen's 1993 guest editorial in the Society for Technical Communication (STC) journal *Technical Communication* reads as an enduring representation of this form. Skelton, then manager of the STC professionalism committee, and Andersen, then assistant to the president of the STC for professional development, wrote a guest editorial on behalf of 21 members of the STC professional committee. The editorial was a statement on the status of the field as a profession. Working exclusively from Wilbert Moore and Gerald Rosenblum's 1970 book, *The Professions: Roles and Rules*, Skelton and Andersen recounted six criteria Moore and Rosenblum provided as a "scale of professionalism": (1) specialized educational preparation, (2) body of knowledge acquired through research, (3) unique and indispensable public service, (4) autonomy in work practices, (5) ethical professional practice according to enforceable codes, and (6) commitment to the values of public service and social responsibility (Skelton & Andersen, 1993, pp. 202–205). Skelton and Andersen reported that technical communication conformed well to principles one and two, as universities provided specialized training and universities and corporations supported and advanced research that informed and was applied to occupational work. They also reported that the field maintained a commitment to "public service and social responsibility" by "using socially responsible language," "facilitating the timely communication of information," and "putting the public good above special interest, i.e., service as public advocates" (Skelton & Andersen, 1993, p. 205).

However, Skelton and Andersen (1993) conceded that, at the time, the field did not yet constitute a unique and indispensable public service. They wrote, "Although technical communication, at this point, is not widely recognized by society as offering a 'unique and indispensable public service,' its value is increasingly recognized by business" (p. 204). Skelton and Andersen also recognized that technical communicators had yet to claim full autonomy over their work. Though, they wrote that they hoped that the advent of "the total quality ethic" in business would create conditions under which technical communicators "should experience increasing autonomy" (p. 204). Finally, they wrote, "Technical communicators do not currently operate according to an enforceable code of ethics unique to the profession" (p. 204).

Although the field fell short in three of their own six criteria, Skelton and Andersen (1993) assumed and thereby asserted that technical communication was a profession (p. 202). They argued that "professionalism ultimately is manifested in the behavior of practitioners," asserting that the occupation is organized as a profession if its practitioners act professionally. Acting professionally here entailed demonstrating (1) commitment to the profession, (2) commitment to a professional calling, (3) commitment to organizing the profession, (4) commitment to education, (5) commitment to a service ethic, and (6) commitment to achieving professional autonomy (Skelton & Andersen, 1993, pp. 205–206).

Skelton and Andersen's choice of *The Professions: Roles and Rules* as their guidebook to the professions provided them with a favorable and relatively diffuse description, as the book does not scale or define occupations as professions but instead largely evaluates characteristics of workers. The book is also concerned not with professionals as independent workers but in locating and defining professional work within organizations. The book's focus is institutionalized professions, an occupational space that would seem particularly suited to technical communication. As Ida Simpson wrote in a 1972 review of the book, "The criteria are not used as a scale to compare the professionalism of occupations but chiefly as rubrics under which attributes of professional roles and their incumbents are described" (p. 408).

Roles and Rules could provide a useful model for how an occupation could emerge (or "evolve" to use the book's metaphor) as a profession while within the confines of a larger institution. However, Simpson (1972), in her review, was critical of the suggestion, arguing that the book never fully connects the activities of presumed professional workers with the unique roles, responsibilities, and accountabilities of their occupations. She wrote,

> But the relations of professional roles to their parent occupations as commonly treated in the literature on professionals are not dealt with. The failure to distinguish explicitly between profession and the roles of professional individuals or to make plain that the scale of professionalism deals chiefly with the latter weakens much of the analysis...." (p. 408)

In other words, simply asserting professional status was not a sufficient condition for recognizing or treating an occupation as a profession.

In an important argument, Simpson (1972) wrote that the book's justification for professional service is convoluted and, in many ways, self-justifying. While somewhat difficult to trace, her point here is worth presenting in full:

> The institutionalization of professional roles is said to be a sequential acquisition of the professional attributes. But in discussing the institutionalization of roles, Moore and Rosenblum appear to start with the assumption that a full-blown profession, including the role expectations which are to be institutionalized in the later stages of the process, already exists. At this point in the analysis, the relations among professional attributes and sequential process stages become difficult to disentangle. The nature of the demand for professional services as described by the authors presupposes that a service orientation has been institutionalized, but the professional role — including its service orientation — is said to be institutionalized only after the demand for professional services has evidenced stable continuity. The service orientation is defined in terms of the very thing that supposedly fosters its institutionalization: the bringing to bear of professional judgment on client's problems. (Simpson, 1972, p. 408)

Simpson's (1972) argument can be equally applied to Skelton and Andersen's (1993) essay as to their theoretical source, as both studies of professional work begin their analysis by assuming "that a full-blown profession" exists and fail to adequately establish the connections between what may be well-intentioned and deeply committed occupational actions and the actual roles, responsibilities, and actions of an actual profession. What we do not get from Skelton and Andersen's essay, to use Simpson's critique, is a "coherent line of reasoning to show systematically the relations of the criteria of professionalism to each other or the process through which professions and professional roles become institutionalized" (p. 408).

This critique is important to current day explications and assessments of what is being articulated as professional communication. Ironically, the foundational assumptions articulated by Skelton and Andersen—that while particular occupational forms or actions do not necessarily fulfill the empirical criteria for professional categorization, desire and ambition to be a profession are sufficient evidence to sustain the argument—have largely held over the more than a quarter century since their argument was published. Similarly, pointing out that a particular rhetoric has an ethical component or is informed by a community's ethics does not necessarily constitute that rhetoric as professional. Ethics is not the sole terrain of the professions. What has been largely elided in these discussions has been a robust articulation of how rhetorical function, intent, and form may be distinctive when discourse enacts a specifically professional purpose.

■ Eliot Krause and the Professions as a Check and Balance

Krause's (1996) articulation of professionalism as a sphere of societal influence and competitive balance within Western democracies has continued to direct my own understanding of the professions' occupational designation and communicative practice. After reviewing the functional, trait, and institutional perspectives of professionalism, Krause constructs a competitive model of the professions that aggregates workplace, economic, and social power into an occupational field. Krause's model emerges from both Freidson's (1970, 1986) studies which showed how medicine used specialized knowledge to create social power and Magali Sarfatti Larson's (1977) analysis of the professions' role in shaping an emergent class based on the simultaneous monopolization and valuation of knowledge. To these models, Krause adds a historical perspective that casts the professions as modern variations of formal artisans' and trade workers' guilds. Putting these approaches together, Krause characterizes the professions as independent guilds that occupy a third-form of social influence and power. Their power is not found in wealth or capital, nor in bureaucratic regulations and institutional hierarchy, but instead is rooted in circumscribing and monopolizing specialized knowledge that societies require for modern life: medicine, education, law, and science, being archetypal, but not exclusive, professional sectors.

Working from a historical perspective, Krause (1996) details the unique occupational patterns and responsibilities that formed conditions for professional occupations in Western economies. As such, he describes the differing conditions for professional work, power, and motivation in the United States versus those in Germany, Italy, and the UK. Engineering and academics, for example, have emerged differently in European economies than in the United States. Academics are nearly exclusively state employees in Europe, and engineers work largely, though not exclusively, for large corporations in the United States. Similarly, an update to Krause's work could consider how different European or Canadian healthcare systems differ in the way physician work is organized versus in the United States. For example, in the United States, over the past two decades, most physicians have become employees of large hospital systems or group practices, and the adjacent occupations of nursing (nurse practitioner) and physician assistant have simultaneously adopted physicians' traditional primary care and public health responsibilities while pushing physicians into greater specialty roles.

Having spent the better part of a career studying the professions, Krause is an advocate for what he envisions as a special contract the professions elicit within their communities according to which professionals "provide service and use their knowledge for economic gain" (1996, p. ix). Importantly, for Krause, the professions are not merely a vehicle for occupational status or a more desirable way to describe a vocation. The professions are necessary, Krause emphasizes, for their efforts to shape, limit, and influence the state and capitalism. "Visualize a triangle," Krause writes, "with the state, capitalism, and the professions at the corners. The state influences and shapes capitalism and professions, capitalism influences and shapes both the state and professions, and the professions act to influence and confront the power of both capitalism and the state" (pp. 1–2). Professionals and, by association, professional communication, are cast as competitive antagonists, methods for eliciting checks and balances against the overreaches of capitalism and bureaucracy.

Krause's ultimate concern appears to be the consumer of professional services. In a question that appears increasingly prescient since 2016, he writes, "[i]f the doctors, the lawyers, the engineers, and the professors lose their power over the delivery of healthcare, legal service, applied science, and knowledge itself . . . and they lose it to capitalism and the state, what will be the implications for all of us?" (1996, p. 2).

Capitalist marketplaces and omnipresent bureaucratic regulation are equivalent functional models for providing consumers with necessary and important services. However, proponents of professionalism argue, left unchecked, capitalism will accelerate the pursuit of efficiency and profit maximization without regard for human life, the environment, or even the system's own long-term sustainability. Similarly, bureaucratic regulations can become oppressive, stifle innovation and investment, eliminate incentives, and legislate without concern for difference, nuance, context, quality, or situation. Here, Krause's (1996) model strikes an important balance among the three sectors and stresses their productive tension. The professions' advocacy for health, safety, and fair pay, for example, mediates capitalist rationalization and its prioritization of profits over the provision of services. Yet, Krause also notes that at times government was required to mediate professional excess and capitalist-leaning monopolization of knowledge. For example, to counter increasing specialization and a lack of access to medical care in the 1960s and 1970s, the U.S. federal government created national Medicare and Medicaid programs and loosened professional physician monopolies by sanctioning alternative practitioners in physician assistants (1965) and nurse practitioners (1974).

These activities occurred as a response to a period from 1930 to the 1960s, during which, while a high point for guild power among U.S. professions, American medicine showed particularly limited concern for the poor, the socially marginal, and at-risk communities (Krause, 1996, p. 284). Thus, the tensions between the different sectors are dynamic. Sectors may over time converge or share interests, and while some historical periods may see a particular sector rise in dominance, like the professions from 1930–1960, other eras may be dominated by government or, as we see in the current era, capitalism.

▪ Capitalizing Professional Communication

When Skelton and Andersen (1993) claimed that, at the time, technical communication could gain professional status because "its value is increasingly recognized by business" (p. 204), they conflated professional and capitalist forms, actions, and interests. Skelton and Andersen do not specify what "value" business is recognizing in technical communication. They do offer that some of this value is found in contributions that "have improved product quality while reducing the time and cost of product development" (p. 204). Improving efficiency, reducing cost, and improving product quality are important and considerable contributions. However, they are contributions that advance the business interests of the corporation. This is an important distinction because it demonstrates how, especially in an American context, the interests of capitalism have been conflated with and have increasingly eroded the independent functions of the professions. When business is able to set the terms that define and value the professions, the power of one sector (capital) has fully subordinated and delegitimized the other.

Krause (1996) discerns this consolidation of professional and capitalist interests well, writing,

> Directly, capitalists are the employers of many professional groups. The characteristically private practice of the American medical and legal professions of 1930 have given way, especially since 1970, to employed physicians and an elite segment of lawyers working directly for big corporations either in legal firms or, increasingly, as "house counsel" within the corporation itself. (p. 35)

Further, he writes,

> Increasingly since the 1970s, though, capitalists have moved to employ professionals including doctors and lawyers, more directly, to take ad hoc action to control the costs created by professionals, and to work with the state toward constraining the remaining guild power of the professions. (Krause, 1996, p. 35)

While American professionals have enjoyed the benefits of enrolling within capitalist work spaces, the financial benefits associated with promoting capitalist enterprises, and the prestige that comes with the "culture of professionalism" (Bledstein, 1976; Krause, 1996, p. 31), such conflation does much to obscure the watchdog and mediating function the professions have historically served in a well-balanced economy. Returning again to Krause (1996), he writes,

> Where state and capitalist power have won out, they and not the profession control the aspects of professional life that we call "the workplace" and "the market" and determine to a large extent how much associational group power the profession has left vis-à-vis the state and capitalism. Subgroups play an important role here—in some cases, the elite remains in some kind of guild control while the mass has succumbed to capitalist or state control, or to a mixture of the two. (p. 22)

When academic researchers conflate the interests of the professions with those of capitalism or those of government, we continue to perpetuate the erosion of professional responsibility and professional power, and we (perhaps unintentionally) promote the interests of capital (or government). In our teaching and our research, we should be more careful to delineate the unique purposes enacted by each sector. Business communication cannot be identified as professional communication because one of the primary purposes of professional communication is to constrain the excesses of business communication. Business communication, by definition, emerges from and promotes capitalist, market-driven, and commercial forms and actions. Its purpose is to generate wealth, commercialize value, and promote the functioning of a market-based economy. Consolidating the business, regulatory, and professional sectors is not only inaccurate, but it obscures the conditions that gave rise to each sector and elides the real and crucial tensions, historical and current, between each sector.

Professional Communication as Functional, Interventional Discourse

Looking back at "Professional Identities" and the literature since, I wonder, anachronistically, if that research could have been more useful if it had switched frames from professional as an occupational category to a communicative function. Despite the baggage associated with the term, perhaps communicative *intent* provides a more productive frame to deliberate what is professional about professional communication than a narrow focus on occupational class, status, or aspiration. There are, of course, occasions when a physician, lawyer, or scientist communicates outside of and in ways unrelated to professional forms and actions.

There are times when physicians make business transactions, scientists are constrained by bureaucracy, and lawyers undoubtedly gossip. Perhaps the question can be restated: What makes professional communication uniquely professional? Alternatively, when does communication deploy a specifically professional action, on what occasion, to what purpose, in what form, with what consequence, with what risk, against what sort of disputant, and toward what cause? Perhaps future research into professional communication can be less focused on occupational status and function and instead can seek to locate specific moments and spaces when communication enacts a specific *professional* activity.

Several years ago, in an effort to improve the institution's reputation, the former provost at an institution where I worked required that peer review activities related to faculty tenure and promotion solicited evaluations from individuals at "top tier universities." The logic here was that passing faculty dossiers through the hands of influential people at more highly ranked institutions would improve the stature and reputation of the university. The evaluators would come to see and associate the university among their own peer group. More pragmatically, when asked to complete reputational surveys for national rankings, these influential people would rank the school higher than they ordinarily might have done. Higher rankings would lead to higher prestige, more undergraduate applicants, higher yields, and students more willing to pay full (or less reduced) tuition.

This effort to conflate marketing with peer review was mostly ignored by committees. However, several faculty members disputed what was seen as the marketization, indeed monetization, of a non-commercial *professional* process. These faculty members also argued that the requirement was misleading: faculty with a teaching-based load (3:3 or 4:4) would be compared with and evaluated by faculty with a research-based teaching load (1:1 or 1:2). Faculty who had no lab space or who were sharing lab space with researchers and graduate students in different disciplines would be compared with faculty with extensive laboratory resources. Faculty in undergraduate-only programs would be evaluated by faculty with graduate students. Asserting that one institution was "peer" to the other was repudiated as a fabrication and a violation of the American Association of University Professors' (1966) ethical standards. Professional standards were eventually reinstated with the appointment of a new provost. When overturning this policy, the new provost explicitly noted that reviews should be obtained by "appropriate" faculty at similar institutions, with similar responsibilities.

On a long weekend in July, the ambulance agency where I volunteer was asked to respond to a two-car motor vehicle accident. After arriving at the accident scene, the responders discovered that a young driver had fallen asleep while driving and veered into the oncoming lane, colliding with another vehicle. Fortunately, there were no life-threatening injuries, but both drivers were taken to the local hospital for evaluation. In his evaluation of the driver who had fallen asleep, the emergency department physician determined that the patient had fallen asleep because the patient was diagnostically morbidly obese and the weight of the patient's neck and chest impeded adequate breathing when the patient was positioned in the driver's seat. The physician determined that the patient represented a threat to public safety and confiscated the patient's driver's license. Several hours later, the patient's father confronted the physician, stating that the patient needed the driver's license for work. The physician was unmoved and stated that the license could be reinstated if the patient lost sufficient weight such that the patient's condition would no longer pose a threat to self or society.

Both of these cases demonstrate moments of professional communication in that they are spaces in which specialists use their objective, knowledge-based positions to confront and mitigate perceived excesses of capitalism. In advocating against capitalism, these actors are also promoting larger conceptual categories (justice, health, knowledge, learning) rather than immediate, materialist, and commercial interests. The patient's ability to drive to work, while a primary concern of business and capital, may have overlapped with but did not add up to the sum total of a physician's domain. That the patient represented a threat to self and society overrode the capitalist's immediate monetary problem. The patient would need to find another way to get to work. Similarly, the administration's ability to monetize its national rankings was a concern unrelated to and separate from the faculty's professional obligations to conduct a fair review of its membership and be truthful with external colleagues. In both cases, professional communication was enacted situationally and deliberately with a clear intent to push back against capitalist incursions.

■ Visualize a Triangle: Movements and Curative Action

"Visualize a triangle," we could write, in a specifically rhetorical version of Krause's (1996) model, with business communication, regulatory communication, and professional communication at the corners. Regulatory discourse influences and shapes capitalism and professions, business communication influences and shapes both the state and professions, and professional communication acts to influence and confront the power of both capitalism and the state. Conflating these practices misrepresents the unique and crucial roles, purposes, and intentions of each sector. Simultaneously, such conflation also subordinates the ethical obligations and social responsibilities of one to the other. Each activity enacts separate and important intentions that could still be better researched, understood, and articulated by rhetorical workplace scholarship.

This is not to suggest that any capitalist-confronting or rule-defying communicative action constitutes professional discourse. The findings of "Professional Identities" continue to be relevant and appropriate even if the social and occupational terrain of the professions may have changed. The professions continue to operate with individual clients, with knowledge-based and conceptual social responsibilities, and with an ethical obligation to uphold their knowledge and the unique functions that knowledge enables. What "Professional Identities" and subsequent work may have overlooked is the layering and integration of the characteristics. While professionals work with distinct and individual audiences (e.g., patient, client, student), the motivation for this activity is the adjudication of the professional concern as it relates to the audience's particular circumstance:

A lawyer's long-term responsibility is the enactment of justice, not necessarily a win for the client. Physicians diagnose disease even if a patient dies. Scientists pursue knowledge, even if that pursuit is disruptive to a student's, community's, or politician's belief. The enactment of societal service takes place through the professional's audience relationship, and the professional is ethically bound to advocating and upholding the concept such work entails. While professional discourse may advocate, not all advocacy or protest brings with it the structural and institutional power the professions wield.

As a correction to capitalist or regulatory overreach, professional communication may include an overt critique or may simply function as a decree. Whether and how this is accomplished; how such decrees are enacted, sustained, and made rhetorically effective; and where and how they inflect capitalism or regulation remains a productive question. The sort of dynamic offered in a rhetorical deployment of Krause's (1996) triangle articulates professional communication as a curative action and a purposeful, even temporary, intervention. It also introduces a certain movement or motion that could turn this discussion away from distinctly occupational frames. For example, a technical communicator could deploy professional discourses to rebalance the power dynamics between users or particular groups and individuals and those who would profit from either the unbounded expansion or the undue restriction of a particular technology (Haas & Eble, 2018). Alternatively, technical communicators could adopt forms of business communication when working to market their products and services, maximize efficiencies, and conduct other actions consistent with marketization of goods and services. Technical communicators adopt regulatory or rules-based communication when creating products that require strict adherence to narrow instructional forms.

Paul Rabinow (2003), in the conclusion of his book *Anthropos Today*, discusses the growing distances between technology, science, and the social and philosophical thought that has attempted to characterize such work. He writes that as "technology was preceding science and achieving a certain autonomy ... this separation and this relative autonomy itself became a phenomenon that required new types of explanation, new narratives, and new metaphors" (p. 135). Rabinow concedes that within such flux, there remains an "impulse" to create a comprehensive narrative and a common account, something to retroactively make sense of where we are and what may have occurred. Yet, Rabinow asks us to resist such an approach, suggesting that such a quest is born of "the reflex to answer old questions" (p. 135). Similarly, attempting to account for a uniquely professional occupational discourse in a fractured, disconnected, and increasingly polarized economy, political culture, and weakened regulatory sector may be seeking answers to questions that are no longer relevant. Rabinow instead offers the metaphors of motion and movement, of a critical practice attuned to what he calls "relations of distance and closeness" (p. 135). To Rabinow's list, I would add modulation. In the example I suggested above, where a technical communicator deploys and moves through situational capitalist, regulatory, and professional discourses, we are presented with new questions of discursive modulation. As Rabinow suggests, such questions entail movement, passage, and rhetorical legitimacy. How might rhetorical agents legitimately pass from capitalist to professional discourses and retain credibility? Might a professional leverage the blunt forces of regulation in order to uphold a commitment to health, justice, or science? How might professional discourses continue to promote core concepts like health, medicine, justice, and science if such concepts are aggregated as equal or contemporary to capitalist and regulatory values? Perhaps the question for a new generation of researchers is how this dynamic is managed, maintained, and modulated by the sorts of new occupations, rhetorical positions, and institutional powers that have emerged over the past 20 years.

This is not to say that we should forget or ignore what we know. Professional communication occupies a distinct purpose apart from, in contrast to, and in competition with other forms of workplace communication and, as such, it is curated in strategic forms and actions within and against these other economic and socially-contested spaces. But movement also permits a certain flexibility. It elides some nostalgia for a discursive order that in actuality may never have been altogether fixed. And, perhaps more importantly, it allows for a renewed appreciation for institutional and disciplinary events that have successfully transformed new discourses into what are now stable acronyms, courses, majors, departments, research journals, and productive, useful work.

References

American Association of University Professors. (1966). *Statement on professional ethics.* https://www.aaup.org/report/statement-professional-ethics (Revised 1987, 2009)

Angeli, Elizabeth L. (2018). *Rhetorical work in emergency medical services: Communicating in the unpredictable workplace.* Routledge.

Artemeva, Natalia & Fox, Janna. (2014). The formation of a professional communicator: A socio-rhetorical approach. In Vijay Bhatia & Stephen Bremner (Eds.), *The Routledge handbook of language and professional communication* (pp. 461–485). Routledge.

Bhatia, Vijay & Bremner, Stephen (Eds.). (2014). *The Routledge handbook of language and professional communication.* Routledge.

Bledstein, Burton. (1978). *Culture of professionalism: The middle class and the development of higher education in America.* W.W. Norton.

Bridgeford, Tracy, Kitalong, Karla Saari & Williamson, Bill.(Eds). (2014). *Sharing our intellectual traces: Narrative reflections from administrators of professional, technical, and scientific programs.* Routledge.

Coppola, Nancy W. (2012). Professionalization of technical communication: Zeitgeist for our age introduction to this special issue (Part 2). *Technical Communication, 59*(1), 1–7.

Couture, Barbara. (1992). Categorizing professional discourse: Engineering, administrative, and technical/professional writing. *Journal of Business and Technical Communication, 6*, 5–37.

Couture, Barbara & Rymer, Jone. (1993). Situational exigence: Composing processes on the job by writer's role and task value. In Rachel. Spilka (Ed.), *Writing in the workplace: New research perspectives* (pp. 4–20). Southern Illinois University Press.

Cox, Matthew B. (2019). Working closets: Mapping queer professional discourses and why professional communication studies need queer rhetorics. *Journal of Business and Technical Communication*, *33*(1), 1–25. https://doi.org/10.1177/1050651918798691.

Del Canale, Marco. (2012). *The impact of changes in the medical environment on physicians' identities* [Master's thesis, Lund University]. http://lup.lub.lu.se/luur/download?func=downloadFile&recordOId=3633612&fileOId=3694087.

Faber, Brenton. (2002). Professional identities: What's professional about professional communication? *Journal of Business and Technical Communication*, *16*, 306–337.

Freidson, Eliot. (1970). *Profession of medicine*. Dodd, Mead.

Freidson, Eliot. (1986). *Professional powers*. University of Chicago Press.

Geisler, Cheryl. (1994). *Academic literacy and the nature of expertise: Reading, writing, and knowing in academic philosophy*. Lawrence Erlbaum.

Haas, Angela M. & Eble, Michelle. (2018). Introduction: The social justice turn. In Angela M. Haas & Michelle Eble (Eds.), *Key theoretical frameworks: Teaching technical communication in the twenty-first century* (pp. 3–19). Utah State University Press.

Henry, Jim. (2000). *Writing workplace cultures: An archaeology of professional writing*. Southern Illinois University Press.

Krause, Elliott A. (1996). *Death of the guilds: Professions, states, and the advance of capitalism 1930 to the present*. Yale University Press.

Larson, Magali Sarfatti. (1977). *The rise of professionalism*. University of California Press.

Mackiewicz, Jo. (2010). The co-construction of credibility in online product reviews. *Technical Communication Quarterly*, *19*(4), 403–426. https://doi.org/10.1080/10572252.2010.502091.

Moore, Wilbert Ellis (with Gerald W Rosenblum). (1970). *Professions: Roles and rules*. Russell Sage Foundation.

New York Times Editorial Board. (2018, December 12). Judges check Trump's immigration cruelty. *The New York Times*. https://www.nytimes.com/2018/12/21/opinion/asylum-jeff-sessions.html.

Ngai, Cindy Sing-Bik. (2018). Professional communication. In Robert L. Heath et al. (Eds.), *The international encyclopedia of strategic communication* (pp. 150–157). John Wiley & Sons.

Petersen, Emily J. (2014). Redefining the workplace: The professionalization of motherhood through blogging. *Journal of Technical Writing and Communication*, *44*(3), 277–296.

Petersen, Emily J. (2016). Reterritorializing workspaces: Entrepreneurial podcasting as situated networking, connected mediation, and contextualized professionalism. *IEEE Professional Communication Conference*. https://ieeexplore.ieee.org/abstract/document/7740516.

Rabinow, Paul. (2003). *Anthropos today: Reflections on modern equipment*. Princeton University Press.

Rice, Jeff. (2016). Professional purity: Revolutionary writing in the craft beer industry. *Journal of Business and Technical Communication*, *30*(2), 236–261. https://doi.org/10.1177/1050651915620234.

Rogers, Jaqueline McLeod & Green, Fiona J. (2015). Mommy blogging and deliberative digital ethics. *Journal of the Motherhood Initiative for Research & Community Involvement*, *6*(1), 31–49.

Rosén, Maria. (2013). From ad-man to digital manager: Professionalization through Swedish job advertisements 1960–2010. *Journal of Communication Management, 18*(1), 16–39. https://doi.org/10.1108/JCOM-04-2013-0038.

Savage, Gerald J. (1999). The process and prospects for professionalizing technical communication. *Journal of Technical Writing and Communication, 29*, 355–381.

Simpson, Ida. (1972). [Review of the book *The Professions: Roles and Rules*. By Wilbert E. Moore in Collaboration with Gerald W. Rosenblum]. *Social Forces, 50*(3), 407–408.

Skelton, Terry & Andersen, Shirley. (1993). Professionalism in technical communication. *Technical Communication, 40*(2), 202–207.

Spigelman, Candace & Grobman, Laurie (2006). Why we chose rhetoric: Necessity, ethics, and the (re)making of a professional writing program. *Journal of Business and Technical Communication, 20*(1), 48–64: https://doi.org/10.1177/1050651905281039.

Sullivan, Patricia A. & Porter, James E. (1993). Remapping curricular geography: Professional writing in/and English. *Journal of Business and Technical Communication, 7*, 389–422.

Chapter 5: Procedural Knowledge and Discourse in Technical Communication: Easy as 1, 2, 3?

Marjorie Rush Hovde
INDIANA UNIVERSITY-PURDUE UNIVERSITY INDIANAPOLIS

Abstract: Within technical communication, understanding the complexities of procedural knowledge and discourse is crucial to creating effective user documentation in many forms. In addition to providing insights into procedural knowledge, this chapter explores differences between descriptive technical discourse and procedural technical discourse that helps people gain procedural knowledge. The chapter also explores several implications of these differences for creating effective procedural discourse, including the importance of usability testing of instructions, followed by a discussion addressing several myths about the creation of and importance of procedural discourse. The chapter closes with implications for future research into procedural knowledge and discourse.

Keywords: procedural discourse, procedural knowledge, instructions, documentation usability

I recently encountered a product whose label proclaimed that setup and use were as "Easy as 1, 2, 3." This phrase appeared to function as a marketing tool to persuade users to purchase a product that would be easy to use. However, in many technical contexts, processes are frequently more complex than "1, 2, 3" and can often frustrate and alienate users who do not know how to complete them. Rich understandings of the complexity of procedural knowledge and its discourse can help technical communicators navigate the challenges that arise when they try to teach users how to use technology or other systems effectively.

Whether technical communicators create stand-alone manuals, online help, training experiences, instructional videos, or other forms of procedural discourse intended for users, they benefit from understanding the complexities of procedural knowledge and its relevant discourse. Because "documentation is a learning medium that can transform the user experience, providing useful and practical information presented in a context-sensitive format" (Hogan, 2013, p. 156), paying close attention to communication can help to develop users' procedural knowledge. In addition, technical communicators are helped by understanding effective processes for creating procedural discourse, processes undergirded by foundational assumptions about the relationships between users and systems (Johnson, 1998).

However, many technical communicators (and their colleagues) may not be aware of the differences between descriptive knowledge and procedural knowledge, and thus they produce discourse that does not help users understand and follow relevant processes (Hovde, 2010). Furthermore, many forms of procedural discourse may be too simple for helping people to function within complex and interlocking systems.

This chapter explores the following:

- perspectives on procedural knowledge and procedural discourse,
- the complexities of thinking about technical communication for non-routine processes in complex contexts,
- practical implications for technical communicators who wish to improve processes for producing effective procedural discourse,
- the role of usability testing of procedural discourse, and
- several myths about the role of procedural discourse.

In addition to exploring my observations and experiences over many years, I draw on insights from a variety of scholars who provide rich understandings for practitioners as well as instructors and scholars of technical communication who wish to understand the richness and complexity of procedural knowledge, understandings that are foundational to creating procedural discourse.

Perspectives on Procedural Knowledge, Procedural Discourse, and Descriptive Discourse

Before exploring the implications of procedural discourse for technical communication, definitions of procedural knowledge, procedural discourse, and descriptive discourse may prove helpful.

Procedural Knowledge

Procedural knowledge exists in action. It typically begins in a situation where a current state is not desirable, includes actions that move toward a goal, and ideally ends when the goal state is achieved (Farkas, 1999). This knowledge "is not just cognitive, but often tactile and visual as well, relying on cues from context on when to act and what to do" (Durack, 1997). Procedural knowledge is a larger category than procedural discourse, but procedural discourse is essential, especially when people are learning to use a complex system.

Procedural knowledge combines "how-to" skills with conceptual knowledge of a system, sometimes called "knowing that." A system is a structure in which users need to work to achieve their goals. It may be a computer system, an organization, a device, an electronic game (deWinter, 2014), or a set of policies. Routine processes are usually easy to learn and remember. For instance, in withdrawing cash from an automatic teller, users insert a card, enter a PIN code, and select an option from the menu on a screen. The processes become slightly more complicated when users wish to check a balance, make a deposit, or transfer money from one account to another. However, even those processes are easy to learn and remember (or figure out if the screen interface is well designed).

Procedural knowledge can be clear cut and uncomplicated when users follow a routine of unvarying steps in a simple system, but it becomes more challenging when users need to follow multiple possible pathways (Albers, 2004; Roochnik, 1996; Swarts, 2014, 2015) to achieve their inter-connected web of goals (Albers, 2004) in complex systems. When routine actions are not possible, owing to contextual factors or combinations of complex systems, users need to think of alternative actions (Farkas, 1999). Goals may shift and emerge as users are trying to create their procedural knowledge. A recent example of insufficiently developed procedural knowledge is the cases of the two Boeing 737 MAX airplanes that crashed because the procedural knowledge of the pilots was inadequate for overcoming problems with new software, primarily because the retraining of experienced pilots proved inadequate for this complex situation (Associated Press, 2019).

People's goals in using a system often spring from their unique contexts, complicated by the fact that users may think in terms that may not be the same as the system's terminology (Mirel, 1993). User goals relate to their contexts and work patterns, involving the "user's mental process" (Albers, 2004, p. 79) more than the possible functions of a system. In addition, users face cognitive, environment, and technology constraints as they work with a system, issues that system designers may not have considered.

People develop procedural knowledge through a variety of approaches. Some users learn through trial-and-error explorations of a system (Mirel, 1993). Others learn through direct instruction combined with practice. Developing procedural knowledge frequently involves multiple senses, according to neuroscientists who argue, "learning and cognition are multi-sensory experiences" (Remley, 2015, p. vii), indicating that multiple parts of the brain are involved. Users may have a variety of strategies for learning—strategies that involve the mind, but also other sensory-motor experiences; procedural knowledge is gained through cognitive, social, and physical means.

Developing procedural knowledge may involve one-on-one interactions with experts or it may involve group training. In the medieval guild system and into the 19th century, apprentices and learners developed procedural knowledge through oral instruction and by imitating what their masters or parents did (Durack, 1998). Early 20th century military training also involved demonstrations, explanations, repetition, and hands-on practice (Remley, 2015, p. 71), so procedural knowledge was transmitted both orally and via practice. Frequently, oral-dominant cultures transmit knowledge differently than literate cultures do (Durack, 1997). (Procedural knowledge also has been considered in some circles to be of a low status and an inferior form of knowledge. This perception continues today—at one university with which I am familiar, students are encouraged to take "knowledge" courses rather than "how-to" courses. For instance, a course in the history of art is acceptable as an elective, but a course in creating art is not acceptable.)

Because time and cost may make synchronous one-on-one or group training prohibitive for helping people develop procedural knowledge, technical discourse can help. However, written documentation and other asynchronous forms of discourse for users also present limitations because their "inherent linearity and rigidity . . . coupled with the necessary reduction of complex situations to sequential units of simple action" may cause users to misunderstand the effective and safe use of a system or tool (Paradis, 1991). Because of the asynchronous nature of much procedural discourse, creators and users may be operating with differing assumptions (van Loggem, 2013). However, despite the limitations of asynchronous procedural discourse, developing procedural knowledge with the aid of discourse is usually more effective than having users learn processes solely through trial and error.

In addition to a person knowing "how" to work within a system, that person's conceptual or descriptive knowledge of a system plays an important role in developing procedural knowledge (Hovde & Renguette, 2017; Swarts, 2018), especially needed when troubleshooting, completing non-routine tasks, or learning new processes. For instance, in the days before graphical user interfaces, I learned of one user who rebooted his computer each time he wanted to escape something. He did not know that a key on his keyboard would allow him to go back to a previous screen, basic conceptual knowledge about a toggle option that would have saved him a great deal of time. Conceptual knowledge, however, is not sufficient for achieving procedural knowledge. For instance, learning music theory can be helpful when learning to play a new instrument, but instruction and practice are needed to produce music. Moving through non-routine processes will require users who possess enough conceptual knowledge to know what to do when conditions shift (Farkas, 1999); conceptual knowledge provides an important foundation when tackling non-routine and/or complex processes.

Procedural knowledge within a person changes over time. When beginning to learn a process, users may need to learn through explicit steps. However, over time and with practice, procedural knowledge becomes internalized and tacit, seeming like second nature to the actor. For instance, a novice may perceive saving a file as several discrete steps, whereas a more experienced user will conceive of the process as a step or two. Although beginners may start with simple, clear-cut procedural knowledge, they often move to addressing problems that are "murky, unpredictable, and uncertain" (Swarts, 2018, p. 38). Complex non-routine processes are more challenging to learn and remember (Albers, 2004; Swarts, 2018), and thus conceptual knowledge plays an important role in learning and memory. For instance, a person moving to a new city may need a map which provides descriptive information that allows for navigation. However, once that person has become familiar with the streets, that person can often figure out the best route, using conceptual knowledge gained through experience and observation.

Overall, procedural knowledge involves knowing how to complete tasks in order to achieve goals. It encompasses both knowing "how" and knowing "that." In addition, procedural knowledge involves possessing enough conceptual knowledge to improvise when non-routine situations arise. Procedural knowledge exists in the doing, so it is difficult to capture solely in discourse. People gain procedural knowledge through instruction and practice, which may include trial and error as well as multiple sensory experiences. Technical communication in many forms at its best functions to "help accommodate technologies and texts to our situated use" (Swarts, 2018, p. 3). Because procedural knowledge exists in the doing and within users' physical bodies and memories, capturing and describing procedural knowledge can be challenging. Although procedural discourse is not the same as procedural knowledge, discourse plays an important role in developing procedural knowledge within individuals and within communities, as discussed in the next section.

Procedural Discourse

Procedural discourse is intended to help people accomplish goals (Farkas, 1999) in relation to a system and to develop their procedural knowledge. The system may be technical, related to an organization, or related to a larger network of resources and actions. For instance,

- Online help can assist users in employing software for their purposes.
- An employee manual can help users figure out how to function within their organization.
- An agricultural manual can help work within "a network of constructed waterways, the knowledge of when and how to irrigate fields, and the entire set of human activities that comprise this method for farming" (Durack, 1997, p. 258).
- Manuals or in-game instructions can assist people in playing electronic games (deWinter, 2014).

Procedural discourse plays an important role in creating larger and more complex procedural knowledge.

Over time, humans have devised a number of forms of procedural discourse which can involve more than words, encompassing a variety of symbol systems including the visual (Remley, 2015; Tenbrink & Maas, 2015). Procedural discourse may take forms such as paper manuals, training sessions, how-to videos, online help, or informal conversations among users, face to face or online. Online forums have the advantage of crowd sourcing, drawing on the resources and experiences of many users to address non-routine uses of a system. This form of user support becomes a conversation or dialog that can adapt to unique needs. These dialogic approaches not only answer questions, but also help users develop abilities to solve future problems (Swarts, 2018, p. 72). Having access to a variety of types of procedural discourse can allow adaptation to unique circumstances and a range of learning styles. Because many technical communicators also design websites that provide content other than online help, understanding procedural discourse assists in creating user interfaces that allow users to move easily through the tasks they need to accomplish on websites.

Whatever forms it takes, procedural discourse is necessarily a simplified version of procedural knowledge and is designed to assist users in learning (Paradis, 1991). Procedural discourse can allow for organizing knowledge and provide a means of sharing knowledge with others distant in time and space. Traditionally, software documentation has been aimed at "the normalization of user behavior . . . to teach the users what the software is capable of doing, how it can be done, and what are the best practices" (Swarts, 2018, p. 100). As they learn, users gradually develop procedural knowledge.

Because the brain changes as it learns new tasks, multimodal instructional materials—"print-linguistic, visual, audio, gestural, and spatial" (Remley, 2015, p. 24)—are crucial and help users learn and remember material because of the reinforcement from multiple senses. Overall, the effectiveness of multimedia may depend on users' learning styles and prior experiences (Remley, 2015, p. 37). Furthermore, if a user has a biological limitation, gaining procedural knowledge may be adapted to take that limitation into account (Albers, 2004; Remley, 2015). Imitation and practice are key to learning new processes, so procedural discourse alone is likely to be inadequate in developing procedural knowledge; nevertheless, the discourse can play a significant role for users.

Effective user documentation has significant social effects because this discourse can "interpret for the lay public the meanings, application, and procedures by which expert products . . . are integrated into the behavioral flow of society itself" (Paradis, 1991, p. 256), thus lowering barriers to access to sophisticated technological systems.

Because discourse can help users attain procedural knowledge, various approaches, especially in written guides, have emerged, some more helpful than others. In looking at approaches to procedural discourse, especially user documentation, one usually encounters several varieties: system-oriented discourse, user-friendly discourse, mixed system and user-task discourse, and user task-oriented discourse, as discussed below.

System-Oriented/Descriptive Discourse

System-oriented/descriptive discourse focuses on describing the features of a system and is most helpful at developing conceptual knowledge but is severely lacking in its ability to develop users' procedural knowledge. The most common format is technical specifications that describe the architecture of a system or product. Descriptions do not include "how-to" information, so users have to extrapolate how to use a system (Hovde, 2010). For example, in the early 1980s, when I was learning to use the word processing program Wylbur on an IBM mainframe, the system-oriented documentation consisted of a ten-foot shelf of papers in no apparent order on one wall of the computer center. At times, users walked over to it and looked up information, but most of us learned to use Wylbur either by asking people near us how to complete tasks and/or through trial and error. The system-oriented guides were useful only to a few people who had appropriate background and who could navigate the materials.

Figure 5.1 illustrates descriptive discourse that focuses on a system. Simply reviewing the table of contents will not let users know what tasks or goals they might accomplish using this publication. Most of the items in the list are nouns or noun phrases, indicating that the documentation describes the system features rather than how to use the system. Such wording is not as helpful to users as verb phrases that indicate user actions (Farkas, 1999, p. 46).

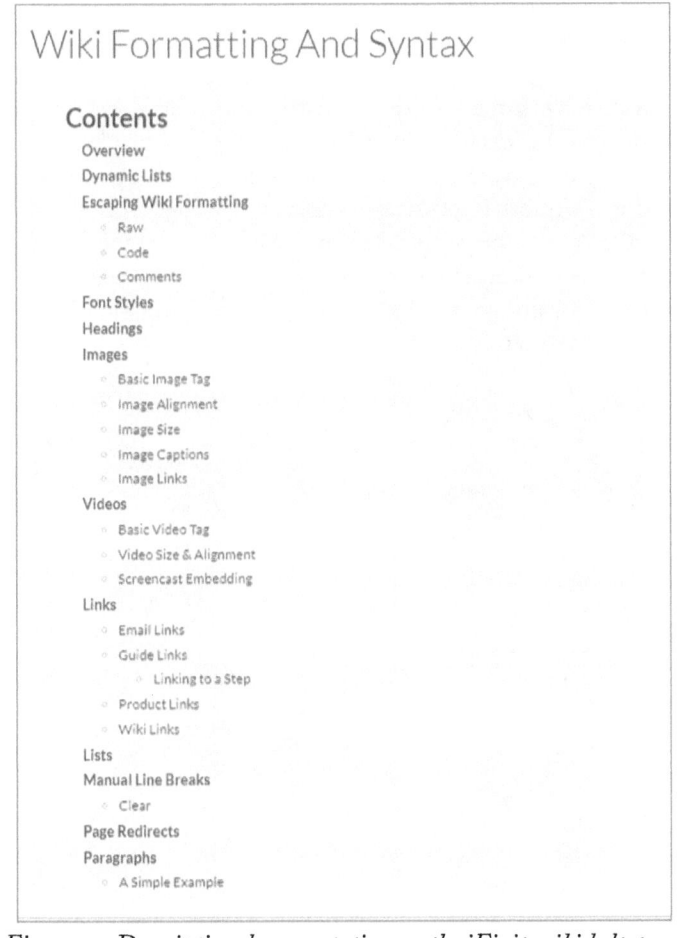

Figure 5.1. Descriptive documentation on the iFixit wiki help page. (Source: https://www.ifixit.com/Help/Wiki_Formatting_And_Syntax by permission of Creative Commons BY-NC-SA 3.0)

One problem with descriptive system-oriented technical discourse is that it may easily become "exhaustive," impeding usability by including information (Remley, 2015) that users may not need. For example, the documentation featured in Michael Salvo et al.'s (2007) study included complete details of a system, but it included much more information than users needed. Figure 5.2 shows an exploded diagram that describes the parts of the system but does not let users know about relevant processes for installing or using the system.

Exhaustive, system-oriented documentation typically ignores users' needs and perceptions, focusing attention on describing a structure (Johnson, 1998). Unsurprisingly, if users see only a static conceptual description, they typically find it difficult to use that system. Some users may try to learn to use a system through trial and error, but most ordinary users do not have the conceptual background, time, or patience to learn through that means.

Another problem with system-oriented documentation is that conflicts between clarity for the reader versus completeness of information about the system may arise because "with the information both hard to find and hard to process, the communication between the interface and the user has broken down and, for all practical purposes, the information doesn't exist" (Albers, 2004, p. 110). Additionally, it is almost impossible to provide complete information about a system; technical communicators need to decide what to include and what to exclude. Ideally, user documentation provides enough information to help users meet their goals but not so much that users become overwhelmed and cannot determine what information is relevant.

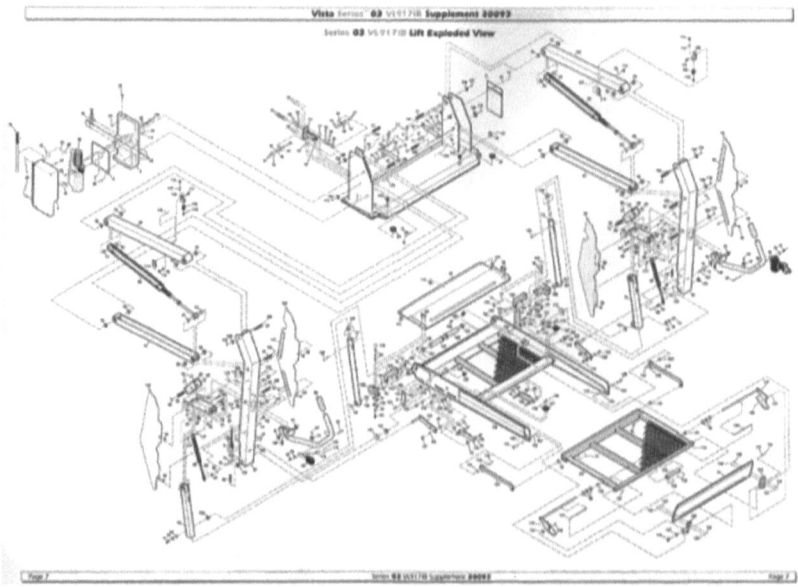

Figure 5.2. Exploded diagram (Salvo et al., 2007, p. 51) describing a system in exhaustive detail. (Used by permission of the Society for Technical Communication.)

Procedural Knowledge and Discourse   145

User-Friendly Documentation

To assist non-specialist users, some creators of technical communication attempt to make documentation user-friendly by using simple, informal language and attractive visual formatting, but "user-friendly" is not the same as user-task-oriented (Johnson, 1998). For example, when Google introduced Chrome in 2008, they released a comic book style explanation of the new browser, using drawings of people with word bubbles and casual, simple language explaining why this browser was unique, as excerpted in Figure 5.3. While the visuals and the language level make the information accessible to an audience with limited technical background, the publication was not focused on how to use the new browser; instead, its content dealt with the logic behind the design and important features of the system.

Figure 5.3. A user-friendly approach is not the same as a user-task-oriented approach, as illustrated by this excerpt from the Google Chrome comic book (McCloud, 2008; Creative Commons Attribution-Noncommercial-No Derivative Works 2.5 License).

Another example comes from a credit union during the earlier days of mobile phones. The user guide consisted of a small booklet that was the size and shape of a mobile phone, as seen in Figure 5.4, and it provided information about the credit union's app for accessing online services. However, while the booklet was user-friendly, it was not user-task-oriented, as seen in its table of contents in Figure 5.4, and thus not overtly helpful for using the bank's online services.

A user-friendly approach often presents a system orientation rather than a user-task orientation (Johnson, 1998). While ordinary language may enhance readability, and personable discourse and visuals are appealing, a publication that is system-oriented but user-friendly does not provide users with a sense of the actions they might complete in relation to the system. In addition, a user-friendly publication may assume that users will read the text in a linear fashion, but learning through doing is often not linear (Johnson, 1998).

One form that user-friendly documentation often takes is a "Frequently Asked Questions" list. While these lists attempt to meet users where they are, such lists are often focused on the system rather than actual user needs (Albers, 2004). In addition, these lists are often poorly organized, forcing users to sift through many questions in order to get to the relevant answer, if they are able to do so at all. Simply providing information is not adequate if it is not shaped and presented in ways that are accessible to users (Albers, 2004). Some creators of documentation realize this limitation and blend system orientation and user-task orientation, an approach that has limitations of its own, as described in the next section.

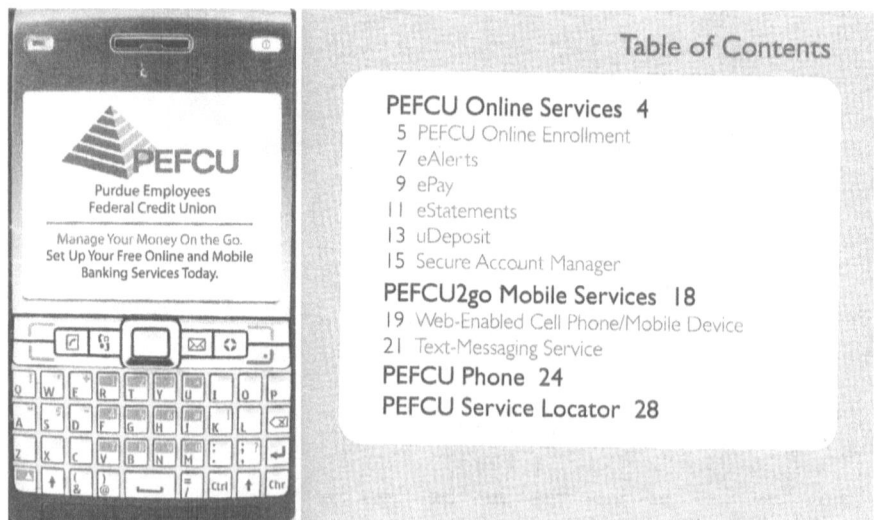

Figure 5.4. Although the cover of this guide employs a user-friendly format, the table of contents remains system oriented. (Purdue Employees Federal Credit Union, 2008. Photo taken by author.).

Blended System Orientation and User-Task Orientation

At times, user documentation mixes system orientation with user-task orientation, as illustrated in Figure 5.5, a table of contents from an older modem manual. The headings sometimes begin with gerunds indicating user tasks, but at other times, headings consist of nouns or noun phrases that occasionally include technical terminology that may be unfamiliar to novices. Furthermore, indentations suggest levels of hierarchy that indicate that the authors may not have been thinking in terms of the users' goals and tasks. Specifically, under the heading "Required Equipment," three user tasks are included at the end of the indented list where one would expect to see equipment items.

Table of Contents
1 **Introduction**
Your New Modem
Protocols, standards and recommendations
The PM1440FX MT Package
Using this Manual
Typographical Conventions
2 **Modem Installation**
Required Equipment
Computer
Serial cable
Telephone
Communications software
Connecting the Modem
Installing the RS232C Cable
Installing the power supply
Testing the Modem
Telephone Connection
Testing the telephone connection
3 **Basic Modem Operation**
Issuing Commands to Your Modem
Command line prefix
Multiple screen characters—Echo command
Setting up the command line
Command buffer
Command acknowledgement
Essential Modem Commands
Sample Command Lines
Using the repeat command
Resetting the modem
Dialing the telephone
Storing telephone numbers

Figure 5.5. The opening of the table of contents of a modem manual with a mixture of procedural information and descriptive information. (From Practical Peripherals, 1993; recreated by author).

This modem documentation mixes system information with procedural information. The system-oriented headings ("Command line prefix," "Command buffer," etc.) do not give the users a sense of what they will learn to do in a given section. I speculate that the creators of this document were not aware of the differences between the two orientations within documentation and were not fully aware of the users' needs.

One line, "Using the repeat command," at first glance looks like a user task because it begins with a gerund. However, that line uses vocabulary focused on the system and does not indicate the goal (guided by context) that the user might have in using the repeat command. According to the manual, the command allows a previously entered command to be repeated, so a more user-task-oriented heading may be worded as "Repeating a previous command." Essentially, the creators probably did not consider that "The task is not in the software, and the user's purpose of interacting with the software is not to engage with it. . . . Instead, tasks live in the world" (Swarts, 2018, p. 29). System-oriented wording is minimally useful to end users.

Although this manual is older, I still frequently encounter this mixed approach in more recent manuals and user documentation, indicating that technical communication still has a long way to go to make sure that user documentation focuses on procedural discourse. (Fortunately, the back cover of this modem manual provides a number to call for tech support.) Overall, system information does not support user action (Albers, 2004). Discourse about a system is needed at times, but it does not meet the needs of people who want to develop procedural knowledge through user-task-oriented discourse, as discussed in the next section.

▪ User-Task-Oriented Discourse

In contrast to the three approaches discussed above, user-task-oriented discourse focuses on how people use a system within their contexts. Technical communicators consider users' purposes and contexts as central to the decisions they make about user-task-oriented discourse. These instructions are typically focused on action adapted to users' situations (Johnson, 1998).

Figure 5.6 provides an excerpt from a user-task-oriented manual's table of contents. Each item begins with a verb, indicating what the user will learn to do in a given section. Although this list seems logical and useful, I rarely encounter this user-task-oriented approach in discourse that is intended to assist users in gaining procedural knowledge. Even if the wording looks user-task oriented, as mentioned in the previous section, the accompanying text may still be focused on the technology rather than the user (Durack, 1998).

Creating user-task-oriented documentation requires more skill than creating system-oriented documentation because in addition to knowing system information, technical communicators also need to know about users' knowledge levels, previous experience, and typical uses of the system relevant to their goals in context. In addition, technical communicators need to know the conventions of procedural discourse that users may expect to see or experience (Hovde, 2010; Paradis, 1991). User-task-oriented instructional discourse needs to "shift from the initial design and manufacturing orientation toward objects to a new orientation toward human thought and behavior" (Paradis, 1991, p. 176). Documentation that relies too much on the system's structure, even if the documentation is user-task oriented, may be too simple for experienced users who want to perform more complex tasks (Mirel, 1993). However, creators of user-task-oriented discourse also assume that users most likely do not need to know every feature of a system in order to use it.

User-task orientation is not inherently better than system orientation within technical communication; each has its function. A technical designer or developer needing technical specifications is unlikely to benefit from task-oriented user documentation, but users who need to know how to use a system are also unlikely to benefit from technical specifications that describe a system. Each approach has a communicative purpose, but when discourse is not designed appropriately for the communication situation, problems arise. User documentation that does not include procedural discourse and relevant conceptual knowledge (Paradis, 1991) may lead to wasted work and negative economic effects for a corporation when technical communicators describe a system in detail but neglect to consider the processes users follow when using the system (Salvo et al., 2007). In addition, poorly created instructional material may affect user safety, leading to liability issues for the organization that produced them (Hogan, 2013; Paradis, 1991; Remley, 2015). Technical communicators ideally aim to create accurate procedural discourse balanced with the level of detail needed by users. This goal is challenging because "there is not a fixed amount of information anyone needs, and different histories can change what a person needs" (Albers, 2004).

Table 5.1 summarizes several differences between user-task-oriented procedural discourse and descriptive discourse.

Editing the Website.. 6
    Create a Backup Copy of Your Website... 7
    Open Dreamweaver... 7
    Open the Desired File... 8
    Edit an Existing Web Page... 9
    Create a New Web Page from Template..11
    Rename a Web Page..12
    Preview Your Web Page...13
    Save Your Web Page...13

Figure 5.6. User-task orientation in a table of contents uses verbs and verb phrases to show users what actions they will learn in each section. (Author created example.)

Table 5.1. Key differences between descriptive discourse and procedural discourse

	Descriptive discourse	Procedural discourse
Purpose	- To describe all of the features of a system	- To teach users how to achieve goal states (Farkas, 1999) and interact with the system or technology - To help users develop procedural knowledge
Scope	- Can be exhaustive (Salvo et al., 2007), describing the system in detail (Albers, 2004)	- Focuses mainly on tasks users need to complete, including conceptual/descriptive details only if they help in completing non-routine tasks
Intended audiences	- Technical designers or developers	- Users, installers, maintenance personnel
Ease of creation	- Relatively easy because a system exists and can be described (Salvo et al., 2007)	- More challenging because creators need to know about subject matter, audience, communication means, organizational constraints, and other situational variables (Farkas, 1999; Hovde, 2010; Johnson, 1998)
Markers of quality	- Accurate - Thorough	- Effective - Easy to use - Memorable - Efficient (Swarts, 2018) - Useful for work beyond the system

Designing procedural discourse to teach users how to complete tasks within interlocked, networked systems is more complex than designing it for completing routine tasks (Albers, 2004; Swarts, 2015, 2018) within simple systems. Hence, technical communicators need to have a good understanding of procedural knowledge so that they can create effective procedural discourse for both routine and non-routine situations. However, they also need to be aware that user-task-oriented discourse is not as helpful "if it does not account for the vagaries of tasks in situ" (Swarts, 2018, p. 27) because user goals typically lie outside the system—in other words, using the system is not typically an end in itself.

Declarative or system knowledge may have a role to play in acquiring and practicing procedural knowledge, but conceptual knowledge works best if it is subordinate to a procedural structure and focus (Farkas, 1999; Karreman, 2004) in procedural discourse. Descriptive discourse such as technical specifications is usually not helpful to end users all by itself because of its focus on conceptual knowledge, which may be useful for troubleshooting or planning non-routine work. In addition, system designers or developers need descriptive discourse so that they can understand a system that they may have to modify or repair. However, not all users need all conceptual information (Salvo et al., 2007). Technical communicators need to understand the differences between conceptual knowledge/discourse and procedural knowledge/discourse and be able to design procedural discourse appropriately for its use. These differing approaches to procedural discourse spring from a history of assumptions and practices in creating user discourse, as discussed in the next section.

History of Assumptions About and Practices of Procedural Discourse Within Technical Communication

Jason Swarts (2018) provides a useful summary of changes in attitudes over the last several decades toward procedural discourse and the ways in which it was presented, and I will summarize that history here. Swarts notes that early 20th century understandings arose of the user manual as crucial to helping non-engineer audiences understand how to operate technology, especially in military contexts. Users of that era were typically not encouraged to vary from the instructions.

With the advent of computers and other advanced technology as early as the 1950s, documentation was frequently system oriented, and the focus remained on efficient use limited to how a system was designed. However, technologies made available to ordinary consumers also created a need for user guides to help them employ those technologies. For example, Figure 5.7 shows the table of contents from an old manual, probably from the 1940s, that includes about eight pages dealing with how to use an electric refrigerator and about 20 pages of menus and recipes, the latter topics no doubt intended for people who wanted to use this device in a well-run household and/or who were moving from an icebox to their first electric refrigerator. (For more on changing relationships between workplace and domestic technology and users, see Durack, 1997.)

This change in audience who had a range of "situated and experiential knowledge," frequently tacit (Swarts, 2018, p. 11), often led to the beginning of users' attempts to adapt technologies in ways that the designers did not intend. Because the goal was efficient use, the conventions of manuals focused on "simple and direct language, short sentences, active constructions, sequentially ordered steps, and a simple focus on one item/task at a time" (Swarts, 2018, p. 12).

In the 1980s, the concept of "Goals-Operations-Methods-Selection" (Mirel, 1993, p. 24; Swarts, 2018) emerged that equated user tasks with system tasks. Later, in the 1990s, understandings developed that user tasks and goals go well beyond system tasks to include "relationships among readers, text, tasks, interface designs, and exploratory types of problem-solving strategies" (Mirel, 1993, p. 25). Notably, "when user needs grew beyond the technology, the documentation served no clear knowledge creation function" (Swarts, 2018, p. 14). Thus, documentation was useful only for a limited set of tasks, but users pushed the boundaries of what software could do as those users became more knowledgeable about software's possibilities and experienced needs beyond those the documentation addressed.

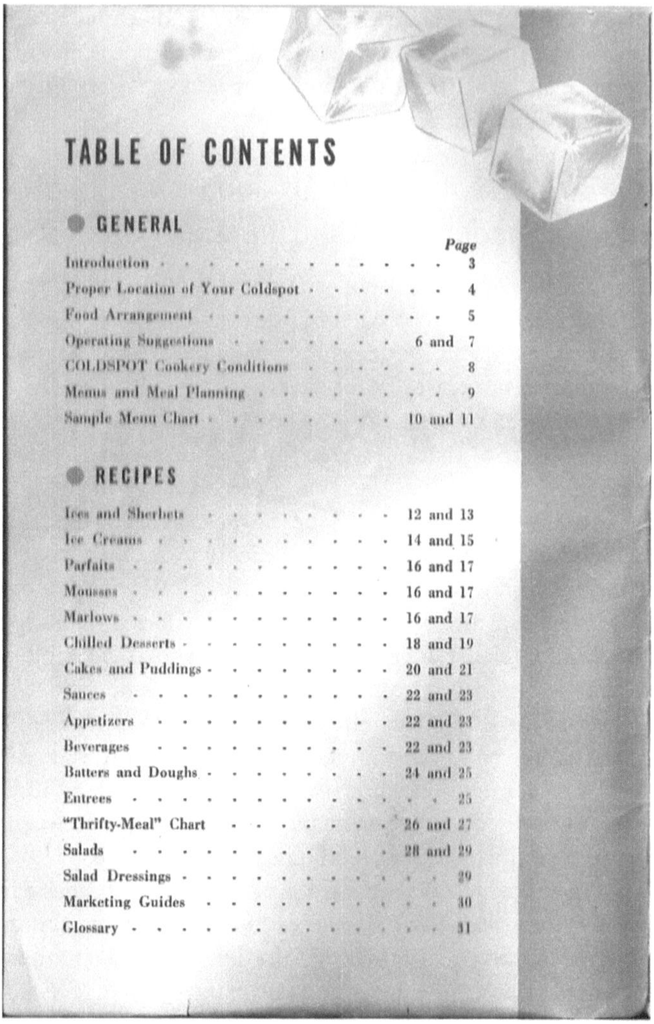

TABLE OF CONTENTS

● GENERAL

	Page
Introduction	3
Proper Location of Your Coldspot	4
Food Arrangement	5
Operating Suggestions	6 and 7
COLDSPOT Cookery Conditions	8
Menus and Meal Planning	9
Sample Menu Chart	10 and 11

● RECIPES

Ices and Sherbets	12 and 13
Ice Creams	14 and 15
Parfaits	16 and 17
Mousses	16 and 17
Marlows	16 and 17
Chilled Desserts	18 and 19
Cakes and Puddings	20 and 21
Sauces	22 and 23
Appetizers	22 and 23
Beverages	22 and 23
Batters and Doughs	24 and 25
Entrees	25
"Thrifty-Meal" Chart	26 and 27
Salads	28 and 29
Salad Dressings	29
Marketing Guides	30
Glossary	31

Figure 5.7. The table of contents from a vintage refrigerator manual includes about eight pages of technical details and about 20 of menus and recipes (Sears, Roebuck and Company, n.d. – photo taken by author).

Also in the 1980s, with the expansion of the availability of computers in workplaces and homes, manuals became more user-task oriented with conceptual information providing users a foundation for understanding the tasks (Swarts, 2018). During this time, controversy arose about whether comprehensive or minimalist manuals were better for learning (Karreman, 2004; Remley, 2015; Swarts, 2018). This minimalist documentation often encouraged users to explore and go beyond what typical documentation offered in using the technology (Swarts, 2018). Various studies indicated that a minimalist manual was preferable for learning by doing, but that users who had conceptual information from more comprehensive manuals did benefit, especially when presented with atypical situations of use (Remley, 2015). However, standard documentation still addressed the basics of operations; fewer means of supporting user learning and action beyond those standard processes were available (Swarts, 2018).

Newer forms of user documentation are currently emerging, but they have not become standardized and may not ever become standardized because of the complex network of user needs in the context of multiple software applications with which users are working. Technical communicators may need to be able to provide "interactive and dynamic help" (Swarts, 2018, p. 19) but also recognize that they are not the only people to create knowledge for users. Given the expanding nature of software and its use, procedural discourse may not be appropriately adapted to these new circumstances (van Loggem, 2013), so new approaches may need to emerge.

Ideally, the history of user documentation would show progress from system orientation to user-task orientation in discourse aimed at users, but I still find far too many examples of system-oriented documentation today. For instance, the documentation included in Figure 5.1 came from a relatively recent source, iFixIt.com, which provides a great deal of technical instructional material, so one would think that organization would understand the importance of user-task orientation. Alas, that is apparently not the case. Technical communicators need to find better means of applying recent advances in thinking about user discourse to actual practice and delivery. Understanding the history of and important concepts about procedural discourse holds many implications for the creation of effective procedural discourse today and in the future. However, additional insights about effective creation processes, as described in the next section, should be useful to technical communicators.

■ Creating Effective Procedural Discourse

In order to create effective, usable, and useful user documentation, creators of procedural discourse in its many forms need to understand not only the differences between system knowledge/discourse and procedural knowledge/discourse but also elements of an effective creation process. If they do not, they may produce unusable system-oriented exhaustive documentation that consumes a great deal of time and resources while being created, is not effective, and does not meet the needs of users. Furthermore, having engineers or marketing personnel (rather than technical communicators) create user documentation may lead to discourse that is not useful or is even unsafe for users (Paradis, 1991). Understanding procedural discourse and knowledge makes it possible to create user-task-oriented documentation (Salvo et al., 2007) that is more likely to help end users. Dávid Farkas (1999) has provided a foundational model of procedural discourse for technical communication on which others have built (Swarts, 2015, 2018), but many technical communication practitioners still struggle with creating "how-to" documentation that helps users complete tasks and processes. Creating effective procedural discourse takes conscious expertise and a context that encourages its creation.

Technical communicators need expertise in several areas in order to create effective procedural discourse. In addition to being aware of good processes, technical communicators also need understandings of the audience of users, authorial image/concerns, content selection, and genre conventions (Hovde, 2010), as well as considering the contextual variables, as explored below.

■ Effective Processes for Creating Procedural Discourse

User-task vs. system-oriented approaches can affect processes for creating technical communication. For instance, creating system-oriented discourse mainly involves getting to know the system well and not necessarily considering end users' needs and practices. If the discourse focuses on user-friendliness, then the creation process focuses on assuring readability. However, if the discourse focuses on user tasks and usability, the creation process begins with understanding users and their situations of use but also includes getting to know the system. Ideally, technical communicators participate early in the technical development process so that they can get to know the system as well as advocate for users when foundational decisions are made about the nature of the interface and user documentation (Johnson, 1998). Formative and summative usability testing of instructional material involving typical users can ensure that the procedural discourse meets its goal of developing procedural knowledge within users.

In addition to understanding users well, technical communicators may find that the role involves becoming "a facilitator or network maker, someone who is skilled at finding the right information and making the right connections and creating the right formats and protocols to meet the users' needs" (Swarts, 2018, p. 150) in relation to complex systems and networked technologies. In these contexts, technical communicators will organize content and make it easy for users to access as well as "managing the process of knowledge creation" (Swarts, 2018, p. 152). This function may go beyond the typical understandings of the roles and natures of technical communicators.

Another element in an effective process is focusing on the usability of the procedural discourse. Technical communicators need to consider more than the tasks that the system implies or is designed for. Instead, they need to consider contextual dynamics of users' work lives to determine if the instructions are useful to users (Durack, 1998; Mirel, 1993). While it is wise to design documentation that is readable, accurate, and has accessible information, technical communicators also need to analyze users' levels of job responsibilities and their desires to adapt systems to their work contexts (Mirel, 1993). Unfortunately, many technical communicators still lack access to rich information about users (Hovde, 2001), so developing these perspectives about users can prove challenging.

Although procedural discourse may be designed with usability in mind, well-designed usability testing of materials plays a crucial role in designing effective user support materials and in ascertaining their effectiveness. Technical communicators who understand the need for user-centered design test their discourse to see if it achieves its aims of teaching users "how-to" knowledge in an efficient and effective manner (Alexander, 2013). Because procedural knowledge is complex, this testing is crucial to help determine if user documentation has reached its goals. Much has been written about the usability of documentation and its testing (Barnum, 2011); a detailed discussion is beyond the scope of this chapter, but testing system-oriented documentation for usability would be futile. Effective usability testing of procedural discourse seeks to ascertain that the discourse "becomes an important part of the understanding process" (Albers, 2004, p. 116) for users. Studying the usability of multimodal instruction can be useful too, especially in determining the most effective mode (video or paper) for instructional discourse (Alexander, 2013).

Usability testing can enhance procedural discourse "to ensure the design contains all the features needed to invoke the proper response and that it is laid out in the manner which users expect" (Albers, 2004, p. 139). Studying how users encounter and seek answers to ill-defined problems is also an important part of a technical communicator's work (Swarts, 2018, p. 64). Content management systems complicate the process of learning about the usability of documentation (Hovde, 2019), but usability remains a crucial part of an effective process that also includes deliberate effort in the complexities of understanding users as well as other communication variables, as discussed in the next section.

Understanding Users' Ways of Learning and Their Uses of Procedural Discourse

Among the communication variables (audience/user, content, author, and format), audience may have the strongest influence in decisions about procedural discourse (Hovde, 2010). Using a rigorous process of understanding users and their needs and goals can make procedural discourse more effective (Albers, 2004; Hovde, 2001), as technical communicators benefit from rich approaches to learning about users rather than speculating about user characteristics (Hovde, 2001). They need to collect "information from a full spectrum of users so the range of knowledge and detail requirements" (Albers, 2004, p. 133) is well understood. For instance, technical communicators might follow conversations in online user forums to discover issues users typically encounter (Swarts, 2018, p. 85). Because "the writer must negotiate the flow of information from the perspective of the user" (Hogan, 2013, p. 157), understanding users is central to making decisions while creating procedural discourse (Hovde, 2010). Whatever the process of developing an understanding of users, those perceptions are crucial to developing effective procedural discourse.

Understanding that users gain procedural knowledge through a variety of modes of instruction combined with practice can lead to innovative forms of materials to help users; visual technical communication in its many forms is especially important because the brain processes visual information more immediately and strongly than linguistic information (Remley, 2015, pp. 28–29). Furthermore, including auditory instruction as it complements other modes of instruction can help users learn (Remley, 2015). Technical communicators need to consider learning theories such as "cognitive load" and "constructivism" (Hogan, 2013, p. 159) when designing task-oriented procedural discourse that adapts to users' learning capabilities. Technical communicators may also need to consider that users often simply scan written instructions rather than reading them in their entirety (Loorbach et al., 2006).

In addition, effective technical communicators need to work within a context in which they have access to information about users and content (Hovde, 2000, 2001, 2002). They need to consider if the typical purpose of the documentation is a tutorial for novices or a reference for experienced users who need to refresh their knowledge (Farkas, 1999)—or a combination of the two. Overall, technical communicators need to understand "how the user thinks and what the user needs so that interface operation, content, and presentation can maximize their respective potentials in communicating with the user" (Albers, 2004, p. x). Of all the communication variables, the user is the most challenging to understand and address in creating procedural discourse.

Technical communicators need to consider users' knowledge levels when designing user documentation. One of the decisions technical communicators need to make is how much detail, especially of conceptual knowledge, to include, whether to create streamlined or detailed documentation. These decisions may affect the ethos or the credibility that users assign to the documentation and the organization that provides it.

Technical communicators may also benefit from the insights of neuroscience as they consider how to assist users in developing procedural knowledge. Specifically, technical communicators need to understand that learning new processes involves more than cognition; it also includes practice/movement to help learn and reinforce that learning. (Remley, 2015, p. 34). Additionally, as they design learning experiences and materials, it is useful to consider the role users' prior experience plays in learning (Remley, 2015). Furthermore, technical communicators need an understanding of the following five principles that apply to how users learn new tasks:

- they prefer to integrate two or more senses as they learn,
- the visual is perceived first and often dominates,
- the timing of when information is received relative to other information affects how it is learned,
- prior experience/learning style affect how they learn, and

- users may focus on one mode of instruction more than others when learning (Remley, 2015, p. 40).

To adapt documentation to users' situations of use, technical communicators perform best when they understand that "people interact with programs differently at various times, depending on their job tasks, their professional approaches to these tasks, and the problems or breakdowns that they encounter during a task" (Mirel, 1993, pp. 25–26). Technical communicators also benefit when they realize that "documented instructions are not rote actions but are interpreted in use, succeeding only so far as that interpretation leads to improved intellective skills, coordination of social interactions and team efforts, and innovative approaches to business processes" (Mirel, 1993, p. 26). Users frequently move quickly beyond routine tasks and make inventive adaptations to suit their workplace or other contexts, especially if they are in roles that require or encourage non-routine usage of the system. These users move beyond an "automated" stance to an "informated" position (Mirel, 1993, p. 37). Users are more interested in acting within a situation, which goes beyond simply acting within a system. If procedural discourse does not include sufficient information, users need to "invent a procedure in the process of applying a tool" (Paradis, 1991, p. 269). Overall, procedural discourse "ought to address the point where user's motivation intersects with technology . . . , a picture of the technology that is inseparable from our situated uses of it" (Swarts, 2018, p. 134). In these ways, documentation can help users develop procedural knowledge.

In teaching people how to use a simple system, technical communicators benefit from task analysis of how users might employ the system. However, in providing instruction in more complex systems, creating user guidance becomes more challenging (Albers, 2004; Swarts, 2015). A simple system and a complicated system are similar in that "various problems can be plotted out and addressed" (Albers, 2004, p. 17), but a complex system is more open-ended with multiplying possibilities for use. Traditional documentation can provide a sense of stability that reflects "assumptions about use, and assumptions about principles that matter most in understanding that technology" (Swarts, 2018, p. 42). Procedural discourse may be able to provide "information with which the knowledge and skills can be built to find their own solutions to their own problems" (van Loggem, 2013, p. 172). However, users' issues may go beyond those situations.

Complex situations are not new. For example, Karen Schriver (1997) provided an example of attempting to set up two VCRs to copy and edit videotapes, a process that also involved a "cable outlet, a converter box, and a TV" (p. 228), all pieces that had to interact to help the users achieve their goals. The creators of manuals for all of these devices did not anticipate such a configuration, so the users invested many hours of trial and error trying to figure out how to make the components work together.

Technical communicators consider users' knowledge levels, needed level of detail, and their ability to process the information cognitively. Furthermore, they need to understand users' "intentions, context, knowledge, skills, and experience" (Albers, 2004, p. 68) to create useful and effective procedural discourse. At times, users may not be able to articulate their needs or their tacit processes (Albers, 2004), so technical communicators need multiple ways to understand users' needs. Overall, "audience and task analysis provides an understanding of the reader's prior knowledge, attitudes, and needs" (Albers, 2004, p. 74), allowing for appropriate design of procedural discourse.

Simply categorizing users as expert or novice may not yield a rich image of user groups and may not take into account the fact that novices may become experts. In addition, a user may be an expert in software, but not in the content relevant to larger tasks. For instance, a user may know how to use a spreadsheet to manipulate quantitative data, but may have no knowledge of principles of accounting. While this user may have an expert level of spreadsheet technical knowledge, the accounting knowledge may be at a novice level, further complicating the task of creating appropriate procedural discourse for that user.

Understanding that users' goals may shift when completing complex processes is a valuable insight for technical communicators (Albers, 2004). In addition, technical communicators creating procedural discourse for ill-defined, complex situations benefit from rigorous methods of understanding users, especially the "mental models" users may possess (Albers, 2004, p. 127), relevant to using that system. When users, especially novices, experience cognitive overload, their mental models cannot account for information, errors increase, and they may omit relevant tasks (Albers, 2004). For procedural discourse to succeed, creators need to understand users' mental models and social contexts when creating it. In addition, users under stress and time pressures in their contexts may not be able to process information well (Albers, 2004). Supplementing their complex understandings of users, technical communicators need to consider format and genre conventions as well as the affordances of various media, as explored in the next section.

■ Understanding Genre Conventions and Media Affordances

In addition to developing a rich understanding of users, their behaviors, their goals, and their ways of learning, technical communicators also need to understand the qualities that various media offer for procedural discourse as well as its genre conventions.

Technical communicators make decisions about the media used for technical communication, especially looking at the "affordances and constraints" of those media as they stimulate learning (Remley, 2015, p. 49). These affordances and constraints may involve the senses the medium employs to aid learning, including various combinations of auditory, visual, and/or tactile experiences. Technical communicators also need to be aware that learning is hampered when too much sensory input is included (Remley, 2015, p. 60) in whatever medium is used.

In thinking about genre conventions, it is crucial to understand that effective procedural discourse typically includes at least information about the system, an implied role for the user, a sense of the context of use, and actions the user will perform (Paradis, 1991, p. 258). Another genre convention to consider in creating procedural discourse is that including a rationale for a specific action can motivate and engage learners (Remley, 2015, p. 76).

To make information salient for users, it needs to be presented in ways that call attention to it and that make it easy to find, helping users to make sense of a situation. The most important content needs to be foregrounded, so technical communicators need to understand how people will use the system in order to understand what to emphasize (Albers, 2004). Effectively designing content provides an "adequate flow of information to the user in a form that makes sense in the situational context" (Albers, 2004, p. 83), helping users interpret meaning and achieve their goals.

Including warnings, cautions, and notes as well as other material may provide conceptual information relating to non-routine conditions. Too much conceptual detail can alienate users, but too little can leave them bereft. Formatting decisions can also help users develop procedural knowledge. However, a bit of deliberate redundancy, especially in making connections between words and visuals, may assist users in being able to understand devices and processes (Tenbrink & Maas, 2015). Visual communication can be especially crucial in procedural discourse, but it needs to be well designed for the audience and the medium (Schriver, 1997). For instance, flow charts showing conditions under which decisions need to be made may help users in non-routine situations (Farkas, 1999).

Technical communicators also need to consider the potential effects of motivational elements in procedural discourse. These motivational elements may include the roles in which the users and authors are cast; the use of non-technical terminology; the usefulness of examples, anecdotes, or metaphors; mentions of users' goals outside of the technology that may lead to specific actions; and the inclusion of testimonials. Furthermore, technical communicators may need to explore ways to balance the inclusion of these motivational elements with the conciseness and the efficacy of the instructions (Loorbach et al., 2006).

Whatever the medium and formats of the procedural discourse, technical communicators need to understand at least the affordances and the genre conventions discussed in this section in order to create effective procedural discourse, which differs in significant ways from other genres. I still encounter far too many examples of procedural discourse that try to explain steps in a paragraph format, that omit crucial visuals, and that do not pay attention to effective design of information. In addition to understanding genre conventions, technical communicators also need to consider how their discourse projects an image about the creators of that discourse.

Understanding the Relations between the Organization and Its Procedural Discourse

Because procedural discourse is often created within an organization, creators need to consider how the context influences its creation (Hovde, 2002) but also how users might perceive the organization's image based on interactions with the procedural discourse it provides. Organizational constraints and resources affect the process of creation; technical communicators may find themselves hampered when colleagues do not understand the nature or importance of effective procedural discourse (Hovde, 2002) and thus do not ensure that the technical communicators have the resources available and/or do not support an effective process for creating usable, useful procedural discourse.

In addition, well-designed procedural discourse can affect the way users perceive the organization. For instance, if an organization provides well-designed, usable, and useful online help, users are more likely to be favorably disposed to that organization. Finally, technical communicators need to think about how they understand and select content for procedural discourse.

Understanding and Selecting Content

Technical communicators ideally select and shape content appropriately so that it is adapted to user's needs. In order to do so, they need access to conceptual knowledge, such as technical specifications (Hovde, 2000), which they combine with their knowledge of the other communication variables (users, content, and organizational/authorship considerations) to create procedural discourse (Hovde, 2010). They need to select the most relevant content for users and ensure its accuracy as well. Including too much detail will overwhelm users (Salvo et al., 2007), and including too little will leave users without necessary guidance. Technical communicators without accurate and rich content knowledge may produce procedural discourse that does not meet user needs and may even lead them astray.

Although technical communicators may experience conditions that allow them to create effective procedural discourse as they consider the variables discussed in this section, several commonly held but misguided ideas may inhibit their work, as discussed below.

Myths About Procedural Discourse and Instructional Materials

Over time, I have noticed that several popular attitudes toward procedural discourse inhibit the creation of effective instructional materials. These myths need to be examined and countered when possible in order to foster the creation of procedural discourse that can empower users.

These myths include "Nobody uses instructions," "Anyone can write instructions," "Technical communicators are merely 'prettying up' technical content," "Good procedural discourse can compensate for a system that was not designed with usability in mind," and "Our system is well designed and intuitive, so user documentation is not needed." This section addresses each of these misconceptions in turn.

■ Nobody uses instructions.

Although procedural discourse/instructional material is often vilified (Johnson, 1998) and many people believe that no one uses that material, research indicates that people do use instructional material (van Loggem, 2014), but in ways that may be unintended by the creators of that material (van Loggem, 2013). For instance, a user may ignore the instructional material initially, but then consult it after reaching an impasse in the use of the system, much as some people only consult a map after they are lost (Mirel, 1993; van Loggem, 2014). One bit of evidence that users do seek procedural discourse can be seen in the popularity of third-party publications such as the *Dummies* and *The Complete Idiot's Guide to . . .* series aimed at teaching people to use software and complete other procedural tasks (van Loggem, 2013).

With the advent of more complex systems and open-ended tasks, users have turned to multiple means of gaining procedural knowledge that are more "interactive, quicker, and can offer more targeted assistance" (Swarts, 2018, p. 6) than traditional documentation. However, a need still exists for technical communicators who understand the dynamics of procedural discourse and how to present it effectively to users in a variety of approaches. Therefore, "If professional users of software are as willing to consult documentation as the findings suggest, then taking pains to design and develop documentation of the highest possible quality is a worthwhile endeavor" (van Loggem, 2014).

Learning to use a system via consulting written instructions is a learned behavior; "One who has learned to do new tasks through demonstration and practice and has never used a print-linguistic document will not understand how to use a manual to learn a new process" (Remley, 2015, p. 26). Hence, offering multiple modes for learning is essential.

Documentation is sometimes devalued within an organization because of the difficulties of measuring return on investment, but "in the long run, misinformed users concluding that a particular software product is useless is even more expensive" (van Loggem, 2014) than creating effective procedural discourse. The complexity of many systems implies that simply improving the interface will not be sufficient for users to learn the system (van Loggem, 2013). Hence, procedural discourse is needed, but it may also need to appear in innovative, user-centered formats.

■ Anyone can write instructions.

Simplifying procedural knowledge into procedural discourse may look easy, but it is actually complex (Johnson, 1998). While it is true that anyone can create some sort of instructions, not everyone can create them to be effective in achieving their goals. In fact, engineers and marketing personnel may create instructions that lead to injury and death (Paradis, 1991). Crucial skills for creating effective procedural discourse include (but are not limited to) knowing how to learn about subject matter (Hovde, 2001), knowing how to learn about users (Hovde, 2000), and knowing how to work within organizational situations to follow a productive process (Hovde, 2002). In addition, technical communicators today need to know how to use content management systems and other tools to create, manage, and distribute procedural discourse in its many forms.

■ Technical communicators are merely "prettying up" technical content.

This myth assumes that presentation can be separated from content, but actually, content does not exist outside of presentation. Instead, technical communicators transform descriptive material based on their knowledge of the technology, the audience, the image that their organizations wish to project, and the best means of communicating procedural discourse to the intended users (Hovde, 2010). Technical communicators select appropriate content for the users' situations of use (Paradis, 1991) rather than offering only exhaustive documentation. Presentation of complex information is crucial to users' abilities to engage with a system and understand it (Albers, 2011). Technical communicators actually serve as knowledge creators (Hovde, 2010) and knowledge managers (Swarts, 2018). Merely "prettying up" content often leads to user-friendly discourse that may be readable and engaging, but does not help users in developing active procedural knowledge.

■ Good procedural discourse can compensate for a system that was not designed with usability in mind.

Even after years of efforts to create usable systems, far too many systems are not designed with principles of effective human-system interaction in mind. Some system designers hold the attitude that training and user documentation can teach people to use a system that is difficult to use (Albers, 2004). However, technical communicators themselves may become frustrated with a poorly designed system and may despair over how to create effective procedural discourse for that system. Indeed, technical communicators may serve as user and/or usability advocates if they are able to participate early in the process of designing a system to be usable.

> The system is well designed and intuitive, so user documentation or procedural discourse is not needed.

In contrast to the previous myth, some interface designers for years have claimed to provide "intuitive" interfaces that do not require user instruction (van Loggem, 2013). However, unless users have undergone the appropriate experiences that lead them to be able to use a system without documentation (deWinter, 2014; Paradis, 1991), creating an intuitive interface is more challenging than designers might think, especially for complex technologies because "access to more complex technologies . . . usually requires a formal framework of explanation . . . that illustrates the contexts and conditions of effective action" (Paradis, 1991, p. 264). Many users lack the mental models needed to comprehend and use a new interface, especially a complex one.

So-called "Intuitive" interface design is typically based on socially constructed experiences and direct instruction rather than the innate features of human nature. For instance, if experienced drivers rent an unfamiliar model of car, they know from years of interacting with automobiles to look for common dashboard controls—headlight switch, wiper control, ventilation controls, etc. Designers of automobile dashboards are also familiar with conventional controls and have usually placed them in accessible places. However, at times, an unfamiliar control is present. For instance, many cars now have a way to turn off "traction control" when one is stuck in mud or snow. However, if drivers are not familiar with this feature, they may not know what the button marked "TC" does and may have to consult the owner's manual, which ideally will provide them with procedural and conceptual knowledge.

"Intuitive" design is thus based on commonly shared experiences and knowledge, which lead to procedural and conceptual knowledge that help users navigate interactions with new systems. These experiences create a mental model that guides how users interact with an unfamiliar and/or complex system. A mental model, built from previous experience, "corresponds to the cognitive layout that a person uses to organize information in memory" and "helps to make connections among disparate bits of information" (Albers, 2004, p. 135). Creating effective procedural discourse benefits from a rich understanding of users' mental models that influence how they learn new information and processes.

An interface that is easy to use generally calls on conventional features and practices, but usually these interfaces are connected to relatively simple processes and systems. In addition, an easy-to-use interface may employ metaphors with which users are familiar. For instance, designers of early graphical user interfaces employed symbols for common office items such as a desktop or a trash can. Users could then take their previous knowledge and transfer it to using the interface. However, when processes and systems become complex, "intuition" may not suffice.

Procedural discourse is part of the user interface (Johnson, 1998; Suchman, 1987), mediating between the intentions of the system designers and the goals of the users, influenced by the technology itself. Because "the lay person is largely isolated from the professional origins of technologies" (Paradis, 1991, p. 257), some form of procedural discourse is necessary for effective use of complex technologies. This procedural discourse "becomes a kind of script for the human-machine interface, in which human physiology is unified with machine action to achieve a utilitarian objective . . . [that can] . . . direct the human-machine interaction so as to deliver the technology to the user's purpose" (Paradis, 1991, p. 268). However, that discourse needs to go beyond simple "how-to" knowledge to help users understand the consequences of their actions (Paradis, 1991, p. 275).

In addition, in an imperfect world, systems are not always thoughtfully created with a focus on users and usability, so user documentation is needed (van Loggem, 2013). Because some systems may need to be versatile and provide a variety of functions, they are necessarily complex. In a complex system, the interface may not be able to provide a rich view of that system to users, but documentation can assist users in understanding the system and its possible uses (van Loggem, 2013). In this sense, procedural discourse is a crucial part of the interface between users and the system (Suchman, 1987).

Because of the complexity of creating effective procedural discourse and the prevalence of the myths discussed in this section, future historical and empirical research is crucial for improving understandings of and the creation of procedural discourse.

■ Areas for Future Research

The nature of procedural knowledge and effective procedural discourse is worthy of further study. The following questions may guide further exploration and inquiry:

1. *What can history teach about principles of effective procedural discourse?* Although some historical research has been done for technical communication in general (Kynell & Kynell-Hunt, 2000; Schriver, 1997; Swarts, 2018), even more insights from the past would be useful for people creating procedural discourse today so that they could understand the effectiveness of a variety of approaches that have been tried over time.
2. *What are best practices for creating procedural discourse for complex processes as we move into the future?* Creating procedural discourse for routine situations is complex enough, and much about this topic has been explored. However, creating procedural discourse for complex, interlocking systems still needs further research (Albers, 2004; Swarts, 2018).
3. *What are the forces that prevent the creation of user-focused procedural discourse? How can those constraints be addressed?* Although much scholarship has focused on the qualities of effective procedural discourse, much of that discourse does not reflect best production practices or the conditions

under which technical communicators created that discourse. Technical communicators work in complex contexts with varying constraints and resources (Hovde, 2002), so future observational studies of influences on the processes of creating user documentation can provide useful insights about the contextual factors that enhance and inhibit the creation of effective procedural discourse.

4. *When and how do users experience procedural discourse?* Although several empirical studies have been completed on this topic (van Loggem, 2013, 2014; Swarts, 2018), much more work is needed to confirm and/or counteract some of the received "wisdom" about procedural discourse. This research should draw on multiple relevant disciplines such as instructional design and cognitive science, which already have rich insights about how learning occurs, so that "the informed design of software documentation demands that the choice for medium and format of the communication, as well as its content, be based on an understanding of the underlying processes of people interacting with software and with documentation" (van Loggem, 2013, p. 176). Results of this research could provide valuable guidance to technical communicators.

5. *How do cultural contexts affect how users access, interpret, and use procedural discourse? What are the effects of procedural discourse on users' access to technology?* Grounded in the current focus on social justice in technical communication (Walton *et al.*, 2019), researchers could explore how technical communication relates to "traditionally marginalized and excluded perspectives, populations, and positions" (Jones *et al.*, 2006, p. 13), including the varied ways members of cultural groups around the world create and use procedural discourse. As technology and technical communication become more globalized, research into cultural and social considerations in procedural discourse will become more crucial.

▌ Conclusion

Procedural discourse works best when it is designed to help users create and carry out procedural knowledge in action; however, it can also provide useful conceptual knowledge to help users address non-routine, complex, and open-ended situations.

Understanding the need for procedural discourse that adapts well to users' situations and needs is central to technical communication. Although scholarship has addressed the dynamics of procedural discourse over several decades, discourse intended to assist users in gaining procedural knowledge is still far too often poorly designed and not tested to see if it meets its goals. With the complexity of technology and other systems increasing exponentially, users need procedural discourse that is well designed to assist them in developing procedural knowledge.

Although a few processes are as "easy as 1, 2, 3," many are not (Swarts, 2018) and thus require thoughtfully created, user-task-oriented discourse in many forms. Technical communicators need to understand the differences between system knowledge/discourse and procedural knowledge/discourse. Additionally, their colleagues who influence the nature of the documentation also need this understanding as technology and its communication become increasingly complex. In addition, technical communicators and their colleagues need to understand processes that enhance the creation of effective learning experiences for users. Well-designed procedural discourse empowers users in multiple contexts as they create and employ procedural knowledge for numerous purposes.

■ References

Albers, Michael J. (2004). *Communication of complex information: User goals and information needs for dynamic web information*. Lawrence Erlbaum Associates.

Albers, Michael J. (2011). Design and usability: Beginner interactions with complex software. *Journal of Technical Writing and Communication, 41*(3), 271–287.

Alexander, Kara P. (2013). The usability of print and online video instructions. *Technical Communication Quarterly, 22*(3), 237–259.

Associated Press. (2019, June 20). Capt. "Sully" Sullenberger slams Boeing for inadequate pilot training on the troubled 737 Max. *Fortune*. http://fortune.com/2019/06/20/capt-sully-sullenberger-boeing-737/.

Barnum, Carol M. (2011). *Usability testing essentials: Ready, set…test!* Elsevier.

deWinter, Jennifer. (2014). Just playing around: From procedural manuals to in-game training. In Ryan Moeller & Jennifer deWinter (Eds.), *Computer games and technical communication: Critical methods and applications at the intersection* (pp. 89–106). Routledge.

Durack, Katherine T. (1997). Gender, technology, and the history of technical communication. *Technical Communication Quarterly, 6*(3), 249–260.

Durack, Katherine T. (1998). Authority and audience-centered writing strategies: Sexism in 19th-century sewing machine manuals. *Technical Communication, 45*(2), 180–196.

Farkas, Dávid K. (1999). The logical and rhetorical construction of procedural discourse. *Technical Communication, 46*, 42–66.

Hogan, Diane B. (2013). Theories that apply to technical documentation. *Connexions: International Professional Communication Journal, 1*(1), 155–165.

Hovde, Marjorie Rush. (2000). Tactics for building images of audience in organizational contexts: An ethnographic study of technical communicators. *Journal of Business and Technical Communication, 14*(4), 395–444.

Hovde, Marjorie Rush. (2001). Research tactics for constructing perceptions of subject matter in organizational contexts: An ethnographic study of technical communicators. *Technical Communication Quarterly, 10*(1), 59–95.

Hovde, Marjorie Rush. (2002). Negotiating organizational constraints: Tactics for technical communicators. *Technostyle, 18*(1), 61–94.

Hovde, Marjorie Rush. (2010). Creating procedural discourse and knowledge for software users: Beyond translation and transmission. *Journal of Business and Technical Communication, 24*(2), 164–205.

Hovde, Marjorie Rush. (2019). Effects of content management and organizational context on technical communication's usability. In Guiseppe Getto, Jack Labriola & Sheryl Ruszkiewicz (Eds.), *Content strategy in technical communication* (pp. 69–88). Routledge.
Hovde, Marjorie Rush & Renguette, Corinne C. (2017). Technological literacy: A framework for teaching technical communication software tools. *Technical Communication Quarterly, 26*(4), 395–411.
Johnson, Robert R. (1998). *User-centered technology: A rhetorical theory for computers and other mundane artifacts*. SUNY Press.
Jones, Natasha N., Moore, Kristen R. & Walton, Rebecca. (2016). Disrupting the past to disrupt the future: An antenarrative of technical communication. *Technical Communication Quarterly, 25*(4), 211–229.
Karreman, Joyce. (2004). *Use and effect of declarative information in user instructions*. Rodopi.
Kynell, Teresa C. (2000). *Writing in a milieu of utility: The move to technical communication in American engineering programs, 1850–1950* (No. 12). Greenwood Publishing Group.
Loorbach, Nichole, Steehouder, Michaël & Teal, Erik. (2006). The effects of motivational elements in user instructions. *Journal of Business & Technical Communication, 20*(2). 177–199.
McCloud, Scott. (2008). *Google Chrome*. https://www.google.com/googlebooks/chrome/.
Mirel, Barbara. (1993). Beyond the monkey house: Audience analysis in computerized workplaces. In Rachel Spilka (Ed.), *Writing in the workplace: New research perspectives* (pp. 21–40). Southern Illinois University Press.
Paradis, James. (1991). Text and action: The operator's manual in context and in court. In Charles Bazerman & James Paradis (Eds.), *Textual dynamics of the professions: Historical and contemporary studies of writing in professional communities* (pp. 256–278). University of Wisconsin Press.
Purdue Employees Federal Credit Union. (2008). *Manage your money on the go: Set up your free online and mobile banking services today*. Purdue Employees Federal Credit Union.
Remley, Dirk. (2015). *How the brain processes multimodal technical instructions*. Baywood Publishing.
Roochnik, David. (1996). *Of art and wisdom: Plato's understanding of techne*. Penn State University Press.
Practical Peripherals. (1993). *PM14400FX MT Modem: Operating Manual*. Practical Peripherals: Thousand Oaks, CA.
Salvo, Michael, Zoetewey, Meredith & Agena, Kate. (2007). A case of exhaustive documentation: Re-centering organizations around user needs. *Technical Communication, 54*(1), 46–57.
Schriver, Karen A. (1997). *Dynamics in document design: Creating text for readers*. Wiley Computer Publishing.
Sears, Roebuck and Company. (n.d.) *Operating suggestions and Coldspot recipes*. Sears, Roebuck, and Company.
Suchman, Lucille A. (1987). *Plans and situated actions: The problem of human-machine communication*. Cambridge University Press.
Swarts, Jason. (2014). The trouble with networks: Implications for the practice of help documentation. *Journal of Technical Writing & Communication, 44*(3), 253–275.

Swarts, Jason. (2015). Help is in the helping: An evaluation of help documentation in a networked age. *Technical Communication Quarterly, 24*(2), 164–187.

Swarts, Jason. (2018). *Wicked, incomplete, and uncertain: User support in the wild and the role of technical communication.* Utah State University Press.

Tenbrink, Thora & Maas, Annika. (2015). Efficiently connecting textual and visual information in operating instructions. *IEEE Transactions on Professional Communication, 58*(4), 346–366.

van Loggem, Brigit. (2013). User documentation: The cinderella of information systems. In Álvaro Rocha, Anna M. Correia, Tom Wilson & Karl A. Stroetmann (Eds.), *Advances in information systems and technologies, Vol. 206* (pp. 167–177). Springer Science and Business Media.

van Loggem, Brigit. (2014). "Nobody reads the documentation": True or not? In *Proceedings of ISIC, the Information Behaviour Conference, Leeds* (pp. 2–5). Information Behavior Conference.

Walton, Rebecca, Moore, Kristen & Jones, Natasha. (2019). *Technical communication after the social justice turn: Building coalitions for action.* Routledge.

Chapter 6: Technical Communication Reimagined Through a Socio-Technical Problem-Solving Lens

Michael J. Albers
EAST CAROLINA UNIVERSITY

Abstract: Designing, writing, and reading a text—of any realistic complexity—is a constant problem-solving and decision-making process. Providing quality content in a complex information environment means providing information for problem-solving within the situation's context. Writing for the socio-technical situation and for problem-solving means positioning the content in terms of the needs of people within that situation and the overall implications of how/why content is needed, used, and how it interacts with other information. A foundational idea of socio-technical theory is that the design of any system can only be understood and improved if both "social" and "technical" aspects are considered together as interdependent elements of a complex situation. The communication situation commonly involves the relationships between people (social systems) and technology (technical systems) and how those systems interact and evolve. Communicating information within a socio-technical environment requires drawing the proper boundaries to make the overall problem manageable and providing the information the reader needs. The socio-technical situation tends to be larger than what is normally considered within technical communication audience analysis and rhetorical studies. For the writer, restructuring the information to meet the needs of the socio-technical environment requires a deep rethinking of how we understand writing, communication, and audiences.

Keywords: socio-technical situation, complex information, decision-making, problem-solving

Jared Spool (2014) tells a story about an auto repair shop and how a person's use of a software estimating application was very different on Friday (with low customer numbers) and Saturday (with high customer numbers). Basically, on Friday, the owner was gushing about how much he loved the application because of the good estimates it provided. On Saturday, he abandoned it for paper because it got in his way.

The difference was not that the software had to be used differently or that the user was different. They were the same task and same person on both days—which reveals the flaw of collecting tasks and audience demographics and calling the analysis complete.

Spool (2014) talks about this response as an example of what he calls service design. Within this chapter, I'll be looking at the same set of issues from a technical communication perspective, not just for a focused application, but for dealing with corporate reports used for decision-making or other non-task-based types of technical communication.

As with Spool's (2014) example, too often, technical communication is written for the ideal situation and then everyone wonders why it collapses so easily and fails to provide useful information (Albers, 2012; Redish, 2007). Or it dumps all the information, and everyone wonders why the reader can't integrate it and use it (Terveen et al., 1995). The basic problem: it failed because it didn't address how both the social aspects and the technical issues of the overall situation—the socio-technical situation—worked as an integrated whole (Trist, 1981; Woods & Roth, 1988.) The main argument of this chapter is the need to bring the socio-technical to the forefront of technical communication analysis.

Both a text's writer and reader confront essentially the same problem. To design and write a text—of any realistic complexity—is a constant problem-solving and decision-making process. To read a text—of any realistic complexity—is a constant problem-solving and decision-making process. In other words, both creating and reading a text can be considered as variations of the same problem. Once a text moves beyond procedural instructions, it must contain information both relevant to the situation and formatted in a way that addresses the reader's needs (Albers, 2004; Wickman, 2014). A trivial-sounding statement, but one which often explains the underlying communication failure of many documents. For concrete examples, see the multitude of "why the document failed" analyses published within the technical communication literature.

All of the reader's information needs, text constraints, and content decisions exist within the situation's problem space. A writer must map that problem space onto the text design space. Both writer and reader must map both problem space and text design space onto the reader's goal space. Taken together, they form a complex socio-technical environment; effective communication within that socio-technical environment requires understanding the integration of people (and their individual response), their social interactions, and technical (technology) aspects. Information and needs within the problem space and goal space shift and change as the situation develops (Cilliers, 1998; Klein, 2014). Sidney Dekker's (2011) book on major failures (airline and major industrial disasters) repeatedly describes the basic problem as thinking about the problem in too narrow of terms with a resulting catastrophic failure.

A foundational idea of socio-technical theory is that the design and performance of any system can only be understood and improved if both "social" and "technical" aspects are considered together and treated as interdependent elements within a complex situation (Trist, 1981). Lisl Klein (2014) considers how socio-technical theory explicitly connects people and technology into an interdependent web—a web where any change to one point ripples out and causes changes to *all* the other points (Albers, 2010). Dekker (2011) did not find single "this failed and caused the disaster" points, but rather, he found long cascades of interdependent events all embedded within a socio-technical context.

Writers face the problem that any complex socio-technical situation has essentially infinite information available. Developing information carries with it a specific representation of that information. Previous work has shown that people define their own tasks and needs in terms which fit their goals (Mirel, 1992). The fit between that specific representation of information and the person's self-defined needs strongly influences its effectiveness. Clearly, developing information to support decision-making requires understanding how they interact within a socio-technical environment (Klein, 2014).

Decision-making within a socio-technical environment requires understanding the relationships within the information (Albers, 2009, 2010). Gary Klein's (1999) naturalistic decision-making model provides the best explanation of how people grasp and use relationships to make decisions (as opposed to the too-common optimized decision matrix methods). Writers tasked with communicating this information must ensure the person knows both the information and how to use/integrate it toward their goals (Robertson et al., 1993; Woods & Roth, 1988). Creating the proper view for the reader requires defining the information boundaries (Laplante & Flaxman, 1995) and knowing how those boundaries affect understanding—boundaries that must be defined by the situation and not by the technical system structure or writer/organizational wants (i.e., providing the easy-to-get stuff; Dekker, 2011).

Writing for decision-making and problem-solving requires understanding the socio-technical situation. However, technical communication is rarely presented through a problem-solving lens suited to working within that complex socio-technical environment. Instead, analysis is defined based on breaking down into single units. Decomposition and analysis of individual pieces works for simple actions and pure technical systems, but fails miserably when people and their social interaction become integral to the situation (Albers, 2009). (Think IKEA instructions versus a five-million-dollar business decision or making a healthcare choice.) Unfortunately, designing for expected or best-case scenarios fails to address the information needs when they move beyond those scenarios (Vicente, 1999).

Within a socio-technical writing situation, technical communication needs to reshape its questions so they are proposed in human-information interaction (HII) terms (Albers, 2012) and focus on defining how the audiences will *interact with the information*, how the audiences will *use it*, and how the various parts of the situation *influence that interaction and use*. Only then will the information work within its socio-technical situation.

Socio-technical research is rapidly developing into its own field, but unfortunately, I don't see technical communication even acknowledging its existence, much less making use of its findings. This chapter strives to begin making the case for considering the socio-technical aspects when creating technical documents.

■ Terms

I begin with defining my terms. Granted, term definition is a standard rhetorical move, but, in this case, it is important because some of these terms are used within technical communication in ways that are not quite how I use them. Obviously, following this chapter's overall argument while using different definitions could prove difficult to impossible. The first part of the chapter—most of it actually—considers the terms we need to define. Each of these terms will be defined and discussed, and then in the later part of this chapter, their interrelationships will be discussed.

The terms to be considered and short definitions are given here. The next sections expand on them.

Complexity	Situations and their information are highly interconnected and any change affects everything.
Writing environment	The environment in which the communication occurs. The type of content—simple or complex—and the reader's use of the information within their situation. (Note that this has nothing to do with how/where the writer produces content.)
Situation	The overall environment in which readers find themselves as they read/research the information. It includes both the technology used to access the information and their overall environment (i.e., corporate directives, what the boss wants, asking others for input, prior knowledge, etc.)
Socio-technical	Communication happens within and depends on an integrated combination of social and technical aspects of the situation.
Decision-making and problem-solving	The ways people make choices to influence the evolution of a situation.

■ Definition of Complexity

Complex information contains lots of ambiguity and subtle nuances within its content. The information interacts with its environment and changes as the situation changes or evolves. Because of these factors, it is impossible to define a "complete set of information" or to completely analyze the situation or provide all paths through it.

Paul Cilliers (1998) described a complex situation by saying,

The interaction among constituents of the system, and the interaction between the system and its environment, are of such a nature that the system as a whole cannot be fully understood simply by analyzing its components. Moreover, these relationships are not fixed, but shift and change. (p. viii)

In a complex situation, the problem will almost always include factors or circumstances not foreseen as part of the original analysis. "As a result, information system design cannot be based solely on expected or frequently encountered situations" (Vicente, 1999, p. 17).

I have previously described complex situations as having six characteristics (Albers, 2004). These factors influence how information must be provided and what information is relevant to a reader.

Characteristic	Explanation
No single answer	There is no single answer or "correct" way to approach a problem.
Open-ended	The proper amount of information cannot be predefined. People collect and analyze information until they are satisfied and then make a decision.
Multidimensional	Multiple factors influence the situation and affect what information is relevant and how the situation will evolve.
History	The previous state of the system influences how the system evolves. Two situations that look identical in a current snapshot, but with different histories, may end up looking very different in the future.
Dynamic	Information does not have a fixed value. It changes as the situation evolves. Likewise, the reader's goals and information needs change.
Non-linear	The overall situation is sensitive to the initial starting conditions, and small changes can result in big differences later.

■ Definition of Writing Environments

At the high level, writing can occur in either highly structured or ill-structured environments. An effective writing methodology and reader expectations are radically different between them.

■ Writing in a Highly Structured Environment

A highly structured environment has clear reader expectations and a clear way of determining if the information is complete and correct. In a highly structured environment—an underlying assumption of most technical communication pedagogy—the reader's basic goal is efficiently completing a task. A step-by-step route can be predefined as the correct path to an answer, and that path can be supported and enforced by a computer system. The high structure means the end result can be judged as a yes/no or correct/incorrect answer.

If the task is to assemble a bookcase, then the writing fits the definition of highly structured. The reader approaches to how to assemble the bookcase are limited and can be fully defined by the writer. The final result can be judged: the bookcase is assembled correctly or not.

Unfortunately, well-defined does not describe most realistic writing situations.

■ Writing in an Ill-Structured Environment

An ill-structured environment lacks the clear-cut answers that were evident in the highly structured environment. The reader's overall goals may be defined, but the paths to achieving those goals and what information is required cannot be fully defined.

In the ill-structured environment—the norm with real-world problems— the reader's goal is one of analysis and problem-solving. The task is not to assemble a bookcase, but to plan next summer's vacation, figure out why sales are down in the west, understand a medical condition, or determine how to improve X (traffic flow, employee morale, course design, etc.). Rather than simply completing a task, the reader needs to be aware of the entire situational context in order to make good decisions. In an ill-structured domain, instead of following a set path, the reader continuously adjusts their path as new information presents itself. As a result, each reader takes a slightly different path and uses slightly different information.

In other words, the writer can't even assume that the information needs are consistent between readers or what information a reader will view before making a decision. Yet, the writer is tasked with creating a design which communicates the information when and how the reader wants it.

■ Situation

The opening definition explained situation as the overall environment in which readers find themselves as they read/research the information. It includes both the technology used to access the information and readers' overall environment (i.e., corporate directives, what the boss wants, asking others for input, prior knowledge, etc.)

The bigger picture can be described by an image (see Figure 6.1) that captures the entire environment. Many technical communication sources seem to work from the view that a person uses one and only one source (the text being currently written) as the information source. But this is rarely true. Instead, a reader uses many sources, only some of which are explicit (documents or asking other people), and some that are implicit (knowledge of "how things are done").

A highlight in Figure 6.1 is that the system—the thing on which most writers and their associated developers focus—is pretty much outside the reader's concern. True, they want it to work smoothly, but they also expect it to just be another source of potential information.

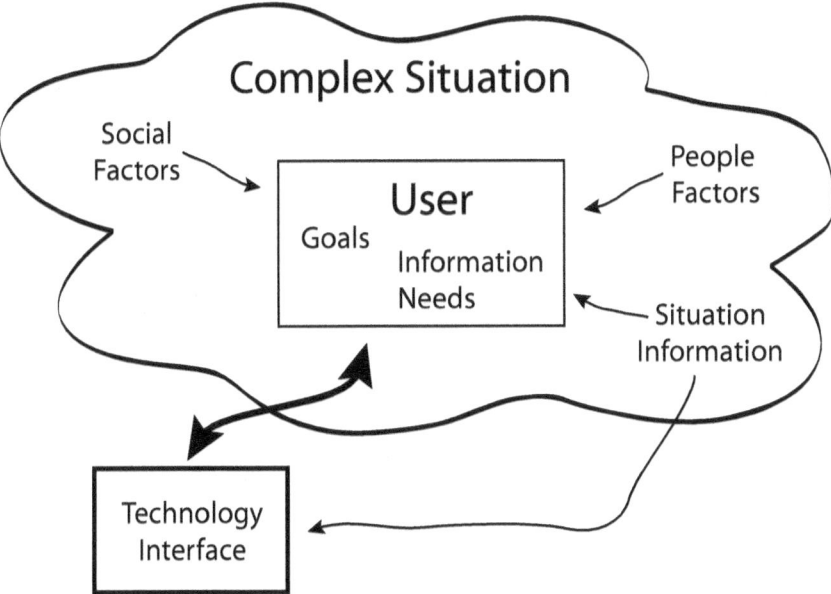

Figure 6.1. Overview of the complex situation. Notice how the system exists almost outside of the situation. Too often design teams place the computer interface front and center while ignoring the rest of the situation (adapted from Albers, 2004).

As a side note, most of the socio-technical literature uses the term *system* and says socio-technical situations operate within a system. That literature, loosely defined, considers system as the entirety of what the reader (and writer) is interested in. However, the word *system* is too easily equated to technology: system equals computer. But that is not what system means in this instance. It is the entire thing—the entire problem space the reader operates within— which a writer must draw boundaries around and within, that matters to the reader. It becomes too easy in discussions of socio-technical systems for participants to start talking past each other because they use different definitions of *system*. Because of that, I use the term *situation*.

■ Definition of Socio-Technical

Klein (2014) sums up the interrelationship of both people and technology:

> Sociotechnical theory makes explicit the fact that the technology and the people in a work system are interdependent. Each affects the other. Technology affects the behaviour of people, and the behaviour of people affects the working of the technology. It is inevitable, it is a real part of the situation, and one therefore needs to take account of how they affect each other. (p. 138)

Most importantly, she emphasizes the relationship. It is flatly impossible to understand either the technology or how people interact with it without considering them together. Any change to one results in a change to the other (which feeds back into a change to the first one . . .).

In 1996, I attended the *HCI International Conference*, and the topic had its own track. The researchers seemed to be totally focused on using the word *socio-technical* in every other sentence. Since then, it has continued to develop into a field with its own research agenda. However, *socio-technical* has had minimal impact within technical communication, much to the determent of technical communication's development as a field.

The idea of socio-technical systems is not new, even back in 1996 when I first encountered it. Russel Mumford (1987) was an early researcher to discuss how having adequate technology without considering the social could still cause poor results.

Let's look at a couple of definitions of *socio-technical* that have been proposed. Wikipedia gives a definition as:

> Socio-technical systems pertain to theory regarding the social aspects of people and society and technical aspects of organizational structure and processes. Here, technical does not necessarily imply material technology. The focus is on procedures and related knowledge, i.e. it refers to the ancient Greek term *logos*. "Technical" is a term used to refer to structure and a broader sense of technicalities. Socio-technical refers to the interrelatedness of *social* and *technical* aspects of an organization or the society as a whole. Socio-technical theory therefore is about *joint optimization*, with a shared emphasis on achievement of both excellence in technical performance and quality in people's work lives. Socio-technical theory, as distinct from socio-technical systems, proposes a number of different ways of achieving joint optimization. (Wikipedia, n.d.)

The Interaction Design Foundation gives a more concise definition but still captures the overall idea.

A socio-technical system (STS) is one that considers requirements spanning hardware, software, personal, *and* community aspects. It applies an understanding of the social structures, roles and rights (the social sciences) to inform the design of systems that involve communities of people and technology. (Interaction Design Foundation, n.d.)

Characteristics of Socio-Technical Systems

In the following passage, Dekker (2011) is talking about failures (major failures; think airplane crashes), but his words also describe the problem with thinking in terms of documenting a narrow topic, rather than considering the entire socio-technical situation:

> The problem with this was that greater complexity leads to vastly more possible interactions than could be planned, understood, anticipated or guarded against. Rather than being the result of a few or number of component failures, accidents involve the unanticipated interaction of a multitude of events in a complex system—events and interactions, often very normal, whose combinatorial explosion can quickly outwit people's best efforts at predicting and mitigating trouble. Interactive complexity refers to component interactions that are non-linear, unfamiliar, unexpected, or unplanned, and either not visible or not immediately comprehensible for people running the system (p. 128)

Two significant characteristics of socio-technical systems are:

> *Non-deterministic:* The same inputs at two different times do not produce the same output. The myriad of subtle (and not so subtle) factors, many of which are not directly accounted for, interact with the situation and prevent it from repeating. At the very basic level, people are involved, and people are highly non-deterministic.

> A situation's history gives it a trajectory and momentum, and although that trajectory might pass through the same point twice, the trajectory itself is different. Thus, the response and results are different.

> *Emergent properties:* The overall performance depends on both the system parts and their relationships, which all operate in a non-deterministic manner. The resultant behavior of a simple system can be predicted based on understanding the parts. Socio-technical systems and their emergent behavior cannot. Emergent properties are bottom-up, highly non-linear, and non-deterministic, which makes them impossible to model (Easterling & Kok, 2002).

"System-level behaviors emerge from the multitude of relationships, interdependencies and interconnections inside the system, but cannot be reduced to those relationships or interconnections" (Dekker, 2011, p. 201).

Examples of emergent properties are things such as a wave at a baseball stadium. No amount of analysis of one person jumping up and sitting down will predict that a bunch of people doing that in a coordinated fashion will produce the appearance of a wave. Likewise, pictures made up of many small images—e.g., pictures of Elvis made of tiny images of his album covers.

Adverse events in complex systems are produced by a complicated combination of events that may never congeal in the exact same way again—the emergent property. Yet, the decision-makers strive to ensure the adverse event will never repeat, which risks making decisions that ripple outward and cause new adverse events. Emergent issues stem not from the event itself, but from the processes that lead up to it. What decisions were made, what events occurred, what assumptions were people working from? What taken-for-granted assumptions were not considered in the decisions?

▍ Decision-Making and Problem-Solving

Decision-making involves analyzing options and making choices. Problem-solving focuses on making a choice to control the trajectory of a situation. Decision-making differs from problem-solving because it focuses on making choices to direct and control a situation, rather than adjusting from undesirable outcomes. On the other hand, they are closely related and can often be used interchangeably without major issues.

In solving the complex problem, the potential choices and reasons for making the choice become of dominating importance. Because people rarely base decisions on simple look-ups (it says 6 here, so the answer is no), the content must support helping them solve a complex problem. Both decision-making and problem-solving tend to be the purpose of information-seeking in complex situations because the reader needs to understand what is happening and make decisions that will support a favorable result.

Fundamentally, decision-making requires integrating the results of multiple queries (Ebert et al., 1997). The question has shifted from a simple "Does this exist?" to much more complex formulations such as "I need to analyze these documents to understand about X. They all discuss X, but which ones contain relevant information? And, more importantly, what is the relevant information for *my* specific needs right now?" That last question is highly pertinent since the relevant information changes as a situation evolves.

Complex situations requiring complex information presentation are a way of life in the modern world. Part of the frustration many people feel searching for information in a computer system arises because the required information they need is hard to integrate into a coherent whole. Loren Terveen et al.'s (1995) work revealed that

> The pragmatics of knowledge use are critical. Simply recording a factor is not enough; issues such as where in the process knowledge is to be accessed, how to access relevant knowledge from a large information space, and how to allow for change also must be addressed. (p. 3)

In other words, socio-technical situations do not lend themselves to the basic task analysis that appears in textbooks. That task analysis is appropriate for step-by-step processes, but fails when the process gets more complex. Instead, communicating technical information through a socio-technical lens requires supporting the way people rapidly assess situations and make decisions based on theories such as Klein's (1999) recognition-primed model rather than classical decision matrix models or simple task analysis (Albers, 1996).

The question concerns not merely whether the readers know some particular piece of domain knowledge, but whether they understand the relationship between different pieces of information. Do they know "that it is relevant to the problem at hand and does he or she know how to utilize this knowledge in problem solving" (Woods & Roth, 1988, p. 420)? People require information that relates to the overall situation, and they need to understand that relationship (Robertson et al., 1993).

Likewise, across multiple studies Barbara Mirel found that users have different conceptions of how to accomplish a task. "In actual work settings, users define their own tasks and task needs according to situational demands, not program design" (Mirel, 1992, p. 15). The design of those systems must encompass a total system that revolves around the goals and information needs of a human and supplies information that makes sense within the person's real-world situation. Felipe Castel (2002) aptly summed up my argument when he said, "Computing does not merely process information, it commits to a certain representation of information" (p. 30). Technical communicators make many of the decisions about that representation; we must make good choices.

Bringing Socio-Technical Reasoning into Technical Communication

In science class, we learned that a rock and a feather fall at the same rate (in a vacuum). Yet, hold a rock and a feather, drop them, and clearly, they fall at different speeds. This obviously means that whatever is attracting them must vary depending on the material—hey, it did to ancient and medieval philosophers, who were adherents of Aristotelian physics. Of course, now we understand the difference is because of air resistance.

The rock and feather example really is relevant to technical communication because too often we risk saying situations are very different because we don't know about/understand the air resistance. In a physical system, air resistance is obvious and easy to measure. In the social sciences, including in technical communication, the stand-in for air resistance may not be obvious. Actually, it probably consists of many different things; some easy to measure, some difficult to measure, some we (erroneously) don't consider worth measuring, and some we don't even know we should measure. But they all define and influence the relationships and, consequently, influence how people understand information and how the overall situation evolves.

▪ Technical Communication Writing Environment

Writing that addresses complex problems and which addresses socio-technical issues is ill-structured. There are too many interrelations within the content for it to be anything else.

The ill-structured environment equates to a wicked problem. Wicked problems—to use Chad Wickman's (2014) term—are a given in technical communication, but we try too hard to reduce them to simple problems. On the other hand, many writers claim they are not really writing in an ill-structured environment, or will acknowledge that the entire process is, but point out that they are working in a small area. They could be better characterized as having rationalized their ill-structured environment into a simple one, a rationalization that proves problematical and which I have discussed on different occasions (Albers, 2004). The decision-making process and information needs for simple (highly structured) and complex (ill-structured) problems are different. We need to acknowledge that difference and provide content differently.

As a field, technical communication has stubbornly refused to move beyond a view of writing as highly structured. This highly structured view permeates technical communication pedagogy, including how we define "what is technical writing."

David Dobrin (2004) put forth a brief definition that "technical writing is writing that accommodates technology to the user" (p. 118). Unfortunately, within the current world, any definition with a strong technology connection must be suspect as too limiting.

Likewise, two of the major introductory textbooks offer these definitions:

> Technical communication encompasses a set of *activities* that people do to discover, shape, and transmit information. . . . The biggest difference between technical communication and other kinds of writing you have done is that technical communication has a somewhat different focus on *audience* and *purpose*. (Markel & Selber, 2019, p. 2)

> Technical communication is a process of managing technical information in ways that allow people to take action. (Johnson-Sheehan, 2005, p. 6)

Both definitions are very writer focused. They describe what a writer must do, rather than focus on communicating information. I also looked for definitions in other major textbooks and found, rather than concise definitions suitable for quoting, longer discussions of what technical communication is and is not. But they still presented those definitions in writer-focused terms. Missing is the acknowledgement about meeting people's information needs when the situation has changed—the Friday and Saturday differences of the opening example.

All of the textbooks' views are tightly tied with the production of artifacts (one or more documents, loosely defined as whatever the audience is expected to read). I'm wondering why we are focused on the production of artifacts. Why are we not focused on communicating the information behind the reason for producing the artifacts? People don't want artifacts; they want information. People do not want a document; they want the information within the document. The document is simply the easiest method of obtaining that information. From a writer's viewpoint, some may consider the document and the information as the same thing, but I think the mindsets of developing an artifact and communicating information are very different. In the one, we are concerned with producing something . . . a something that gets tweaked for the sake of being a good artifact. Whether or not that tweak is meaningful with respect to its communication value can get lost. These types of problems make me think of the book *The Design of Everyday Things*, where Don Norman (2002) disparagingly described many deeply flawed designs with "probably won a prize," because many flawed designs he critiques did, in fact, win design awards.

From a technical communication perspective, along with the standard issues such as audience analysis, defining the socio-technical situation involves understanding the relationships between the information elements and defining the boundaries of interest. These two issues, relationships and boundaries, are typically ignored in both practice and within technical communication pedagogy. Yet, together, they make or break the text's ability to effectively communicate its information. We must understand their importance, determine them during the analysis, and create content that reflects how we defined them.

■ Relationships

Relationships form the foundation on which people understand complex information (Albers, 2009, 2010). It is not the pieces of information but the relationships between them that provide the understanding. The analysis must capture both the information and the relationships. In capturing the relationships, the analysis captures how people understand and interpret the information. That understanding and interpretation is not about the information per se but about the relationships within the information. Understanding relationships forms a simplified explanation of why an experienced person can look at a collection of data and know what's happening while an inexperienced person can recite all the data but still lacks an understanding on which to base decisions. Being able to quickly understand the relationships is a major aspect of naturalistic decision-making (Klein, 1999).

Much of a reader's comprehension exists in their understanding of the relationships between and within pieces of information. The reductionist approach of breaking problems into smaller pieces breaks up those relationships and interactions. After understanding the smaller pieces, the analysis must then work back outward or risk failing because it failed to capture the relationships and interactions which make up the situation. It fails because it fails to capture the essential elements needed to understand a situation.

Thus, information relationships are not just a nice-to-know thing. The information understanding exists within the relationships, not with the individual text elements. Without understanding the relationships, people cannot make good decisions. Thus, writing from a socio-technical lens means understanding

- how those relationships form,
- what makes them form,
- how changes to the relationships propagate through the system,
- the biases people exhibit in understanding them,
- how the relationships change as the situation changes, and
- how they differ between related situations.

Unfortunately, too often an attempt at an analysis measures the easy-to-measure and disregards the rest. And often jumps right in to measure the easy-to-measure and doesn't try to define what should be measured. The result describes the overall situation very poorly, and the idea of deep analysis gets a bad reputation. The problem was not in the data collection or in the analysis but in what data was collected.

Relationships come in two major types: functional and non-functional. Functional relationships are directly connected— Such as, if we increase X, then we know Y will change. Non-functional relationships are more situation dependent—Such as, "we can't put a new parking lot there because it encroaches on a natural wet area and we risk an environmental lawsuit." Some information elements have nothing to do with building a parking lot, but the overall social aspects build a relationship between environmental groups and parking lot location. Clearly, non-functional relationships can have a major impact on decision-making, but they are easy to ignore since they rarely appear in system block diagrams. At first glance, they appear outside of the problem scope, or they never get mentioned to the people doing the analysis.

Relationships and Feedback Loops

The analysis leading up to content development needs to consider the entire situation at multiple levels. Relationships form and exist for both macro- and micro-levels of both social and technological interactions.

Relationships form a two-way integration, and changes to a piece of information ripple out; the resultant change can ripple back, again changing the original information. In other words, the relationships within the situation are part of the feedback loops that control and (de)stabilize the situation.

The feedback loops within relationships allow the system to adapt. As part of the change, an information element itself may/may not change, but its relationship to other elements will change. With the overall web formed from the relationships, the strength and type of changes are very difficult to predict. Consequently, how the socio-technical situation will react is very difficult to predict.

A bunch of blocks connected with springs and sitting on a surface act as a metaphor for the socio-technical situation (see Figure 6.2). They must be on a surface because it represents the internal friction and unknowns within the situation.

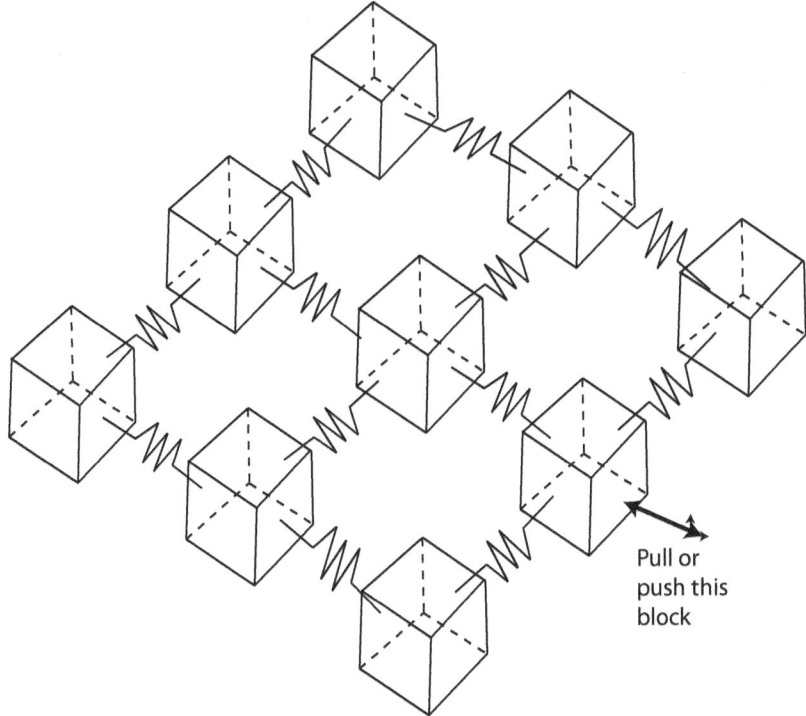

Figure 6.2. Blocks connected by springs. Movement of the marked block makes all the other blocks move. Thinking in terms of a larger number of irregular and varied sized blocks makes the concept more realistic for visualizing the issues of communicating complex information.

The overall readjustment occurs because a dynamic stable system is kept in equilibrium though a set of feedback loops of information and control. Each block movement affects other blocks, which changes the spring tension (relationships) between them. Each move results in the entire system readjusting itself to a new position that minimizes the overall tension. The proper level of analysis is not individual components and how they break but system constraints and objectives.

The friction element introduces the non-linear response. If they were suspended from a frame (or any other way that minimizes friction), displacing a block and then moving it back would result in the overall system returning to the previous point. Once friction is introduced, then it will not return to the starting position but some position different from both the starting and pre-return position. This is the critical factor ignored in too many decisions made with the belief of "if it doesn't work, we'll just go back to what we had before."

> It is impossible to move a block without the change rippling through the system. Complex systems operate under conditions far from equilibrium. Inputs need to be made the whole time by its components in order to keep it functioning. Without the constant flow of actions, of inputs, it cannot survive in a changing environment. The performance of complex systems is typically optimized at the edge of chaos, just before system behavior will become unrecognizably turbulent. (Dekker, 2011, p. 138)

One way to think about the system operating far from equilibrium is to think of the spring diagram with most of the springs stretched to the point where any less/more tension will cause the block to move. It also means the overall system is not just hanging there, steady, waiting for the block to move. Instead, the dynamic nature of the situation is constantly slightly moving different blocks, and the overall system is in a state of constant readjustment.

From a design perspective, this means you can't understand the entire situation or predict the effect of a change. It might seem like it will have minimal effect, but if combined with some other random changes, it risks tossing the system into violent gyrations before reaching a new equilibrium point—with no guarantee that the new point will be desirable or expected.

On the other hand, this block system (and complex systems in general) tends to be highly resilient to the loss of any one part (remove a block); it will adapt and reach a new equilibrium point which does not include the part. The non-symmetry aspects of a complex system mean that if the part is reintroduced, rather than returning to the old equilibrium point, it will rebalance itself from the current point and will end up with a new equilibrium point. Decisions cannot be simply reversed.

Technical Communication Reimagined   185

■ Boundaries

We must consider how and why we are drawing the boundaries at the beginning of any design process. The boundaries *define* the system of interest. They should, of course, be based on the reader's information needs, which, almost by definition, make defining the boundaries a non-trivial problem. The relationships and the potential ripple effects help to define where the boundaries should be drawn.

Based on how a boundary gets defined, both the relevant information and its presentation change. Misdefining a boundary redefines how the person views and understands the situation (Laplante & Flaxman, 1995; Robertson et al., 1993). Contrary to the common practice, writing with a socio-technical lens means acknowledging that the area inside the boundary includes both the system and the social situation in which the system is embedded. "Define the boundaries not by the system itself, but by the purpose of the description of the system" (Dekker, 2011, p. 139).

The old style of writing manuals that provided a menu option by menu option description (start at File-new and write through Help-about) drew the boundaries too small (see Figure 6.3). Here, the relevant information is just information about the operation of one menu option. No connection to other menu options; no connection to the tasks in which people would use it. It's easy to write up each item, but people rarely address problems with just one menu option. The narrow boundary ignores the actual problem embedded within a socio-technical situation and, instead, simplifies it into a straightforward technical description problem.

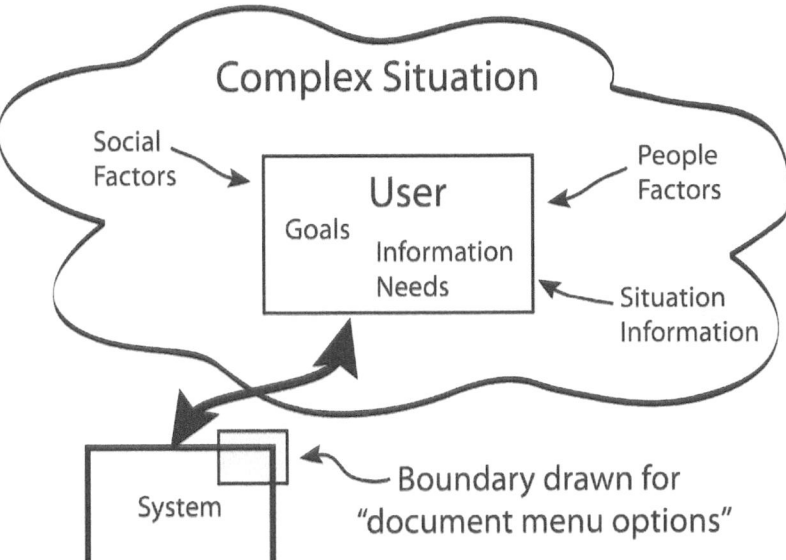

Figure 6.3. Drawing too small boundaries. To make the document highly structured, the boundaries get defined with respect to the system. The result is correct and usable, but generally useless, documentation.

I've said multiple times in different writings that people have no desire to use any software system and that they don't care about the menu options (Albers, 1996, 2010). What they do care about is accomplishing a task, and the software system happens to be the easiest route. By drawing the wrong boundaries, technical communication has managed to have minimal impact by not helping people accomplish the task.

Expanding beyond simple system interface explanations requires looking at the bigger picture and much deeper understanding of both audience and the socio-technical situation that is common (see Figure 6.4). It requires understanding what decisions the audience wants to make and how they tend to go about making them. Note that I'm not talking about a fancy artificial intelligence (AI) system that makes the decisions. If the system provides sales information for both sales staff and management, it cannot make decisions about how to focus sales. However, the upfront analysis and design should consider how the readers go about looking at various pieces of information and synthesizing them. In this case, the goal of a good socio-technical communication would be to help with the synthesis and resulting decisions. In any writing or designing situation, we must define boundaries. Draw them too small and we only look at isolated tasks/events and don't see the big picture. Draw them too large and we get buried in an exponentially growing collection of relationships, most of which are irrelevant.

One often-voiced complaint with any view of drawing inclusive boundaries, including a socio-technical view, is that determining the boundaries is impossible. The spring diagram in Figure 6.2 can correspond to the boundaries. As changes ripple out, they should get less and finally make little change. Defining the boundaries to match the decrease in the ripple provides a workable boundary in terms of both containing effects and providing information.

Too often, complaints focus on the edge cases, with comments such as "Well, yes, this is good, but I have *one* client where it doesn't apply; therefore, the entire thing is crap." In the end, dealing with people ends up being probabilistic, and the 80/20 rule applies. It is impossible to make 100 percent of the readers happy. The analysis and writing must focus on the 80 percent, where most of the information needs are. When people enter the situation, 100 percent will never happen.

Writing (Reading) Within a Complex Socio-Technical Environment

Looking at writing (reading) within a complex socio-technical environment, we encounter a disconnect. Technical communication and technical communication pedagogy are rarely presented through a problem-solving lens suited to working within that complex socio-technical environment. Providing content in a complex information environment means providing information for decision-making or problem-solving within the situation's context.

Technical Communication Reimagined 187

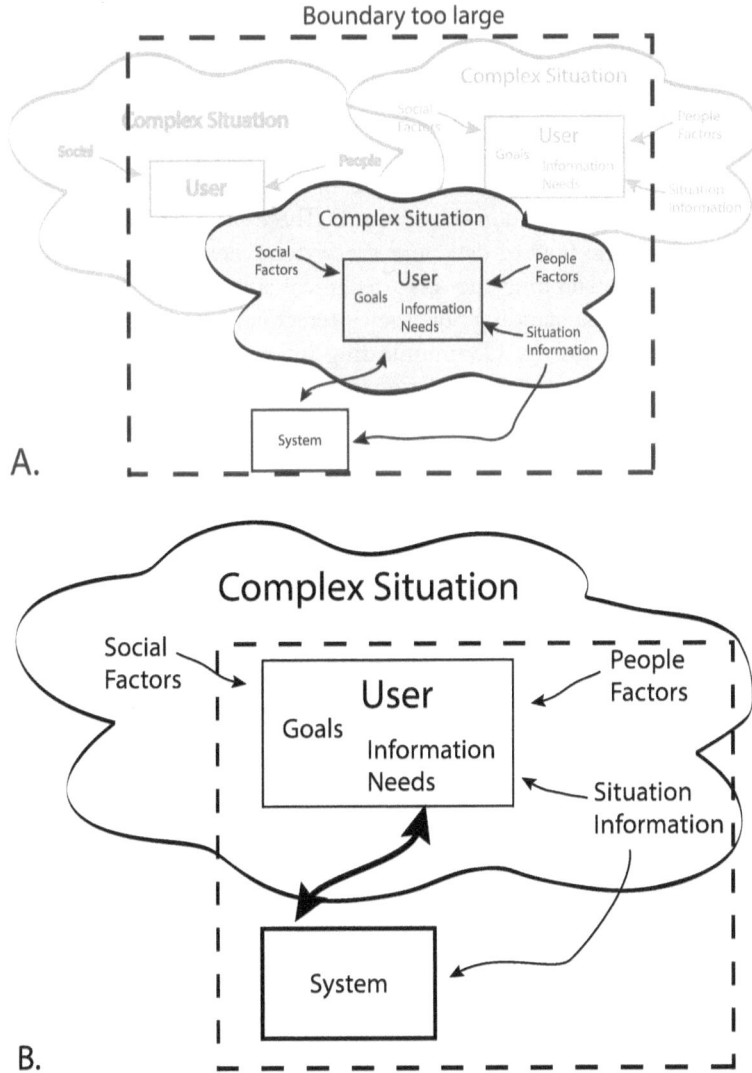

Figure 6.4. Drawing boundaries. A. has pushed the boundaries too far out and is typically what the person who claims they can't draw boundaries outside of the system box views. B. does not include the entire complex situation, but does bring the boundaries into a manageable size.

Writing for decision-making and problem-solving means positioning the content in terms of the socio-technical situation. Unfortunately, too much information is provided as a data dump and not as information structured to assist the readers in achieving their goals. Clearly, a writer needs to use critical thinking skills to reflect on the needs of people within that situation and the overall implications of how/why content is needed and used. For the writer, restructuring the information to meet the socio-technical environment needs requires a deep rethinking of what we mean by understanding, communicating, writing, and audiences.

We have seen that the system boundaries and the information relationships drive the overall problem space of what is defined as relevant. That problem space of any socio-technical system is made up of a set of interacting sub-systems (which may themselves contain sub-systems). These sub-systems contain people, the environment, the flow of data, and the systems delivering that data within the situation. Understanding the socio-technical situation in which the person operates requires understanding how they interact with, how they trust, and how they use those sub-systems. Communicating information within that situation requires drawing appropriate boundaries across those sub-systems.

Sub-systems must be identified as part of the early content analysis, and, more importantly, their interactions must be defined. There are also the information influencers that are not a specific source but which influence how a source is understood and used. Examples of influencers are bosses that insist certain information be considered (even if it is more/less irrelevant), power structures that affect how people perceive the information source, or technical system interactions that strongly control data flow (poor interface design, cumbersome approval/release of information process, strong and restrictive data security, etc.).

The analysis of a communication situation aims to provide a foundation for an optimized communication. Socio-technical system optimization requires a joint optimization. Although it is easier to optimize either the social or technical sides (or some subset of factors within each), the result will not be a highly efficient or effective system. Too many sources (for instructors, students, and practitioners) describe technical communication as being about understanding the audience and explaining the material clearly—a view which oversimplifies reality. This view gives a starting point but fails to produce texts that provide a high-quality reader experience. The initial pre-writing analysis can begin at this simplified starting point, but it then needs to transition into the messier view of how communication really happens. The analysis now needs to consider the common factors within the relationships between people (social systems) and technology (technical systems) and how those factors interact and evolve.

Many post-failure reports bear out how problems cascaded through a system where some parts worked wonderfully, yet the system as a whole failed miserably. Documentation failure can often be traced to a focus on one aspect, typically a narrow view of the technical aspects (as marked in Figure 6.3). It could be described as written from a view of "I'm describing the system. How it gets used is not my concern." The document fails because the analysis and content failed to address the complex interdependencies that exist within the reader's socio-technical environment. The entire complex situation cloud in Figure 6.1 got ignored during the writing process.

Within that socio-technical writing situation, technical communication needs to reshape its questions in human-information interaction (HII) terms (Albers, 2012) and focus on defining how the audiences will *interact with the information,* how the audiences will *use it,* and how the various parts of the situation *influence that interaction and use.* All three factors must be considered equally.

■ All Decisions Are Local and All Implications Are Global

Receiving information which properly delineates the relationships and boundaries can help improve the decisions. With few exceptions, decision-makers are not trying to sabotage a system/process; they want to make good decisions, and they want the overall project to succeed. Even for those that are trying to sabotage it (or don't care if they run it into the ground in the long term), the decisions make sense within their agenda and priorities. In other words, decisions exist as a snapshot of a single instance and are made with the belief that they are the best possible (best compromise) decision reflecting the decision-maker's goals. The future may reveal them as horrid or wonderful, but when they are made, they are considered good.

Although I doubt many people take issue with the previous paragraph, it hides a significant issue that has a major impact on the decision-making process: decisions are local; implications are global. Decision-making research has concluded that essentially all decisions are local. Unfortunately, decision implications are global—the rippling that occurs because of any, even minor, change. "Behavior that is locally rational, that responds to local conditions and makes sense given the various rules that govern it locally, can add up to profoundly irrational behavior at the system level" (Dekker, 2011, p. 159). It is only later, when viewed with hindsight and viewed from a larger viewpoint, that we can see the flaws in the logic and the poor decision path that the people followed. The post-failure analysis that discusses the lack of information or poor presentation of proper information can trace the failure back to designs that never connected information to the larger picture. The analysis must consider and expect that people make choices and decisions in isolation based on considerations of individual parts. In addition, the content developers must acknowledge that although they may have a view of the overall relationships, the people making the decisions will not.

People try to make decisions as if they are adjusting to a static system and forget that they are adjusting to a dynamic, highly interconnected system (see Figure 6.2). They think in terms of a simple system and take a highly local view of the change (see Figure 6.5). The thought process follows along the lines of "We are only making a minor change. It will never affect anything else." But the springs are all under tension and the other blocks move. That movement of the other blocks, the ripples through the overall situation, may push it over the edge of stability and cause profound changes. Effective writing about the situation needs to bring the dynamic nature front and center to remind the decision-makers that there will be ripple effects and the potential ripples must be considered.

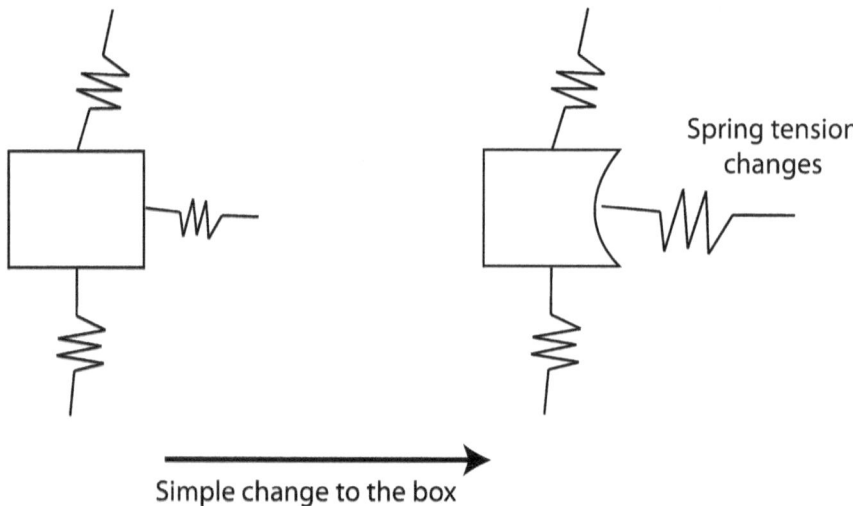

Figure 6.5. Making a change to just one box. It is easy to have an attitude of "pay no attention to that spring (relationship), for we only care about this box." But the spring tension (relationships) change and, consequently, the entire situation changes.

Figure 6.5 shows the problem of a decision-maker only worrying about their one small box. (High-level decision-makers—executive-level—may be making decisions about multiple boxes, but they still make decisions about their group of boxes.) Their decision is local. These decisions are all local decisions, made to optimize the current local point. They are all sensible and logical within the current local conditions. Discussions which raise questions about interconnections are often shut down with comments along the lines of "Yes, that may be important, but today we are only concerned about . . ." Consideration of the longer-scale dependencies is deferred to another day or, more realistically, deferred forever. The boundaries are redrawn small, and the relationships/ripple effects of the decision are ignored.

Complexity, essentially by definition, means a huge number of interacting parts that give rise to unpredictable outcomes. Each new component or layer of organization creates an explosion of new relationships and a myriad of new ways to draw boundaries. A problem is that the analysis tends to rationalize that it's analyzing a simple system with only a few parts; we see only the parts directly of interest to us *now*. When viewed small piece by small piece, then, yes, it might look simple, but that is like examining the fuel pump on a car and forgetting that it connects to the rest of the engine. We rationalize the small view and ignore the large view. We focus on the block and forget the springs.

Technical communicators need to draw proper boundaries to reveal the influence and potential ripple paths. Writing through the socio-technical lens captures the complexity of the situation and helps to force readers to consider how their item of interest sits within and affects a bigger situation. The end goal of developing content is to best communicate it to the readers. Making sure that communication supports making complex decisions requires using a socio-technical lens.

■ Example: Wolves in Yellowstone

One compelling example of the interconnections can be seen by looking at the reintroduction of wolves back into Yellowstone. Some researchers claim they significantly changed the landscape. It's a long sequence that looks obvious in retrospect, but since there were essentially an infinite number of ways their introduction could have gone, it was not predictable (the single path is traced out in Figure 6.6).

The compressed version: Before wolves, deer ate the trees at stream banks so nothing grew to stabilize the ground, and streams became fast moving and eroded their banks. Wolves pushed the deer away from the streams (but didn't actually eat many of them); ground cover came and stabilized the soil, trees grew, beavers came, the stream got dammed and turned into slow-flowing streams, amphibians and wetland reptiles/mammals came; stream banks overgrew; erosion stopped. Obviously, wolves have no direct effect on a stream, but introducing the wolves started a chain of ripples that changed fast, free-flowing, eroding streams to slow, meandering streams.

Some decisions have minor ripples and may cause the situation to bounce back to almost the starting position (adding beavers when there are not enough trees would not fix the problem). Others cause an avalanche of changes (adding the wolf) that may be far removed from the initial goal and totally not a concern of the decision-maker (the group who added wolves). The key item which can cause the avalanche may not appear remotely related to later developments.

This example itself may or may not be true (Fong, 2018; Kuhne, 2019). But it still shows a potential train of relationships rippling through a complex system—interactions that would never be predictable at the beginning of the process. They cannot be predicted because the non-linear aspects of the situation make it sensitive to the initial conditions, and uncontrolled factors also have an influence. Afterwards, it looks like a straightforward chain of events, but the reality is that a large event tree is constantly pruned down to give "what really happened" (see Figure 6.6).

This example talks about wolves, but a similar string of events can occur whenever major decisions are made—when a new software system is implemented, major hiring policies are changed, or a company's focus shifts between products. These are all company-wide decisions and affect the entire corporate environment. But, likewise, decisions made at a much lower level can ripple through and have profound effects on a specific unit.

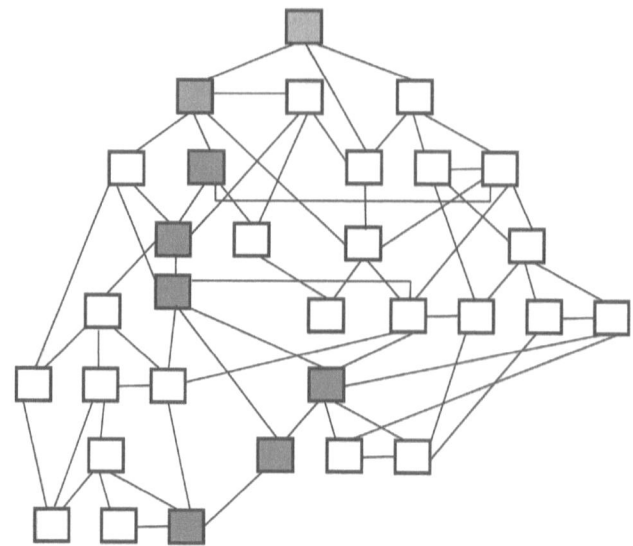

Figure 6.6.: Chain of events. All of the outcomes make up the large tree, but only a single chain actually occurs (the gray boxes). Many socio-technical factors combine to drive which specific chain occurs for a process.

As technical communicators, when we develop content, we need to ensure it provides the reader with a good view of the potential consequences; the content must capture the socio-technical situation. It is very easy for expectation bias to consume the reader. They only see what they expect to see and foresee that their decisions will unfurl as they expect. Potential unintended consequences are ignored.

■ Looking Ahead

■ Research Needed

While research often looks at decision-making in isolation, in reality, decision-making occurs as part of larger tasks and makes up only a single element in achieving a larger goal (Orasanu and Connolly (1993). Decisions occur within a cycle which "consist[s] of defining what the problem is, understanding what a reasonable solution would look like, taking action to reach that goal, and evaluating the effects of that action" (Orasanu & Connolly, 1993, p. 6). Although they were talking about decision-making research, the statement applies equally to much technical communication research. The analysis and documentation focus on one component and miss that they are embedded within and constantly react to a much larger framework.

Complex situations and their complex information presentation needs have become the norm. Yet, as people search for and interact with information in a computer system, they feel frustrated because the information they need is hard to integrate into a coherent whole. Back in 2007, Janice Redish (2007) discussed how the usability of complex systems is not the same as the usability of simple systems, but a fundamentally different beast. Yet, over ten years after Redish made her call for the development of complex usability methods, little has changed. It appears her message has been lost. We seriously need to begin the research needed to handle information communication within a complex situation.

Current technical communication analysis methods are focused on how to communicate about the one component—in other words, dealing with simple systems with their right/wrong answers. Our task analysis methods are good for what they are designed for but do not go far enough. They fail to capture the bigger picture in which the information need is embedded—note the term *information need* and not *task*. When the communication goal moves beyond tasks and into decision-making, we have essentially no methods that go beyond high generalities of "understand your audience." A true statement, but not one that provides methods for determining the socio-technical needs of the audience within a decision-making environment. The audience, the environment, and the technical systems all interconnect and interact with each other. We need methods of how to consider those relationships when creating information. Returning to the Spool (2014) story which started this chapter, we need methods to help distinguish the different information needs on different days and how to address them.

■ Pedagogy

Moving to a problem-solving view is not simply a teaching problem but, rather, a mindset of shifting from providing information to asking why the person needs the information in the first place and how they are going to use it, and what problem they are solving. It is about understanding the entire situation. An important question that must be answered at a deep level is what drives why people are looking at the information in the first place. The need for information should not be viewed as a simple need.

Instead of dealing with the real goals, too much technical communication takes the simple view and writes about "how to retrieve the information" or "how to perform X." What it means, how to interpret X, and how X connects to other information are defined as outside of scope. The complex problem of the reader's situation has been redefined to a simple problem. And then the writers wonder why no one values their work. Within the classroom, we need to discuss these bigger issues and ensure the students understand that a simple "I'll just write up all the information about it" or "I'll write 14 different sections, one for each audience" will never communicate the information in a way that fits the reader's needs.

We also need to make students realize that complex situations cannot be broken down into individual pieces—this works wonderfully in the hard sciences and in computer science but fails miserably in the social sciences. The deconstruction approach fundamentally changes the problem because it loses the relationships (Albers, 2009). Once, in a discussion of complex relationships, a student was adamant that everything within them could be broken down into a series of simple actions that could be understood and/or documented. Not really. Yes, each identified action/problem can be solved, but within the complex problem space, the problems that need to be solved change. It is not the same sequence each time. And often, those simple actions become too soft and have no real specific action. Not unlike a command to "now research your topic." True, but it doesn't give guidance on how/what to research, or how to know if the results are valid and/or sufficient.

The ideas of defining relationships or drawing boundaries within a situation seem to be completely off the technical communication radar. Task analysis teaches us to collect the steps or required information. It doesn't explicitly contain the additional factor of defining the relationships, which is essential to understanding a complex situation. Rhetorical analysis has too much desk work (sitting at a desk and thinking/reflecting) rather than interacting with people to collect data.

∎ Conclusion

Transiting technical communication to a problem-solving view using a socio-technical lens is not simply a teaching problem. Rather, it's a change in the discipline-level mindset that requires shifting from providing information—typically in a step-by-step fashion of highly structured writing— to asking why the person needs and uses the information in the first place, how and where they are going to use it, and what problem they are solving, and *then* providing them with the information presented in a manner relevant to their needs (see Figure 6.1). Technical communication needs to reshape its basic conception of communicating information to one that privileges problem-solving and decision-making rather than simply providing information or procedures. The writing goal of being clear, complete, and correct becomes much fuzzier within this world. None of these three terms has clear, complete, and correct answers, something which causes cognitive dissidence for everyone—students, instructors, and practicing professionals.

A foundational idea of socio-technical theory is that the design and performance of any system can only be understood and improved if both "social" and "technical" aspects are considered together and treated as *interdependent* elements within a complex situation. Although at this point in this chapter, the previous sentence should be obvious, the socio-technical situation tends to be larger than what is normally considered within audience analysis, much of technical communication, or studies of the rhetorical situation. These two sentences sum up the essence of this chapter, and I hope they become the most quoted lines. Fully understanding them requires understanding what needs to change to create content through a socio-technical lens. Within a complex situation, the social and the technical are highly interdependent. Creating content requires teasing out those interdependencies and the relationships that form them. Audience analysis must be more than basic demographics and must get at the fundamentals of how/why the audience needs the information. As part of that defining of how/why, the problem space must be defined. Boundaries must be drawn to make the problem manageable for the writer but still relevant for the reader. The boundaries must include the important relationships, which can give the impression of expanding the problem beyond the "I need this information" view. Yes, the writer must provide that information, but they must also ensure it remains within the context of the entire socio-technical situation. Without that context, the reader lacks the full information required to make a high-quality decision.

Final Thoughts

Looking back, I realize that when I wrote the books *Communication of Complex Information* (2004) and *Human-Information Interaction* (2012), I had too strong of a focus on the writer conveying information to a reader aspects of the communication situation. In other words, assuming in Figure 6.1 that the arrowhead size denotes significance, the arrows pointing at the reader were much larger than the arrows going outward from the reader. But thinking of this in socio-technical terms requires rethinking their relative size. In fact, they must be either the same, or, perhaps, the outward arrows are actually larger.

Why larger? Because the reader will be taking the information, making decisions, and affecting the situation. Technical communication through a socio-technical lens is not about providing information to a reader but about understanding how that reader will be influencing the situation. Yes, there are feedback loops with information coming to and from the reader, but, ultimately, it's how the reader changes the situation that matters.

The technical communicator's job is to provide the information needed to allow the reader to make decisions that change the situation in a manner they desire, and to monitor that the changes are processing as expected. Accomplishing this task requires understanding the situation in which the information is used—what information is relevant, how it interconnects, where it comes from, how both the information and the relationships evolve as the situation develops, and how to draw the boundaries to define the situation.

References

Albers, Michael J. (1996). Decision making: A missing facet of effective documentation. In *Proceedings of the 14th Annual International Conference on Systems Documentation: Marshaling New Technological Forces: Building a Corporate, Academic, And User-Oriented Triangle (SIGDOC '96) (pp. 57–65)*. Association for Computing Machinery. https://doi.org/10.1145/238215.238256.

Albers, Michael. J. (2004). *Communication of complex information: User goals and information needs for dynamic web information*. Erlbaum.

Albers, Michael. (2009). Information relationships: The source of useful and usable content. In *SIGDOC '09: Proceedings of the 27th ACM International Conference on Design of Communication (pp. 171–178)*. Association for Computing Machinery. https://doi.org/10.1145/1621995.1622027.

Albers, Michael J. (2010). Usability and information relationships: Considering content relationships when testing complex information. In M. J. Albers & B. Still (Eds.), *Usability of complex information systems: Evaluation of user interaction.* (pp. 109–132) CRC Press.

Albers, Michael J. (2012). *Human-information interaction and technical communication: Concepts and frameworks*. IGI Global.

Castel, Felipe. (2002). Ontological computing. *Communications of the ACM, 45*(2), 29–30.

Cilliers, Paul. (1998). *Complexity and postmodernism: Understanding complex systems.* Routledge.

Dekker, Sidney. (2011). *Drift into failure: From hunting broken components to understanding complex systems.* CRC Press.

Dobrin, David. N. (2004). What's technical about technical writing? In J. Johnson-Eilola & S. A. Selber (Eds.), *Central works in technical communication* (pp. 107–123). Oxford University Press.

Ebert, David, Zwa, Amen, Miller, Ethan, Shaw, Christopher & Roberts, D. Aaron. (1997). Two-handed volumetric document corpus management. *IEEE Computer Graphics and Applications, 17*(4), 60–62.

Easterling, William E. & Kok, Kasper. (2002). Emergent properties of scale in global environmental modeling—Are there any? *Integrated Assessment, 3*(2–3), 233–246.

Fong, Maria. (2018). *Wolves change ecosystem and geography in Yellowstone*. Tufts University. http://sites.tufts.edu/tuftsgetsgreen/2018/02/27/wolves-change-ecosystem-and-geography-in-yellowstone/.

Interaction Design Foundation. (n.d.) *What are socio-technical systems?* Retrieved January 28, 2019 from https://www.interaction-design.org/literature/topics/socio-technical-systems.

Johnson-Sheehan, Richard. (2005). *Technical communication today.* Pearson.

Klein, Gary. (1999). *Sources of power: How people make decisions*. MIT.

Klein, Lisl. (2014). What do we actually mean by "sociotechnical"? On values, boundaries and the problems of language. *Applied Ergonomics, 45*, 137–142.

Kuhne, Michael. (2019). *Scientists debunk myth that Yellowstone wolves changed entire ecosystem, flow of rivers*. AccuWeather. https://www.accuweather.com/en/weather-news/scientists-debunk-myth-that-yellowstone-wolves-changed-entire-ecosystem-flow-of-rivers/70004699.

Laplante, Phil & Flaxman, Harvey. (1995). The convergence of technology and creativity in the corporate environment. *IEEE Transactions on Professional Communication, 38*(1), 20–23.

Markel, Mike & Selber, Stuart. (2019). *Practical strategies for technical communication* (3rd ed.). Bedford/St. Martin's.

Mirel, Barbara. (1992). Analyzing audiences for software manuals: A survey of instructional needs for 'real world tasks'. *Technical Communication Quarterly, 1*(1), 15–35.

Mumford, R. Enid. (1987). Socio-technical systems design: Evolving theory and practice. In Gro Bjerknes, Pelle Ehn & Morten Kyng (Eds.), *Computers and democracy* (pp. 59–77). Avebury; Aldershot.

Norman, Donald A. (2002). *The design of everyday things*. Basic Books.

Orasanu, Judith & Connolly, Terry. (1993). The reinvention of decision making. In Gary Klein, Judith Orasanu, Roberta Calderwood & Caroline Zsambok (Eds.), *Decision making in action: Models and methods* (pp. 3–20). Ablex.

Redish, Janice. (2007). Expanding usability testing to evaluate complex systems. *Journal of Usability Studies 2*(3), 102–111.

Robertson, George, Mackinlay, Jock & Card, Stuart. (1993). Information visualization: Using 3D interactive animation. *Communications of the ACM, 36*(4), 57–71.

Spool, Jared. (2014). Service design: Pushing us beyond the familiar. *Center Centre – UIE*.https://articles.uie.com/service_design/.

Terveen, Loren, Selfridge, Peter & Long, M. David. (1995). Living design memory: Framework, implementation, lessons learned. *Human-Computer Interaction, 10*(1), 1–37.

Trist, Eruc. (1981). The evolution of socio-technical systems—A conceptual framework and an action research program. In Andrew Van de Ven & William Joyce (Eds.), *Perspectives on organisational design and behaviour.* (pp. 19–75) Wiley Interscience.

Vicente, Kim. (1999). *Cognitive work analysis*. Erlbaum.

Wickman, Chad. (2014). Wicked problems in technical communication. *Journal of Technical Writing and Communication, 44*(1), 23–42.

Wikipedia. (n.d.) *Sociotechncial system*. Retrieved January 28, 2019 from https://en.wikipedia.org/wiki/Sociotechnical_system.

Woods, David D. & Roth, Emilie. (1988). Cognitive engineering: Human problem solving with tools. *Human Factors* 30(4), 415–430.

Chapter 7: Applied Rhetoric as Disciplinary Umbrella: Community, Connections, and Identity

Jennifer R. Veltsos
MINNESOTA STATE UNIVERSITY, MANKATO

Matthew R. Sharp
EMBRY RIDDLE AERONAUTICAL UNIVERSITY

Jacob D. Rawlins
BRIGHAM YOUNG UNIVERSITY

Ashley Patriarca
WEST CHESTER UNIVERSITY

Rebecca Pope-Ruark
GEORGIA INSTITUTE OF TECHNOLOGY

Abstract: This chapter argues that many of the existing names and boundaries in use around professional communication create artificial separations among research, pedagogy, theory, and action related to the practice of rhetoric in contemporary society. Scholars working in this area teach and conduct research across a variety of disciplines, but we share a rhetorical foundation and a concern for the practical application of that theory. This combination of classical rhetoric and public action provides a way to move our work beyond the confines of the academy and actively engage in rhetorical work within the communities where we work, live, and research. We argue that *applied rhetoric* is an overarching term that more accurately describes the interdisciplinary work used by scholars, teachers, and practitioners in diverse areas of communication who work to clarify ideas that help people accomplish goals, to explicitly connect research to teaching, and to be a force for good in the world.

Keywords: applied rhetoric, praxis, disciplinarity, identity

The modern study of rhetoric, in all its forms and functions, spans a wide range of disciplines. Rhetoric scholars identify as researchers and practitioners of professional and technical communication, rhetoric and composition, organizational rhetoric, the rhetoric of science, the rhetorics of health and medicine, public rhetoric, or civic rhetoric, among others. While the result of this ever-expanding specialization may be an increased influence of rhetoric across a range of disciplines, professionally we have been fractured into niches that use seemingly arbitrary boundaries to distinguish our work from others. This fracturing arguably began in 1914, when public speaking teachers left the National Council of Teachers of English to form what would become the National Communication Association. This split between rhetorical scholars of written and spoken work was the first of many separations where rhetoricians working in specific contexts would set out to find specialized audiences and collaborators for their work. While this specialization is seen as useful by some for the purposes of deepening expertise, it does have a downside (Harlow, 2010).

This ever-increasing disciplinary specialization and its concomitant specialization of discourse (Russell, 2002) makes cross-disciplinary collaboration simultaneously (and paradoxically) more necessary and more difficult (Harlow, 2010). Furthermore, individual scholars may find themselves feeling somewhat lost—between rhetorical traditions, research methods, and pedagogies—not fully at home in their own discipline but not completely accepted by their peers in other disciplines. Historically, these disciplinary divides have caused theoretical scholarship to be separated from and privileged over scholarship on pedagogy (Leff & Lunsford, 2004; Zarefsky, 2004). Scholarship with a more traditional, humanist approach is often separated from and privileged over that which examines the social utility and practical application of rhetorical theory (George & Trimbur, 1999; Mountford, 2009). Furthermore, rhetoricians are experiencing "an erosion of their influence" (Mountford, 2009, p. 407) even within their long-standing disciplinary homes of English and speech communication.

With these problems in mind, in June 2018, an interdisciplinary group of scholars gathered at the inaugural meeting of the Applied Rhetoric Collaborative to discuss characteristics that link their work and how to cross the deepening disciplinary lines within our field. Together, 25 attendees who specialize in technical communication, design thinking, environmental communication, classical rhetoric, engineering communication, and communication studies discussed ways to cross those artificial boundaries. What emerged from the inaugural symposium (and two follow-up symposia) was a clear desire to connect our teaching and scholarship with our communities, to promote the application of rhetoric in a variety of situations and purposes outside of academia, and to develop conversations and collaborations across our current disciplinary lines.

This was not the first time that scholars have crossed disciplinary lines to attempt a reunification of rhetoric's progeny. According to Diana George and John Trimbur (1999), the Conference on College Composition and Communication, founded in 1949, lists among its original goals the unification of teachers of composition and communication. By bringing the instructors of those disparate courses together, many assumed the so-called "communication approach" to the first-year course—combining instruction in speaking and writing—would take hold. But the inclusion of "the 4th C" in both the organization and in the course's curriculum turned out to be "a brief affair, characterized by mutual attractions and misgivings, that proved unable to imagine a future for itself" (George & Trimbur, 1999, p. 682). Next was the Wingspread Conference in January 1970, which had the laudable goal of "finding a suitable definition of rhetoric and a common goal for future study suitable for interdisciplinary alliances" (Mountford, 2009, p. 407). Although the conference's proceedings, titled *The Prospect of Rhetoric*, were optimistic about this newfound interdisciplinary mission (Mountford, 2009), Thomas O. Sloane (2010) claims that "it is less visionary or prophetic about the future of rhetoric than it is diagnostic of its present condition. It offers a prospectus for lines of inquiry needed to take our discipline into the future" (pp. 3–4). And, while the direction provided by such a prospectus would have been a necessary step to accomplishing the mission of the conference, little more came of it. The Alliance of Rhetoric Societies (ARS), founded in 2003, however, showed more promise. An "organization of organizations" (Clark, 2004, p. 5), ARS was intended to unite the study of rhetoric across traditions in response to

> the difficulty rhetoric scholars have experienced in learning about each other's work, in sharing insights with those who are working on similar projects but in different traditions, in making their collective voice heard by granting agencies, and through an absence of coordination among their respective scholarly organizations (Clark, 2004, p. 5).

Resulting in a special issue of *Rhetoric Society Quarterly* in 2004, the calls to action from ARS prioritized the necessity of rhetorical education and an understanding of rhetoric as agentive and action-oriented (Clark, 2004; Geisler, 2004; Hauser, 2004; Leff & Lunsford, 2004; Zarefsky, 2004). Unfortunately, David Zarefsky's (2004) warning against the "fatal flaw" of these recommendations proved true: with no clear implementing agent (individually or organizationally), they were largely "left on the shelf" (p. 37). And, finally, the "Rhetoric In/Between the Disciplines" seminar at the 2013 Rhetoric Society of America Institute led to "The Mt. Oread Manifesto on Rhetorical Education" (2014). The manifesto lamented the separation of writing and speaking instruction and encouraged rhetoricians to "cross departmental and disciplinary lines and collaborate to design and implement an integrated curriculum in rhetorical education (p. 3). Though encouraging that scholars of rhetoric still consider an integrated curriculum a worthy goal, to our knowledge, this manifesto, like the many before it, has not led to any great revolution in the curriculum or, for that matter, in the study of rhetoric.

Most of these past efforts for reunification of rhetorical traditions have focused on the idea of education and pedagogy as the rhetorician's "birthright" (Hauser, 2004, p. 52). Accordingly, they framed the rhetorician's responsibility and contribution to society in terms of educating our students and preparing them with the rhetorical skill required for civic life (Geisler, 2004; Hauser, 2004; Leff &

Lunsford, 2004; Mountford, 2009; The Mt. Oread Manifesto, 2014; Rood, 2016). This desire to use our rhetorical expertise to "make a difference in the world" through our pedagogical context is admirable, but as encouraging as these past efforts are, very little has come from their optimism for the reunification of rhetorical education and scholarship.

Enter the Applied Rhetoric Collaborative, which expands that desire to make a difference through rhetoric from being entirely rooted in our pedagogical endeavors to also include other aspects of our professional and nonprofessional lives. In other words, while we share "a commitment to rhetoric as action," we see our ability to encourage "a society that grants [rhetorical] agency more broadly" as expanding beyond the classroom (Geisler, 2004, p. 15). Applied rhetoric, as we define it, includes using our rhetorical expertise in innumerable contexts to effect positive change in the world, including but not limited to our classrooms.

In our definition, applied rhetoric is a combination of classical rhetoric theory, professional practice, and public action. It uses rhetoric to solve complex problems at work, in our classrooms, in our communities, and in our public and private lives. In this chapter, we explore this definition of applied rhetoric as the thread that crosses existing disciplinary lines to connect business, technical, scientific, and professional communication. The combination of classical rhetoric and public action provides a way to move rhetorical work beyond the confines and disciplinary divisions of the academy and to actively engage in the work of rhetoric within the communities where we work, live, and research. Applied rhetoric (as an organization, a discipline, and a professional identity) is uniquely positioned to clarify ideas that help people accomplish things, to explicitly connect research to teaching, and to be a force for good in the world.

■ Our Shared Rhetorical Roots

Applied rhetoric is by no means a new term, although it has not been used consistently in scholarship or in pedagogy over the past five decades. One of the earliest scholarly references to applied rhetoric emerged in the field of linguistics. Robert Kaplan's 1970 "Notes Toward an Applied Rhetoric" focuses on supporting advanced English learners as they learn how to analyze and create common discourse patterns. While promising, *applied rhetoric* quickly fell out of favor as a term in linguistics scholarship, replaced with *contrastive rhetoric* and other similar terms.

More recently, applied rhetoric has emerged as a key term in two scholarly tangents: rhetoric of science and rhetoric of economics. In 2013, Carl Herndl and Lauren Cutlip announced the foundation of an Institute for Applied Rhetoric of Science and Sustainability at the University of South Florida, which would focus on "science policy, citizen participation, modeling, and data visualization" (p. 5), four areas that sit at the intersection of theory and practice. Four years later, Herndl (2017) described an "applied RSTEM" (rhetoric of science, technology, engineering, and medicine) and wrestled with moving the field into the post-critical age. Herndl suggested the common theme we seek, but he warned that the term could suggest a resistance to theory as well. Locke Carter (2005) defines applied rhetoric as covering the broader fields of technical communication, business communication, and rhetoric and composition, as well as mass and speech communication. In Carter's view, applied rhetoric relies on a market of ideas in which professional communicators (and their academic counterparts) argue for the value of their work. A key takeaway of Carter's work is the continued emphasis on the real-world practicality of rhetoric instead of solely focusing on the more instrumental and critical approaches that tend to take priority in academia. The shared values developed at the three Applied Rhetoric Collaborative symposia—connecting academics with communities and practitioners and developing deliverables to help them accomplish their goals—echoed Carter's work.

The term *applied rhetoric* has also been incorporated into program and course descriptions. Although publicly available program and course descriptions do not always convey what actually occurs in those programs or courses, they provide a window into what a particular program values. As Lisa Melonçon and Sally Henschel (2013) note, the presence of a course—or a program itself—within a course catalog lends it authority. Thus, the existence of these courses and programs suggests that applied rhetoric as a disciplinary umbrella is a concept many of us already acknowledge in our pedagogy and program administration.

The heart of applied rhetoric is the theory and practice of rhetoric itself. From the earliest records, rhetoric was a public practice, whether in arguing the merits of the law or influencing the decisions of a purely democratic Greek society. Over the millennia of rhetorical discussion, the definitions and practices of rhetoric have expanded and shifted until the modern concept of rhetorical scholarship includes public speaking, composition and writing, professional communication (in all its varieties), social construction and organization, and materialist studies (among many more fields and subfields). Ironically, while modern law and politics have evolved to become their own action-oriented fields, they remain dependent on the practice of rhetoric but have distanced themselves from its theoretical foundations to focus on the practical, professional applications of rhetoric. Rhetorical scholarship, on the other hand, has expanded to include a wide variety of approaches, methods, and topics but often downplays the public actions that defined classical rhetoric.

The expanding definition of rhetoric has been coupled with a dispersion of rhetorical scholars across the colleges and departments of modern universities. Geographically, rhetoricians are scattered throughout institutions. Many reside in English departments, where they are housed with literature, writing studies, linguistics, and creative writing while they teach courses like composition, technical communication, business communication, rhetorical studies, usability, visual communication, and proposals. Some are in communication studies departments, teaching public communication and often focusing more on oral rather than written genres. Some are in business schools, where they may teach written or oral business communication courses. Others are embedded within technical disciplines like engineering, and yet others work on their own as consultants in writing centers and other academic support services. Even outside of these organizational units, still other colleges and departments may hire individual rhetoricians to teach writing or speaking in service courses (see Harlow, 2010 and rhetmap.org).

This dispersion of scholars allows for academic specialization and focused instruction but can also lead to practical problems: territorial disputes; competition for limited funding; competition for students; and confusion in defining disciplines for colleagues, administration, and students. It can also discourage collaboration between scholars who, while relying on the same rhetorical traditions, are seeking to meet different disciplinary standards for presenting and publishing research, including separate conferences and publication venues.

Although the dispersion of rhetorical scholars across universities can be a source of tension and conflict, we also see it as an opportunity for interdisciplinary collaboration and action. By focusing on our shared rhetorical roots and our desire for practical action, the term *applied rhetoric* provides a unifying umbrella to connect scholars with disparate interests and academic homes. The term explicitly returns our focus to our shared roots in classical rhetoric. Applied rhetoric is not associated with a particular subject matter but rather with supplying arguments. For example, to a casual observer, the programs of the symposia held by the Applied Rhetoric Collaborative may seem eclectic. Speakers have discussed modern sophistry, story maps, internet comments, Wikipedia, recipe books, the designation of national monuments, new materialism in the workplace, veteran's studies, engineering communication, the Chicago Statement, faculty development centers on campus, pet rescue adoption policies, expert witnesses in murder trials, and Martha Stewart's product lines. By design, the symposia programs have not been "identifiable with knowledge of any specific subject" (Aristotle, 1358a/1991), yet there has been a remarkable cohesion in the presentations because of their focus on rhetorical research and practice in public spheres. By applying our "distinct abilities of supplying words" (Aristotle 1358a/1991), the small group has proven that different theoretical foundations, research methods, topics, and applications can coexist and even speak to each other through applied rhetoric.

The connection of matter and language (Cicero, III.v/2001) and their effect on persuasion, effectiveness, and ethicality have been at the heart of rhetorical studies for centuries. By organizing ourselves around the concept of applied rhetoric, we expand Cicero's definition to include the matter, the language, and the application. This combination of matter, language, and application extends Lloyd F. Bitzer's (1968) foundational rhetorical triangle to more fully reflect the practice of rhetoric in its many forms and forums.

Technical and Professional Communication as Applied Rhetoric

Although applied rhetoric scholars have a variety of academic homes, a number of us have historically located ourselves within the disciplines of technical communication, business communication, or an uneasy combination of the two. The history of technical communication has been well documented (see Connors, 1982; Kynell, 1999; Kynell & Tebeaux, 2009; Moran & Tebeaux, 2011, 2012; Staples, 1999). Less well documented is the history of business communication, but it, too, traces its roots to rhetoric (Carbone, 1994; Reinsch, 1996). Yet even with 2,000 years of history and philosophy, both have something of an identity crisis. Who are we? What do we do? These questions are asked and answered again and again as changes in technology and communication practices expand our boundaries. More, the act of claiming to be a scholar of one or the other can be tricky; simply choosing the names within this section sparked an ongoing discussion about where and how to use which term. It can also be fraught: perceived disciplinary turf issues in some departments may mean that declaring ourselves a scholar in one of these "practical" forms might exclude us from also identifying as a scholar of rhetoric.

Complicating matters is the interdisciplinary nature of both technical and business communication. Each exists in relation to other disciplines and other workplace problems. Through these interactions, we have collectively become something of an intellectual magpie:

> Technical communication shares and borrows methods, theories, and even content areas with design communication, speech communication, and rhetoric and composition as well as with psychology, education, and computer science. These fields share questions about usability, Web-site design, and information management. What makes technical communication distinct and recognizable? (Rude, 2009, p. 175)

Rachel Martin Harlow (2010) concurred, describing technical and professional communication as a "third culture discipline" that uses our relationships with other disciplines to synthesize ideas and methods that meet our needs. As if to confirm this notion, the journal *Business Communication Quarterly* became *Business and Professional Communication Quarterly* in 2014. James Dubinsky (2014) explained the change as a move toward interdisciplinarity and to more accurately reflect the shared "intellectual and methodological roots" of its authors.

Being a mashup discipline means we spend precious time creating lines of demarcation, sometimes arbitrarily. For example, what is the difference between technical communication courses and business communication courses? Their locations within institutional structures suggest the difference would be significant. Technical communication programs (and, thus, the courses) are overwhelmingly located within humanities-focused departments, such as English or stand-alone technical communication departments (Melonçon & Henschel, 2013). Business communication courses are usually located in a center within the business school, within a particular department in the school, or in the business school in general (Sharp & Brumberger, 2013).

Given these institutional separations, one could reasonably expect an equally significant difference in the content and structure of these courses. In practice, however, the distinctions are harder to identify. Several curriculum audits have identified the typical content of business communication courses from the perspectives of both employers and teachers (Moshiri & Cardon, 2014; Russ, 2009; Wardrope & Bayless, 1999). More recently, Kristen Lucas and Jacob D. Rawlins (2015) proposed five core business communication competencies: professional, clear, concise, evidence-driven, and persuasive. Sally Henschel and Lisa Melonçon (2014) identified essential conceptual skills for technical communication as rhetorical proficiency, abstraction, experimentation, social proficiency, and critical systems thinking. Yet no curriculum audits are publicly available for technical communication service courses. When researchers do examine the technical writing service course, they address specific facets of the course, such as the inclusion of intercultural communication components (Barker & Matveeva, 2006; Matveeva, 2007, 2008) or effective conversion of the course into online formats (Battalio, 2006), rather than its curriculum. Coppola (2010) described the Society for Technical Communication's effort to build a body of knowledge that would help establish technical communication "as a fully mature profession" (p. 12). Yet several years after its founding, Ray Gallon (2016), a former member of the Society for Technical Communication (STC) Board of Directors, acknowledged that the project was still largely incomplete.

More than 25 years ago, Nancy R. Blyler (1993) suggested that the curricular separation may be based on differing intents: business communication is persuasive, whereas technical communication is instructive and informative. But, she argues, if technical communication is rhetorical, then thinking in terms of persuasion vs. instruction is a moot point. Blyler also suggested that the separation seemed to be based on the documents that students will write in the workplace. Business communication consists of annual reports, sales, advertisements, and proposals; technical communication consists of reports, instructions, descriptions, manuals, and specifications. While anecdotal evidence suggests that these perceived differences are still in place, a recent survey of business and STEM faculty at mid-sized public universities in the northeast and midwest United States indicated that the differences are now negligible (Patriarca & Veltsos, 2017).

If we were wrong about the curricular separations, could we be wrong about other separations that divide us into increasingly small niches of research (Russell, 2002)? Could we be more similar than we realize? Saul Carliner (2012) traced the confusion that is caused when we define ourselves too narrowly, citing an STC study in which participants offered hundreds of working titles, including product information specialist, documentation specialist, business analyst, and information developer. Other titles in use include writer, editor, usability specialist, content strategist, and content developer. The term *technical communicator* was intended to be more inclusive, but it may still alienate those of us who do not work in technical fields, develop technical content, or teach students how to communicate in technical disciplines. The revised term *technical and professional communication* (TPC) is sometimes used to reflect the various forms that our work can take, yet it is wordy and somewhat redundant (which doesn't reflect well on our own abilities as communicators). Furthermore, as the work associated with technical communication expands (e.g., usability, project management), we must continually redefine what technical communication means and does.

It's not just the work of TPC that is changing. Natasha N. Jones, Kristen R. Moore, and Rebecca Walton's (2016) antenarrative identified several threads within technical and professional communication scholarship that stretch the "pragmatic identity" (p. 213) of our field beyond its usual focus on efficiency and problem-solving. They noted issues of feminism and gender; race and ethnicity; international and intercultural communication; community and public engagement; user advocacy; and diversity, social justice, and inclusion. In widening the scope of TPC, they recenter our attention on the human impact our work has on society.

We go one step further and connect our work not only through its human impact but also through its use of rhetorical theories and strategies. Like the Sophists, we resist classification based on location, subject of study, or methods (Harlow, 2010). Instead, we embrace a shared identity as practitioners of rhetoric rather than shared practices of work, research methods, pedagogical methods, or subject matter expertise. Applied rhetoric provides the flexibility and adaptability that Teresa Henning and Amanda Bemer (2016) suggest are "fundamentally linked to a technical communicator's power" (p. 325) and important factors in career satisfaction and career health. Ironically, the shift towards applied rhetoric also provides a measure of stability for a scholar's career: our theory, our pedagogy, our methods, and our boundaries may shift, but the underlying theme of using rhetorical theory and strategies to solve problems remains constant. If we acknowledge the rhetorical thread that connects our work, we can more easily see how our research might intersect or how findings in one area, like rhetoric of science and medicine, might help practitioners or teachers in another, like business communication. Rhetoric is the mother tongue that we use to talk about our work.

■ Applied Rhetoric as *Doing* Rhetoric

Beyond the flexibility and adaptability inherent in the term, applied rhetoric emphasizes that rhetorical theory must be brought to bear onto something else—some activity, reality, or materiality that may or may not appear to be rhetorical at first glance—for a specific purpose. This differentiation between the theoretical and the practical is reminiscent of a debate in the field of technical communication that caused quite a stir in the 1980s. It began with Carolyn Miller's (1979) article "A Humanistic Rationale for Technical Writing" and Elizabeth Harris' (1979) article "Applications of Kinneavy's *Theory of Discourse* to Technical Writing," both in the same issue of *College English*. Those essays provoked a response from Elizabeth Tebeaux (1980), in which she claimed that the theoretical approaches to curriculum design advocated by Miller and Harris "ignore the purely pragmatic topics and problems that must be emphasized in the course" (p. 823). Both Miller (1980) and Harris (1980) responded to Tebeaux's criticism by claiming that she overemphasized the role that industry practice should have in the development of course material and underappreciated the role that rhetorical and linguistic theory should have in that development. And so began (or continued) the debates between the practical and the theoretical—the industry and the academy—in technical writing.

While these debates still exist within technical writing and other related disciplines, rhetorical theory has often been used to try to bridge the gap. Miller's (1989) "What's Practical About Technical Writing?" offers the beginnings of an answer in the Aristotelian concept of praxis, which she identifies as a middle ground between theory and practice, informed by both and, hopefully, working to shape both. Patricia Sullivan and James Porter (1993) define praxis as "a 'practical rhetoric' focused on local writing activities (practice), informed by as well as informing general principles (theory)" (p. 226). Similarly, J. Blake Scott and Lisa Melonçon (2017) propose the concept of techne—the combination of theoretical principles and practical knowledge in a stable, yet highly contingent foundation for ethical, rhetorical conduct—as a way to guide the development of the discipline. And, Robert R. Johnson (2010) combines the practical and theoretical by acknowledging the dual telos of techne: the end product as well as the use of that product. In fact, he claims that the products produced through techne "are essentially inert until they are placed into use" (p. 677). Each of these approaches emphasizes application of rhetorical theory to specific, unique contexts and phenomena. Through that application, both theory and practice develop and evolve.

By embracing the moniker of applied rhetoric, we embrace this idea of praxis and continue to extend it beyond curricular concerns. Miller (1989) argues that theory should inform practice through the curriculum by training students to be critically aware professionals, but that's not the only (or even the most effective) way we can engage with practice. The various disciplines that applied rhetoric covers already have a long history of reaching beyond the academy in attempts to merge theory and practice. For instance, scholars in technical and professional communication have often partnered with or engaged with industry as sites of research (see Faber, 2002a; Spinuzzi, 2003; Winsor, 2003; Zachry & Thralls, 2007), and the scholarship of civic and public rhetoric often engages directly or indirectly with community-based programs (see Ackerman & Coogan, 2010; Blythe et al., 2008; Deans et al., 2010; Flower, 2008; Grabill, 2006; Simmons, 2007). The inaugural issue of *Rhetoric of Health and Medicine* (*RHM*) acknowledges that the purpose of that discipline is "to engage and inform other fields and extra-academic practices" (Melonçon & Scott, 2018, p. v), so much so that the journal makes room for outward-facing scholarship such as persuasion briefs and dialogues with stakeholders outside of the academy. By analyzing and engaging with the actual application of rhetoric, *RHM* is applied rhetoric. Thinking of these various discourse communities as applied rhetoric suggests (perhaps even demands) that these collaborations across departments, across disciplines, and especially with practitioners outside of the academy become more intentional and more central to our work.

In the past decade, technical and professional communication has also seen a turn towards the "wicked problem," a poorly defined, complex problem that cannot be solved with a simple response; indeed, the problem is often redefined as new solutions are offered (Rittel & Webber, 1973). Though the concept was originally developed for public policy planning work, it has been adopted within technical and professional communication as a response to communication challenges involving audiences with multiple, often competing, needs. The continual redefinition required by wicked problems also aligns with cultural studies approaches to technical communication, particularly Slack et al.'s (1993) argument that technical communicators continually make and remake meaning within the deliverables they create. More, wicked problems frequently require the perspectives of scholars and practitioners from multiple disciplines (Rittel & Webber, 1973).

The concept of wicked problems has continued to frame several topics within technical and professional communication. Jeffrey M. Gerding and Kyle P. Vealey (2017) incorporate work in entrepreneurship, civic and public rhetorics, and technical communication to argue for what they call "hybrid solutions" to wicked problems that appeal to investors' need for stability and address the evolving nature of the problems (p. 303). More recently, Brock Carlson (2019) described the situations facing community organizers in Appalachia as wicked problems that can best be addressed using local knowledges and nonstandard communication strategies. Most often, though, the wicked problems framework has been applied to issues related to environmental communication, including the 2010 Deepwater Horizon spill along the U.S. Gulf Coast (Wickman, 2014), a reconceptualization of scientists as *audiences* for science communication (McKiernan & Steinbergs, 2016), and a community-focused study of how individuals can be persuaded to believe and act in accordance with climate science (Shirley, 2019).

Despite the possibilities in the turn towards wicked problems, we argue that the concept is rooted too deeply in the problems themselves to be a useful umbrella for the work we do in our field. As its name suggests, it is also deeply rooted in *complex, ever-changing* problems. The umbrella of applied rhetoric, however, allows a focus on *resolution* or, at least, *mitigation* to any problems that are posed, rather than on the problems themselves, and the ability to tackle problems without worrying about their complexity or scope. Though many scholars are indeed focused on wicked problems, many prefer to focus on smaller, local problems that can be solved through rhetoric: resolving the contradictions posed by local pet rescue policies, advocating for environmental changes through rhetorical story maps, or leading a faculty development center at a university. Using applied rhetoric as an umbrella includes those scholars working on either type of problem, or on the multitude of problems that could exist between the two.

As a disciplinary umbrella, applied rhetoric allows us to focus on the cyclical engagement between theory and practice in the way that the necessarily practical discipline of rhetoric should. Jeffrey T. Grabill (2006) claims that practice-driven rhetorical research requires that "usefulness become a primary epistemological, ethical, and political value" (p. 162). Prioritizing the usefulness of rhetoric at a disciplinary level allows us to learn from each other's successes and failures at engaging our students; our colleagues, disciplines, and institutions; our corporate, nonprofit, and government partners; and our communities and publics in the work of rhetoric. Together, we move within and outside of the academy to improve the ways that rhetoric is used for its various, nearly unlimited ends. Herndl and Cutlip (2013) say this kind of move toward praxis allows the rhetoric of science, technology, and medicine to "flourish as a participant in interdisciplinary research projects in which rhetoric functions as a significant contributor to research, outreach, and policy formation" (p. 4), allowing the discipline's scholars to "move from talking about science to doing science" (p. 7). As a framework, applied rhetoric thus creates a space for scholars to move from *talking* about rhetoric to actually *doing* rhetoric.

By focusing on the doing of rhetoric, we allow our work to move beyond the academy and positively affect the communities we study, and we allow them to affect us, as well. Here, too, we're not the first to make this argument. The disciplines within rhetorical studies have a long history of engaging with various publics. Ellen Cushman (1996) argues that "in doing our scholarly work, we should take social responsibility for the people from and with whom we come to understand a topic" (p. 11) by contributing the resources of our positions to help "people disrupt the status quo of their lives with language and literacy" (p. 13). And, David J. Coogan and John M. Ackerman (2010) argue that

> communities can benefit from the increased attention of rhetoricians in pursuit of democratic ideals, but rhetoric can also benefit from community partnerships premised on a negotiated search for the common good—from a collective labor to shape the future through rhetoric in ways that are mutually empowering and socially responsible. (pp. 1–2)

Sometimes called "participatory action research" (Sullivan & Porter, 1997), this type of mutually beneficial partnership between scholars and communities is one example of the kind of work we think about when we envision a future for applied rhetoric. Stuart Blythe, Jeffrey T. Grabill, and Kirk Riley (2008) envision another in their "critical action research" where they conduct research "on behalf of citizens rather than with them" (p. 276). In either case, by engaging with communities within academic, civic, and other public contexts, applied rhetoric is the more "expansive, collaborative, and consequential way" of thinking about rhetoric that Caroline Gottschalk Druschke (2014) encourages for the rhetoric of science, but applied rhetoric allows us to think that way about the rhetoric involved in nearly any discipline or context because rhetoric is a "transdisciplinary, emplaced, engaged field by its very nature" (Druschke, 2014, p. 6).

That kind of extension of work, though, comes with a certain responsibility. Carolyn R. Miller (1989) references the concept of phronesis, or practical wisdom, as integral to praxis. In Aristotelian terms, the focus of praxis is on good, effective conduct, so the reasoning appropriate to praxis is that which "necessarily concerns both universals and particulars: it applies knowledge of human goods to particular circumstances" (Miller, 1989, p. 22). Johnson (2010) goes as far as to say that the ancients would consider it unthinkable to remove techne from cultural and ethical contexts. Applied rhetoric thus requires us to inhabit a middle ground between an overreliance on either theory or practice to guide our judgment. We must analyze each situation as a unique opportunity for rhetoric, a unique opportunity for "arguing in a prudent way toward the good of the community" (Miller, 1989, p. 22). Many scholars are already doing this difficult work, and across quite different areas. For example, research that seeks to understand how and why vaccine refusal communities find anti-vaccination rhetoric persuasive can be used to develop strategies for public practitioners and scholars for compassionate, effective communication with these communities (Campeau, 2019; Lawrence, 2020; Scott et al., 2015). Scholars can work with government entities to improve the credibility of their websites through usability testing (Youngblood & Youngblood, 2013; Youngblood, 2018), improve public planning processes with rhetorical listening strategies (Moore & Elliott, 2015), and improve communication among deployed service members (Mallory, 2019).

This work does not always show up in peer-reviewed scholarship, though. We see examples of public practice in social media posts that popularize the idea of students going to faculty office hours and share examples of how to reach out to faculty if students need help (e.g., Wise, 2020); in offering free resume review services through the local library for community members; in helping a university better communicate to its students via usability testing; in leading the way on organizational policies that support activist movements and civil rights; and in many, many more public-facing situations.

Applied rhetoric can thus "inform and ameliorate" practice (Melonçon & Scott, 2018, p. v). While all rhetoric is practical in the sense that the methods of rhetoric must be brought to bear onto something else, applied rhetoric is a useful term for our work because it allows rhetoric and rhetors the opportunity to make a difference within our communities. Our efforts lead to practice, particularly public practice. In this way, we are returning to the earliest roots of rhetoric in that our efforts are outwardly focused to influence the societies and communities in which we live.

■ Conclusion

It's clear that we are not the first nor the only group to use the term applied rhetoric, nor even the first to attempt to establish a disciplinary umbrella for scholars with ties to rhetoric. Given the history and scholarship we have reviewed here, one might simply think that we're joining a Greek chorus. (Pardon the pun.) But through these symposia, this organization, and even this chapter, we hope to help bring the term into widespread use. We view applied rhetoric as being uniquely positioned to clarify ideas that help people accomplish things, to explicitly connect research to teaching, and to be a force for good in the world. Not all scholars or practitioners of technical or business communication may identify as rhetoricians, but it is clearly the tie that binds us together.

Rhetoric is an action, one that scholars and practitioners alike perform because all communication is inherently rhetorical. Miles A. Kimball (2017) proposed that the skills we often associate with technical communication are essential skills for . . . well, everyone. As Blyler (1993) suggested, the crucial knowledge is to be able to identify and respond to contextual issues of workplace documents, understand how documents express communal values and expectations, and adapt messages and strategies to a variety of situations. Effective communication isn't the result of hunches, habits, talent, or luck. Rhetorical theory legitimizes the rationale for decisions about what works and why (Hart-Davidson, 2001), and it is not limited to those who identify as technical communicators. In fact, while we're talking about what technical communication is (or is not), the rest of the world is just doing it. They are writing product reviews (Mackiewicz, 2011), answering questions on message boards (Frith, 2014), creating YouTube tutorials (Chong, 2018), and pitching new businesses (Roundtree, 2016).

Through our work with the quasiprofessional Applied Rhetoric Collaborative, we are casting a wide net to create "a community of like-minded people who share professional interests but also enjoy one another's company" (Carliner, 2012, p. 62). By including the broad fields of technical and business communication, communication studies, and specialized areas like rhetoric of science, rhetoric of economics, risk and crisis communication, social media rhetorics, and more, applied rhetoric is the most accurate and inclusive term for our field because it references our shared rhetorical foundations and allows for a breadth of topics and methodological approaches. After all, to expand our own horizons, we should routinely interact with people who are doing other things.

Rhetoric is, by its nature, a practical art, an applied method. We use theory. We question theory. We develop theory. But we do those things by examining rhetoric as it is applied in various contexts. We hope this perspective will work to inspire conversation and innovate rhetorical practice across those various contexts. Let that be the legacy of applied rhetoric—a cross-disciplinary revival of rhetoric's ancient practical purpose.

References

Ackerman, John M. & Coogan, David J. (Eds.). (2010). *The public work of rhetoric: Citizen-scholars and civic engagement*. University of South Carolina Press.

Aristotle (1991). *On Rhetoric: A Theory of Civic Discourse* (G. A. Kennedy, Trans.). Oxford University Press. (Original work published ca. 322 BCE)

Barker, Thomas & Matveeva, Natalia. (2006). Teaching intercultural communication in a technical writing service course: Real instructors' practices and suggestions for textbook selection. *Technical Communication Quarterly, 15*(2), 191–214. https://doi.org/10.1207/s15427625tcq1502_4.

Battalio, John T. (2006). Teaching a distance education version of the technical communication service course: Timesaving strategies. *Journal of Technical Writing and Communication, 36*(3), 273–296. https://doi.org/10.2190/D86G-UGCH-BFX8-10EY.

Bitzer, Lloyd. (1968). The rhetorical situation. *Philosophy & Rhetoric, 1*(1), 1–14.

Blyler, Nancy R. (1993). Theory and curriculum: Reexamining the curricular separation of business and technical communication. *Journal of Business and Technical Communication, 7*(2), 218–245. https://doi.org/10.1177/1050651993007002003.

Blythe, Stuart, Grabill, Jeffrey T. & Riley, Katherine. (2008). Action research and wicked environmental problems: Exploring appropriate roles for researchers in professional communication. *Journal of Business and Technical Communication, 22*(3), 272–298.

Campeau, Kari L. (2019). Vaccine barriers, vaccine refusals: Situated vaccine decision-making in the wake of the 2017 Minnesota measles outbreak. *Rhetoric of Health & Medicine, 2*(2), 176–207.

Carbone, Mary T. (1994). The history and development of business communication principles, 1776–1916. *The Journal of Business Communication, 31*(3), 173–193.

Carliner, S. (2012). Three approaches to professionalization in technical communication. *Technical Communication, 59*(1), 49–65.

Carlson, Erin B. (2019). *"There is wealth in the struggle": Unearthing and embracing community knowledges through organizing work in Appalachia* [Doctoral thesis, Purdue University]. https://doi.org/10.25394/PGS.8292149.v1.

Carter, Joyce Locke. (Ed.). (2005). *Market matters: Applied rhetoric studies and free market composition*. Hampton Press.

Chong, Felicia. (2018). YouTube beauty tutorials as technical communication. *Technical Communication, 65*(3), 293–308.

Cicero. (2001). *Cicero on the Ideal Orator (De Oratore)* (J. M. May and J. Wisse, Trans.). Oxford University Press. (Original work published ca. 55 BCE)

Clark, Gregory. (2004). Introduction. *Rhetoric Society Quarterly, 34*(3), 5–7. https://doi.org/10.1080/02773940409391285.

Connors, Robert J. (1982). The rise of technical writing instruction in America. *Journal of Technical Writing and Communication, 12*(4), 329–351.

Coogan, David J. & Ackerman, John M. (2010). Introduction: The space to work in public life. In John M. Ackerman & David J. Coogan (Eds.), *The public work of rhetoric: Citizen-scholars and civic engagement* (pp. 1–16). The University of South Carolina Press.

Coppola, Nancy. (2010). The technical communication body of knowledge initiative: An academic-practitioner partnership. *Technical Communication, 57*(1), 11–25.

Cushman, Ellen. (1996). The rhetorician as an agent of social change. *College Composition and Communication, 47*(1), 7–28.

Deans, Thomas, Roswell, Barbara & Wurr, Adrian J. (2010). *Writing and community engagement: A critical sourcebook*. Bedford/St. Martin's.

Druschke, Caroline G. (2014). With whom do we speak? Building transdisciplinary collaborations in Rhetoric of Science. *Poroi, 10*(1), Article 10, 1–7. https://doi.org/10.13008/2151-2957.1175.

Dubinsky, James. (2014). Letter from the executive director of ABC. *Business and Professional Communication Quarterly, 77*(1), 3–4. https://doi.org/10.1177/2329490614524173.

Faber, Brenton D. (2002a). *Community action and organizational change: Image, narrative, identity*. Southern Illinois University Press.

Faber, Brenton. (2002b). Professional identities: What is professional about professional communication? *Journal of Business and Technical Communication, 16*(3), 306–337.

Flower, Linda. (2008). *Community literacy and the rhetoric of public engagement*. Southern Illinois University Press.

Frith, Jordan. (2014). Forum moderation as technical communication: The social web and employment opportunities for technical communicators. *Technical Communication, 61*(3), 173–184.

Gallon, Ray. (2016). Four years on the STC board—A review. *Rant of a Humanist Nerd*. https://humanistnerd.culturecom.net/2016/06/21/four-years-on-the-stc-board-a-review/.

Geisler, Cheryl. (2004). How ought we to understand the concept of rhetorical agency? Report from the ARS. *Rhetoric Society Quarterly, 34*(3), 9–17. https://doi.org/10.1080/02773940409391286.

George, Diana & Trimbur, John. (1999). The "communication battle" or whatever happened to the 4th C? *College Composition and Communication, 50*(4), 682–698.

Gerding, Jeffrey M. & Vealey, Kyle P. (2017). When is a solution not a solution? Wicked problems, hybrid solutions, and the rhetoric of civic entrepreneurship. *Journal of Business and Technical Communication, 31*(3), 290–318. https://journals.sagepub.com/doi/10.1177/1050651917695538.

Grabill, Jeffrey T. (2006). The study of writing in the social factory: Methodology and rhetorical agency. In J. B. Scott, B. Longo & K. V. Wills (Eds.), *Critical power tools: Technical communication and cultural studies* (pp. 151–170). State University of New York Press.

Harlow, Rachel M. (2010). The province of Sophists: An argument for academic homelessness. *Technical Communication Quarterly, 19*(3), 318–333. https://doi.org//10.1080/10572252.2010.481530.

Harris, Elizabeth. (1979). Applications of Kinneavy's *Theory of Discourse* to technical writing. *College English, 40*(6), 625–632.

Harris, Elizabeth. (1980). Elizabeth Harris responds. *College English, 41*(7), 827–829.

Hart-Davidson, William. (2001). On writing, technical communication, and information technology: The core competencies of technical communication. *Technical Communication, 48*(2), 145–155.

Hauser, Gerard A. (2004). Teaching rhetoric: Or why rhetoric isn't just another kind of philosophy or literary criticism. *Rhetoric Society Quarterly, 34*(3), 39–53. https://doi.org/10.1080/02773940409391289.

Henning, Teresa & Bemer, Amanda. (2016). Reconsidering power and legitimacy in technical communication: A case for enlarging the definition of technical communicator. *Journal of Technical Writing and Communication, 46*(3), 311–341. https://doi.org//10.1177/0047281616639484.

Henschel, Sally & Melonçon, Lisa. (2014). Of horsemen and layered literacies: Assessment instruments for aligning technical and professional communication undergraduate curricula with professional expectations. *Programmatic Perspectives*, *6*(1), 3–26.

Herndl, Carl G. (2017). Introduction to the symposium on engaged rhetoric of science, technology, engineering, and medicine. *Poroi*, *12*(2), Article 2, 1–15. https://doi.org//10.13008/2151-2957.1259.

Herndl, Carl G. & Cutlip, Lauren. (2013). "How can we act?" A praxiographical program for the Rhetoric of Technology, Science, and Medicine. *Poroi*, *9*(1), 2151–2957. https://doi.org/10.13008/2151-2957.1163.

Johnson, Robert R. (2010). Craft knowledge: Of disciplinarity in writing studies. *College Composition and Communication*, *61*(4), 673–690.

Jones, Natasha N., Moore, Kristen R. & Walton, Rebecca. (2016). Disrupting the past to disrupt the future: An antenarrative of technical communication. *Technical Communication Quarterly*, *25*(4), 211–229. https://doi.org/10.1080/10572252.2016.1224655.

Kaplan, Robert B. (1970). Notes toward an applied rhetoric. In Robert Lugton (Ed.), *Preparing the EFL teacher: A projection for the '70s* (pp. 45–74). The Center for Curriculum Development.

Kimball, Miles A. (2017). The golden age of technical communication. *Journal of Technical Writing and Communication*, *47*(3), 330–358.

Kynell, Teresa. (1999) Technical communication from 1850–1950: Where have we been? *Technical Communication Quarterly*, *8*(2), 143–151. https://doi.org//10.1080/10572259909364655.

Kynell, Teresa & Tebeaux, Elizabeth. (2009). The Association of Teachers of Technical Writing: The emergence of professional identity. *Technical Communication Quarterly*, *18*(2), 107–141. https://doi.org//10.1080/10572250802688000.

Lawrence, Heidi Y. (2020). *Vaccine rhetorics*. The Ohio State University Press.

Leff, Michael & Lunsford, Andrea A. (2004). Afterwords: A dialogue. *Rhetoric Society Quarterly*, *34*(3), 55–67. https://doi.org/10.1080/02773940409391290.

Lucas, Kristen & Rawlins, Jacob D. (2015). The competency pivot: Introducing a revised approach to the business communication curriculum. *Business and Professional Communication Quarterly*, *78*(2), 167–193. https://doi.org/10.1177/2329490615576071.

Mackiewicz, Jo. (2011). Epinions advisors as technical editors: Using politeness across levels of edit. *Journal of Business and Technical Communication*, *25*(4), 421–448.

Mallory, Angie. (2019). *Interpersonal communication in war zones: The U.S. Marines' use of rhetorical listening as a communication behavior* [Doctoral dissertation, Iowa State University]. ProQuest. https://search.proquest.com/openview/6cac51338be81c76f2436119a98e504d/1.pdf?pq-origsite=gscholar&cbl=18750&diss=y.

Matveeva, Natalia. (2007). The intercultural component in textbooks for teaching a service technical writing course. *Journal of Technical Writing and Communication*, *37*(2), 151–166. https://doi.org/10.2190/85J8-2P74-1378-2188.

Matveeva, Natalia. (2008). Teaching intercultural communication in a basic technical writing course: A survey of our current practices and methods. *Journal of Technical Writing and Communication*, *38*(4), 387–410. https://doi.org/10.2190/TW.38.4.e.

McKiernan, Katherine R. & Steinbergs, Andra. (2016). Scientists as audience: Science communicators as mediators of wicked problems. In Jean Goodwin (Ed.), *Confronting the challenges of public participation: Issues in environmental, planning and health decision-making* (pp. 103–108). CreateSpace. https://doi.org/10.31274/sciencecommunication-180809-9.

Melançon, Lisa & Henschel, Sally. (2013). Current state of U.S. undergraduate degree programs in technical and professional communication. *Technical Communication*, *60*(1), 45–64.

Melançon, Lisa & Scott, J. Blake. (2018). Manifesting a scholarly dwelling place in "RHM". *Rhetoric of Health & Medicine*, *1*(1–2), i–x. https://doi.org/10.5744/rhm.2018.1001.

Miller, Carolyn R. (1979). A humanistic rationale for technical writing. *College English*, *40*(6), 610–617.

Miller, Carolyn R. (1980). Carolyn Miller responds. *College English*, *41*(7), 825–827.

Miller, Carolyn R. (1989). What's practical about technical writing? In B. E. Fearing & W. Keats (Eds.), *Technical writing: Theory and practice* (pp. 14–24). Sparrow.

Moore, Kristen R. & Elliott, Tim J. (2015). From participatory design to a listening infrastructure: A case of urban planning and participation. *Journal of Business and Technical Communication*, *30*(1), 59–84. https://doi.org/10.1177/1050651915602294.

Moran, Michael G. & Tebeaux, Elizabeth. (2011). A bibliography of works published in the history of professional communication from 1994–2009: Part 1. *Journal of Technical Writing and Communication*, *41*(2), 193–214. https://doi.org/10.2190/TW.41.2.f.

Moran, Michael G. & Tebeaux, Elizabeth. (2012). A bibliography of works published in the history of professional communication from 1994–2009: Part 2. *Journal of Technical Writing and Communication*, *42*(1), 57–86. https://doi.org/10.2190/TW.42.1.e.

Moshiri, Farrokh & Cardon, Peter. (2014). The state of business communication classes: A national survey. *Business and Professional Communication Quarterly*, *77*(3), 312–329. https://doi.org/10.1177/2329490614538489.

Mountford, Roxanne. (2009). A century after the divorce: Challenges to rapprochement between speech communication and English. In A. A. Lunsford, K. H. Wilson & R. A. Eberly (Eds.), *The SAGE handbook of rhetorical studies* (pp. 407–422). Sage.

Patriarca, Ashley & Veltsos, Jennifer R. (2017, October). *Investigating the curricular differences of technical and business communication service courses* [Presentation]. Association for Business Communication 82nd Annual International Conference, Dublin, Ireland.

Reinsch, Jr., N. Lamar. (1996). Business communication: Present, past, and future. *Management Communication Quarterly*, *10*(1), 27–49. https://doi.org/10.1177/0893318996010001003.

Rittel, Horst W. J. & Webber, Melvin M. (1973). Dilemmas in a general theory of planning. *Policy Sciences*, *4*(2), 155–169. https://doi.org/10.1007/BF01405730.

Rood, Craig. (2016). The gap between rhetorical education and civic discourse. *Review of Communication*, *16*(2–3), 135–150. https://doi.org/10.1080/15358593.2016.1187456.

Roundtree, Aimee K. (2016). Startup weekends: Invention as process for proto-entrepreneurs. In *2016 IEEE International Professional Communication Conference (IPCC)* (pp. 1–14). IEEE. https://doi.org//10.1109/IPCC.2016.7740506.

Rude, Carolyn. (2009). Mapping the research questions in technical communication. *Journal of Business and Technical Communication*, *23*(2), 174–215. https://doi.org//10.1177/1050651908329562.

Russ, Travis L. (2009). The status of the business communication course at U.S. colleges and universities. *Business Communication Quarterly*, *72*(4), 395–413. https://doi.org/10.1177/1080569909349524.

Russell, Davis R. (2002). *Writing in the academic disciplines: A curricular history* (2nd ed.). Southern Illinois University Press.

Scott, Jennifer, Kondrlik, Kristin, Lawrence, Heidi, Popham, Susan & Welhausen, Candice. (2015). Rhetoric, Ebola, and vaccination: A conversation among scholars. *Poroi*, *11*(2). https://doi.org/10.13008/2151-2957.1232.

Scott, J. Blake & Melonçon, Lisa. (2017). Writing and rhetoric majors, disciplinarity, and "techne." *Composition Forum*, *35*(35), 1–14.

Sharp, Matthew R. & Brumberger, Eva R. (2013). Business communication curricula today: Revisiting the top 50 undergraduate business schools. *Business Communication Quarterly*, *76*(1), 5–27. https://doi.org/10.1177/1080569911247187.

Shirley, Beth J. (2019). *Adapting environmental ethics and behaviors: Toward a posthuman rhetoric of community engagement* [Doctoral dissertation, Utah State University]. USU Digital Commons. https://digitalcommons.usu.edu/etd/7513/.

Simmons, W. Michele. (2007). *Participation and power: Civic discourse in environmental policy decisions*. SUNY Press.

Slack, Jennifer D., Miller, David J. & Doak, Jeffrey. (1993). The technical communicator as author: Meaning, power, authority. *Journal of Business and Technical Communication*, *7*(1), 12–36.

Sloane, Thomas O. (2010). Prologue: *The Prospect* as prospectus. In Mark J. Porrovecchio (Ed.), *Reengaging the prospects of rhetoric: Current conversations and contemporary challenges* (pp. 1–4). Routledge.

Spinuzzi, Clay. (2003). *Tracing genres through organizations: A sociocultural approach to information design*. MIT Press.

Staples, Katherine. (1999). Technical communication from 1950–1998: Where are we now? *Technical Communication Quarterly*, *8*(2), 153–164. https://doi.org/10.1080/10572259909364656.

Sullivan, Patricia A. & Porter, James E. (1993). Remapping curricular geography: Professional writing in/and English. *Journal of Business and Technical Communication*, *7*(4), 389–422. https://doi.org/10.1177/1050651993007004001.

Sullivan, Patricia & Porter, James. (1997). *Opening spaces: Writing technologies and critical research practices*. Ablex.

Tebeaux, Elizabeth. (1980). Let's not ruin technical writing, too: A comment on the essays of Carolyn Miller and Elizabeth Harris. *College English*, *41*(7), 822–825.

The Mt. Oread manifesto on rhetorical education 2013. (2014). *Rhetoric Society Quarterly*, *44*(1), 1–5. https://doi.org/10.1080/02773945.2014.874871.

Wardrope, William J. & Bayless, Martha L. (1999). Content of the business communication course: An analysis of coverage. *Business Communication Quarterly*, *62*(4), 33–40. https://doi.org/10.1177/108056999906200404.

Wickman, C. (2014). Wicked problems in technical communication. *Journal of Technical Writing and Communication*, *44*(1), 23–42.

Winsor, D. A. (2003). *Writing power: Communication in an engineering center*. State University of New York Press.

Wise, B. [@wisebeck]. (2020, November 9). *Reading reflections today and had a little string of "I wish I'd asked for help but I didn't know how"* [Thumbnail with link attached] [Tweet]. Twitter. https://twitter.com/wisebeck/status/1325992759614824449.

Youngblood, N. E. & Youngblood, S. A. (2013). User experience and accessibility: An analysis of county web portals. *Journal of Usability Studies*, *9*(1), 25–41.

Youngblood, S. A. (2018). Site identity, artifact duplication, and disambiguation in Alabama local emergency management agencies (LEMAs). *Communication Design Quarterly Review*, *6*(1), 9–15. https://doi.org/10.1145/3230970.3230972.

Zachry, M. & Thralls, C. (2007). *Communicative practices in workplaces and the professions: Cultural perspectives on the regulation of discourse and organizations.* Baywood.

Zarefsky, D. (2004). Institutional and social goals for rhetoric. *Rhetoric Society Quarterly*, *34*(3), 27–38. https://doi.org/10.1080/02773940409391288.

Part Three: Reassembling with Emerging Relationships

Chapter 8: New Ways of Reading: Making Sense of Complex Biomedical Writing Using Existing Guidelines

Lisa DeTora
HOFSTRA UNIVERSITY

Abstract: Technical communication scholarship often seeks to critique or intervene in powerful medical and scientific discourses. Yet differences in what Carolyn Miller referred to as the communal rationality of scholarly fields may require new ways of reading to make such work possible. This chapter examines guidelines for regulatory documentation that make visible the intellectual framing of biomedical research. Regulatory documentation includes an array of materials that health authorities and government agencies use to authorize and evaluate biomedical research as well as the technical aspects of developing and manufacturing medicinal products. Publicly available guidelines illustrate how those who compose and evaluate regulatory documentation constitute communal rationality within their various specialty areas. Technical communication scholars can use such guidelines to examine the strengths and limitations of the discourses prescribed therein. The author outlines the current place of regulatory documentation relative to technical communication scholarship and offers methods for interpreting these complex discourses using theoretical framing from rhetorics of science, health, and medicine.

Keywords: scientific writing, medical writing, regulatory documentation, rhetoric of science

A Discursive Disjunction

A recent exchange in *Technical Communication Quarterly* highlights a disjunction between technical communication scholarship and regulatory documentation, a legally mandated and complex biomedical discourse. Cathleen O'Connell (2020), a pharmacist and pharmaceutical product labelling expert, corrected Molly Kessler and S. Scott Graham (2018) regarding a newly minted acronym, PDL (prescription drug labels), that conflated multiple discrete documents intended for distinct expert and nonexpert audiences, as legally mandated worldwide. O'Connell sees Kessler and Graham (2018) as failing to identify accurately how prescription drug labelling documentation creates material danger for patients who diverge from prescribed practice, missing sites for authentic rhetorical intervention. O'Connell notes that many dangers to patients from the labelling that accompanies prescriptions arise from product naming conventions (p. 92), but she does not articulate the rhetorical stakes of her corrections in detail. Kessler and Graham's (2020) somewhat nonplussed response to O'Connell reiterates the value of their initial argument, which hints that the very fact that "PDL" looked like a logical category to two highly educated adults is likely a problem requiring some rhetorical intervention. The authors "applaud" (O'Connell, 2020 p. 92) and "thank" (Kessler & Graham, 2020 p. 2) each other, yet seem unable to engage in a genuine exchange. This type of well-meaning but stalled communication is one impetus for the current chapter.

Technical communication studies have paid relatively little attention to biomedical regulatory documentation like the product labelling in the above example. Regulatory documentation generally refers to the many different materials that must be submitted to government agencies for review to authorize medical research studies or to market drug products. My essay "Principles of Technical Communication and Design Can Enrich Writing Practice in Regulated Contexts" (2018) outlines how technical communication knowledge can be brought to bear on two specific regulatory genres: lay summaries of clinical study results and integrated discussions of benefits and risks identified across clinical studies. One shortfall of that paper is a lack of grounding in the then-existing technical communication scholarship, such as Gregory Cuppan and Stephen Bernhardt's (2012) work on reviewer practices for clinical study reports. A subsequent entry, co-authored with Michael J. Klein (2019), situates a brief but complex element of the clinical study report, the patient safety narrative, as a form of intercultural communication. We suggest that recognizing our own limits and identifying the scope of potential interventions can help technical communicators better understand how to communicate across these discourses. I expand on that idea here and call on technical communication and rhetoric scholars to consider biomedical regulatory documentation as representing a series of different intellectual cultures, each of which constitutes itself rationally, as described in its accepted guidelines.

Beliefs among technical communications scholars about positivist discourses in the sciences may impede true exchange with practitioners of regulatory discourses in biomedicine, the term used to describe the various scientific fields that ultimately contribute to drug development and medical practice. Since Carolyn Miller (1979) presented the humanistic rationale for technical communication, questions lingered regarding the "communal rationality" (p. 617) of scientific discourses, which may appear to present only "contextless logic" (p. 617). Charles Bazerman (1988) critiqued as irresponsible a structured scientific format—introduction, methods, results, discussion—that leaves readers to piece information together themselves. Bazerman values the sort of genre conventions Alan Gross (2019) later linked to a scientific sublime, exhibiting elegant writerly characteristics largely absent from structured scientific prose, which makes specific demands on readers. In an essay on layered literacies in technical communication pedagogy (DeTora, 2020a), I cite Allen Renear and Carole Palmer (2009), who describe scientific reading as "simultaneously to search, filter, scan, link, annotate, and analyze fragments of content" (p. 828) from many different texts. Sam

Hamilton (2014), a biomedical writing expert, also describes this type of reading as routine for regulatory reviewers. O'Connell (2020), too, writes in this register, organizing information about multiple documents into tables to identify their distinct audiences and contents. A further point of note, as I argued in the *International Journal of Clinical Practice* (2017), is that biomedical discourses routinely use the same word (like *safety* or *labelling*) to mean different things, even within a single sentence, which further compounds reading demands. I contend that these differences do not signal a lack of communal rationality and that technical communication scholars may need new ways of reading to effect real change in, or even comprehend, these complex discourses.

What technical communication scholars might understand as the rhetorical stakes of regulatory documentation are made visible in guidance documents prepared by and for biomedical writers and reviewers. I see these guidelines as forming a metadiscourse—a way of writing about writing—that can provide useful information to technical communication scholars, not the least of which is an insight into the communal rationality of biomedical discourses. For instance, these guidelines are a means of differentiating the audiences and conditions of production that O'Connell (2020) saw Kessler and Graham (2018) mistakenly conflating. Increased knowledge about structured documentation and its production and reception, including settings like the Food and Drug Administration (FDA) advisory committees that Graham and coauthors (2018) examine, may be gained by reading these existing guidelines. In fact, the metadiscourses of biomedical experts, their thinking and writing about what makes good documentation, ultimately reveal multiple sites for technical communication and rhetorical interrogation.

■ Regulatory Documentation of Clinical Studies

Much biomedical research is regulated by law (in the US, Title 21 of the Code of Federal Regulations [C.F.R.] applies to the FDA), and permission must be obtained to begin a clinical study or to market a medicinal product, which is a broad term used to describe drugs, vaccines, and biologicals used to prevent or treat diseases or other physical conditions. In this context, regulatory documentation must be submitted to a health authority for review before, during, and after each clinical study and when seeking to market a medicinal product (see Table 8.1). Health authorities are groups like the FDA in the United States or the European Medicines Agency, which are charged by governments to help protect public health by regulating medicinal products. Clinical studies generally test an investigational medicinal product in human volunteers and contribute to an overall clinical program designed to support specific claims made (or intended to be made) on a product's labelling. Scientific evidence collected via laboratory and animal research, which also must be documented, is used to justify the initial clinical studies of any product (see Benau, 2020; DeTora, 2020b; Hamilton, 2014; O'Connell, 2020). As observed by groups like the International Committee of

Medical Journal Editors (ICMJE; 2019), results of clinical studies should also be published in a peer-reviewed biomedical journal.

Biomedical research generates all the scientific information needed to evaluate a medicinal product, and regulatory documentation extends to into various intellectual domains, including chemistry, cell culture and other laboratory research studies, clinical studies, and statistical meta-analyses (Benau, 2020; DeTora, 2020b; Wood & Foote, 2009). Scientific subject matter experts in these fields often have only a passing familiarity with regulatory documentation or publication requirements, which creates a need for experts to educate authors and reviewers (see Battisti et al, 2015; Clemow et al, 2018; Cuppan & Bernhardt, 2012; Hamilton, 2014; Winchester, 2017). Regulatory and medical writers are called on to fill this need, and the intellectual demands of their work has continually increased over time (see Benau, 2020; Clemow et al, 2018; Gillow, 2015; Hamilton, 2014; Winchester, 2017). In fact, the complexity of individual regulatory documents, like those listed in Table 8.1, means that medical and regulatory writing professionals may specialize in one specific documentation type, scientific discipline, and/or therapeutic specialty area (see Benau, 2020; Clemow et al, 2018; DeTora, 2020b; Hamilton, 2014).

Each of the documents listed in Table 8.1 must meet specific legal requirements, some of which apply worldwide. However, regulations explain what must be done, not how to do it. Hence, guidelines are published by groups such as the International Council on Harmonisation of Technical Requirements for Pharmaceuticals for Human Use (ICH)[1], the International Committee of Medical Journal Editors (ICMJE), the Regulatory Affairs Professional Society (RAPS), and other experts (Benau, 2020; DeTora, 2020b; Hamilton, 2014; ICMJE, 2019; Wood & Foote, 2009). These guidelines are a rich source of information about how biomedical audiences view and understand not only documentation but also the biomedical research endeavor more generally. Health authorities also require complex submissions, like the Investigational New Drug application (IND) in the US or the Investigational Medicinal Product Dossier (IMPD) in the EU, which mandate a certain organization so that reviewers can find the information they need. The most common format for these filings is described in ICH M4 (R4) Organisation of the Common Technical Document for the Registration of Pharmaceuticals for Human Use (2016; see Figure 8.1, Table 8.2). The Common Technical Document (CTD) is organized hierarchically so that critical discussions are supported by reference documents of increasing granularity. Although, on occasion, a dossier will be built around a single pivotal study to meet a specific medical need (see DeTora, 2020b), usually each study report is a more minor element of a dossier. Next, I will discuss the disjunction between technical communication scholarship on clinical study reports and how these reports are understood within regulatory discourses.

---

1. The ICH guidelines are reproduced verbatim in various national guidance documents with different effective dates; thus, citations tend to include the ICH alphanumeric designation, a convention hereafter followed in this chapter.

Table 8.1. Clinical studies: An overview of some key documentation steps

Document	Function
Study protocol	Outlines the rationale and details the methods for conducting a clinical study
Investigator's brochure	Reviews known information about an experimental medicinal product, such as chemistry, as well as effectiveness and safety in animals and humans
Background packages and meeting outcomes or minutes	Set out the goals (or outcomes) of meetings with health authorities, key questions, and any background information needed to allow regulatory evaluation
Investigational new drug application (IND)/Investigational medicinal product dossier (IMPD)	Supports a request to investigate or continue investigating a product for a specific indication in part by organizing the documentation that reports and explains all known information about a product into defined formats so that health authorities can approve or deny permission to start or continue clinical studies.
Trial and results registries	Publicly provide information about the design of clinical studies seeking participants, their locations, and eligibility criteria for participants and/or the results of completed trials
Annual reports and periodic updates	Provide required safety or other information to a health authority during a specified period
Expedited safety report	Notifies health authorities of certain adverse events, deaths, and hospitalizations, within a specified period (e.g., 24 or 72 hours)
Amendments	Describe changes to a study protocol, investigator's brochure, or other previously completed document
Statistical analysis plans	Specifies planned statistical analyses and their methods
Study report	Briefly reviews the study methods and rationale and presents the results
Lay summary	Presents a high-level overview of study design and results in plain language intended for the general public
Product labelling	Characterizes a product, its approved conditions of use and storage, and presents key clinical data in a complete context for specific audiences of prescribers, regulators, and patients. Comprises patient package inserts and other types of documents.
Marketing application	Requests permission to market a medicinal product for a specific use. These dossiers collect and organize the documentation that reports and explains all known information about a product into a defined format so that health authorities can approve or deny permission to market the product.
Publications	Communicate study designs or findings in peer-reviewed journals or professional meetings

References: US 21 CFR § 314.50; Benau, 2020; Consultation, 2018; DeTora, 2020b; EU 536/2014; Gillow, 2015; ICH E3, 1995; ICH E6 (R2), 2016; ICH E9, 1998; ICMJE, 2019; O'Connell, 2020; Wood & Foote, 2009

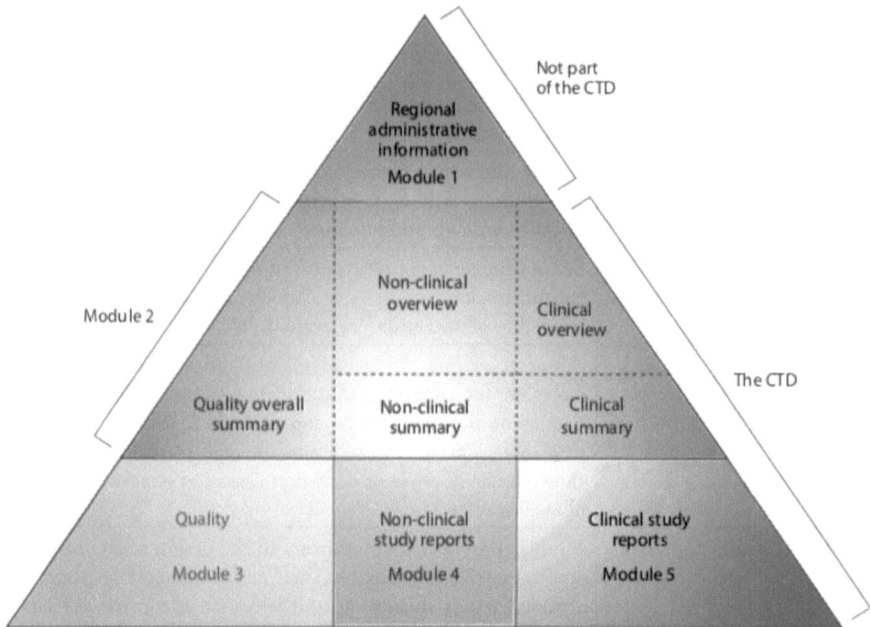

The CTD triangle. The Common Technical Document is organized into five modules. Module 1 is region specific and modules 2, 3, 4 and 5 are intended to be common for all regions.

Figure 8.1. The Common Technical Document triangle as shown in ICH M4 (R4). Image is in the public domain.

Table 8.2. Specific Common Technical Document components that present clinical study results

Document name (Guideline)	Source documents	Functions
Clinical Overview (ICH M4E [R2])	- Clinical Summary - Nonclinical Overview - Quality Overall Summary - Clinical study reports Published literature	Provides an overall critical analysis and interpretation that justifies the proposed label indications and explains whether the benefits of the investigational product outweigh the anticipated and known risks in the intended setting, based on the expert opinion of the authors
Clinical Summary (ICH M4E [R2])	- Clinical study reports - Supplemental statistical analyses Integrated summaries - Data tables of combined analyses across studies - Published literature	Summarizes all studies and analyses done in the clinical development program, which may include combined statistical analyses across multiple studies but not opinion or interpretation

Document name (Guideline)	Source documents	Functions
Integrated summaries (US 21 C.F.R. § 314.50 (d)(5)(v))	• Clinical study data • Statistical analysis plan	Summarize statistical analyses and other examinations of data across different studies or a clinical program based on U.S. Food and Drug Administration requirements
Clinical study reports (ICH E3)	• Clinical study data • Study protocol(s • Statistical analysis plan • Investigator and site information • Published literature • Investigator brochure	Briefly review the rationale and methods for a clinical study and then present its results

Definitions of acronyms: C.F.R.: Code of Federal Regulations; ICH: International Council on Harmonisation of Technical Requirements for Pharmaceuticals for Human Use; E: efficacy (refers to clinical studies); M: multidisciplinary

Clinical Study Reports and Technical Communication Scholarship

George Lakoff and Mark Johnson (2003) describe how cognitive frames help people understand new information in the light of their existing habits of thinking, while incompatible framing impedes understanding. The opening example I used is an apt example of conflicting frames that also may be seen in technical communication views of clinical study reports, which have tended to follow Miller's (1984) interpretation of reports as a call to action. Elizabeth Angeli's (2019) overview of technical communication scholarship also suggests that all reporting is intended to persuade others to act. The most sustained attention to clinical study reports in technical communication scholarship has been by Stephen Bernhardt, Gregory Cuppan (both respected medical writing consultants; see 2012), and various coauthors. These authors describe clinical study reports as making various arguments, a claim that is not consistent with guidelines for writing these reports in part because the word *report* does not always mean the same thing in regulatory documentation contexts as it might in technical communication scholarship. In fact, a study *report* is most likely to present or restate information.

In the CTD format, the clinical study report appears in Module 5, where it should function as a reference for the Clinical Summary, which factually summarizes, but does not critically interpret, data across multiple studies. The Clinical Summary should support the Clinical Overview, which provides a critical analysis and interpretation of available clinical data and the benefit-to-risk assessment needed to justify using a product in medical practice (see Table 8.2 and Figure

8.1). Thus, study reports are a supporting reference for the arguments intended to support actions, like approving a product for marketing (Benau, 2020; DeTora, 2020b; ICH M4E [R2], 2016; Wood & Foote, 2009). As Hamilton (2014) observes, regulators read across multiple documents *and* multiple dossiers, making it critical that clinical study reports be consistent and concise. In this context, study reports serve essentially as vehicles for data rather than arguments or calls for action.

The most widely used guidance for clinical study reports, ICH E3 Content and Structure of Clinical Study Reports (completed in 1995), requires a brief recap of the study protocol and statistical analysis plan, any changes to the plan that occurred during the study (like changing study formats due to COVID-19), and the study results, highlighting any novel or unforeseen findings, especially unanticipated safety outcomes (see Table 8.2). Crucially, ICH E3 allows authors to omit an overall critical analysis because an effective scientific discussion, such as that in a Clinical Overview, generally requires results of multiple studies to justify claims about benefits and risks (see ICH M4E [R2], 2016). Bringing additional regulatory guidance documents into the mix reveals more differences between a clinical study report and the idea of a report as a call to action. These subtleties do not impact the value of Bernhardt and colleagues' work (see Cuppan & Bernhardt, 2012), which educates reviewers to attend to audience needs rather than personal preference; the differences I describe become more important when considering regulatory documents and dossiers as rhetorical entities.

Cuppan and Bernhardt (2012) and Bernhardt (2003) describe statistical analysis as a way of extrapolating general findings from individual results. Yet, regulatory guidelines like ICH E9 Statistical Principles for Clinical Trials (1998) explicitly advise against extrapolating the general from the particular, instead recommending that statistical analyses be made on a population basis to limit variability. Planned sample size and statistical power, hence the relationship between the individual and the general, are determined before a study begins to prevent errors or bias from skewing results. Cuppan and Bernhardt (2012) also describe study reports as presenting a case for whether a clinical study is adequate and well controlled. Yet both the ICH E6 (R2) update to Good Clinical Practice (2016) and ICH E10 Selection of Control Groups (2000) guidelines specify that clinical studies must be demonstrably adequate and well controlled *before* they begin. For instance, as described in the Declaration of Helsinki (2018) and ICH E10 (2000), any new product must be compared with any existing standard treatments and not just placebo. Hence, the clinical study report described in ICH E3 (1995) only reiterates earlier reasoning and does not make new persuasive or critical claims, which must be located elsewhere. The reading culture within regulatory settings, as noted earlier, is highly tolerant of fragmented narratives and simultaneous reading across different documents, which makes balancing multiple guidance documents appear natural to insiders even as it appears alien to those who share Bazerman's (1988) or Gross's (2019) textual sensibilities.

Rhetorical Models for Understanding Regulatory Documents

Nathan Stormer's discussions of *taxis* (2004) and *mnesis* (2013) offer means for understanding how argument and persuasion might operate in regulatory documentation. In an essay on articulation and *taxis*, Stormer (2004) describes how the gaps between words and things that might appear to be a natural consequence of the relationship between language and the material world are actually historically and culturally constructed performances. Seeing rhetorical articulation as a performance may be helpful when considering the construction of regulatory documents and dossiers because this understanding calls on readers to attend to the arrangement of words and things in different intellectual cultures. Rhetorical *taxis*, or the arrangement of elements within a text, creates pathways through scientific documents (and, in the setting of regulatory documentation, larger dossiers of documents and groups of guidelines) that invite specific types of readings, in this case, across multiple texts rather than through a single work. Seeing the arrangement of documents in regulatory dossiers as a mode of rhetorical *taxis* can allow for new cross-disciplinary strategies and theoretical frameworks for analyzing texts. Expanding the idea of articulation to include the guidelines that explain requirements for documents and studies could also help address the metadiscursive concerns of technical communication (for example, the need to understand how texts operate to address user needs or occupy a space within an intellectual culture).

Recognizing regulatory documentation as a representation of intellectual cultures in biomedicine may help rhetorical scholars come to terms with what might otherwise appear to be flawed discourses. A potential added benefit might be seen in Jennifer D. Slack and colleagues (1993) interpretation of articulation in the context of social and cultural theories, such as Stuart Hall's (1973) concept that communication occurs through successful decoding of encoded information. In regulatory discourse, readers may be assumed to have found the "key" to specific documents by familiarizing themselves with guidelines. Slack and coauthor's model of articulation also considers how power relations inherent in different subject positions impact communication. Since power structures like government or corporate hierarchies constrain regulatory discourses, technical communicators might identify areas for interpretation and invention, as Cuppan and Bernhardt (2012) do, by examining relationships within organizations. While the technical communicator might have little power or influence in regulatory documentation, they may exert considerable influence in creating adaptations for general use, a role likely to become more important with a current move toward data transparency (see ICMJE, 2019; Regulation EU 536/2014; Tomlin, 2008). Technical communication scholarship is also well situated to examine the differences between peer-reviewed publications and regulatory documentation of the same study.

Stormer (2004) also examines how bodies of knowledge can be brought to bear on one another through the idea of prosthesis, or the augmentation of textual or material entities. Prosthesis can be helpful in understanding individual regulatory documents and the dossiers in which they appear, as well as the relationships between documents and audiences that O'Connell (2020) elucidated. Guidelines, for instance, can be seen as augmenting individual documents and dossiers. Regulatory documents also rely on the presence of figures and tables, some of which are legally mandated and prescribed, to impart information. These visually dense textual augmentations might be framed for humanities and social science readers via Thierry Groensteen's (1999/2011) who explains how meaning may exist outside argument (or even language) and shows readers how to construct pathways through visually complex texts, like graphic narratives. His semiotic analysis of images, text, and spaces on a page shows how seemingly disparate elements may operate simultaneously to impart greater meaning than any element alone. This mode of reading parallels the scientific reading practices Renear and Palmer (2009) and Hamilton (2014) describe as routine in the sciences, but may be more readily understood by readers trained in social sciences or the humanities. By recognizing the prosthetic function of various types of unspoken knowledge in regulatory documentation, technical communicators might, as suggested by Lisa Melonçon's (2017) discussion of user experience design, more easily situate themselves in the space of the reviewer or other end user. Such skill could be critical when adapting regulatory materials (or other biomedical research data) for new audiences or when examining the role of published literature that may be referenced within regulatory documents.

Recursivity, as discussed by Stormer (2013), is also a useful model for understanding regulatory documents and guidelines. Stormer suggests that the functions of memory and forgetting (*mnesis*) are inextricably linked in many rhetorical activities because an understanding of the prior state is essential to the value of the current reality. On a pragmatic level, regulatory dossiers and guidelines are intended to be updated as new information comes to light, which makes regulatory documentation necessarily recursive (see Benau 2020; Clemow et al, 2018; DeTora 2020b; Wood & Foote, 2009). In biomedical discourses, Stormer's rendering of rhetorical recursivity and its connection to *mnesis* provides a model for replacing outdated or incorrect information with new, more reliable data or for medical inquiry that seeks to limit undesirable signs and symptoms. An essential recursive function could, for example, link undesirable disease symptoms before and after treatment, which is a core aim of both clinical study reports and peer-reviewed manuscripts. Recursivity is also helpful for understanding the continually shifting landscape of guidelines, which are routinely updated to address new discoveries or unmet medical needs.

The rhetorical models just discussed provide a vantage point for unpacking the fragmentation of scientific argument Bazerman (1988) sees as problematic but scientific like Hamilton (2014) view less critically. The arrangement of textual elements in regulatory documents may rely on an interplay between language and other elements (like figures and tables) that work together to communicate information about a current state simultaneously with past knowledge. Or, these elements may be expected to work across multiple texts, allowing reviewers to find information by providing consistent visual and textual cues. Crucially, the combination of visual elements and language can convey meaning even in the absence of argument. Rhetorical theory, thus, can help explain the contexts and functions of regulatory and other complex biomedical discourses by considering the arrangement of objects within these texts (or texts within compilations or dossiers) as a sort of performance that varies by a document's type and broader context, such as the way it is intended to serve its readers. Layering considerations, such as power relations and guidelines or visual rhetoric, into this milieu offers new sites for interpretation, integration, and theorization.

▪ Regulatory Metadiscourses and Textual Production

One obstacle to understanding any biomedical discourse is the sheer volume of available guidance documents. While ICH guidelines are widely accepted, all regulatory agencies provide additional guidance to explain their expectations. Other groups like professional societies and research centers attempt to interpret this wealth of information in targeted ways for authors; some examples are shown in Table 8.3. Guidelines help educate authors about technical requirements for research conduct and reporting as well as how to write documents, like informed consent forms, protocols, publications, or even advertising to recruit study participants. Documentation guidelines function much like how-to books, identifying the basic needs of a highly educated core readership, such as minimum content requirements, in a prescribed order of presentation, while presuming that their users and readers are familiar with scientific and regulatory requirements. Innovation and creativity are discouraged in this context. Although, as Hamilton (2014) notes, minor adjustments may be made to some regulatory documents, the reasons for these changes must arise from scientific logic rather than textual preferences.

Regulatory documentation requirements are backed by the force of law, which can make writerly innovation not only unwelcome but dangerous. One effect of scientific reading practices and the genre conventions of regulatory reports is that many documents, including clinical study reports, are compiled by combining elements that either existed previously, like study methods, or are understood as "generated" rather than written in a humanistic sense (Benau, 2020; Clemow et al, 2018). Hamilton's (2014) discussion of study report authorship concentrates on combining elements following a logical progression and does not mention concepts like persuasion or argument. Similarly, even in noting a move away from a mechanistic model of medical regulatory writing, Rita Tomlin (2008) signals a need for added scientific knowledge to manage increasingly complex content rather than skills in persuasion or argument. The demand for scientific expertise obviates discussions about whether this knowledge carries intellectual value, even as it elides argument (Benau, 2020; Clemow et al, 2018; DeTora, 2020b, Hamilton, 2014; Winchester, 2017).[2]

Table 8.3. Guidance documents for biomedical writing

Document type	Definition	Applicable examples for clinical studies
International Council on Harmonisation of Technical Requirements for Pharmaceuticals for Human Use (ICH) guidelines	Consensus documents developed by an international group of regulators, academics, and industry experts to determine appropriate standards and reporting	*ICH E3 Content and Format of Clinical Study Reports* *ICH E6 (R2) Good Clinical Practice* *ICH E8 Clinical Trials* *ICH E9 (R1) Statistical Analysis Plans* *ICH M4E (R2) Common Technical Document*
Ethical guidelines	International and country-specific guidelines for the appropriate treatment of human beings enrolled in research studies	*Declaration of Helsinki* *Belmont Report*
Regulations	Rules of law established by governments in order to regulate health authorities and manufacturers	*Food and Drugs Title 21 Code of Federal Regulations* *Regulation (EU) No 536/2014*
Professional society guidelines	Guidelines and requirements for different professional groups such as medical writers, medical publications professionals, and regulatory affairs professionals	*Regulatory Affairs Professional Society Fundamentals* *Good Publication Practice Guidelines*
Journal guidelines		
Standards established by medical journal editors for the quality and integrity of publishable work as well as ethical practices of authorship, peer review, and editorial responsibilities *International Council of Medical Journal Editors Recommendations* *Committee on Publication Ethics Guidelines*		
EQUATOR Network	Guidelines for the minimum appropriate information to report for various types of studies	*Consolidated Standards for Reporting the Results of Randomized Trials (CONSORT) Guidelines*

---

2. A huge body of research in biomedicine examines the meaning of authorship in biomedicine and is outside the general scope of this paper.

The rhetorical articulation of regulatory documentation is further complicated by requirements for conciseness. ICH guidelines for all documents state that information should not be repeated between text, tables, and figures: text should provide a high-level characterization of data, highlight noteworthy observations, or present a concise discussion and analysis (Hamilton, 2014; ICH E3, 1995; ICM M4 (R2), 2016). This ethos creates reading burdens as well as obstacles for authors. Since structured scientific formats require the reader to successfully decode hybrid visual and textual elements, the initial arrangement of these elements can be challenging, even for experts. Although reviewers are expected to have enough scientific acumen to actively decipher these documents, some sites for rhetorical intervention still exist. Hamilton (2014), for example, asks medical writers to consider the balance between necessary data to support regulatory review, visual clarity, and the possibility that electronic conveniences, like linking, can present obstacles to reviewer experiences, suggesting another space where Melonçon's (2017) work on user experience design might be brought to bear. Making interventions will require a deeper understanding of the material conditions under which regulatory documentation is produced as well as its rhetorical limitations.

■ Calls to Action and Discursive Contexts

Although clinical study reports do not convey the type of call to action that technical communication scholarship has tended to seek in them, it does not follow that such calls to action do not exist in regulatory discourses. The sites for such calls may be located using guidance documents. For instance, pharmacovigilance, a specialized discipline, monitors safety and side effects associated with medicinal products and may lead to specific actions, as described in the constellation of guidelines under ICH E2A-E2F Pharmacovigilance (1994–2014). During clinical studies, safety problems may require that researchers stop or pause a study or remove a product from the market. These problems may be too urgent to delay until a study report can be written; hence rapid or expedited networks use short reporting forms (see ICH E2A Clinical Safety Data Management: Definitions and Standards for Expedited Reporting, 1994). Although these forms may be considered a report of sorts, they generally lack the type of narrative information that would provide a meaningful context for readers operating outside of regulatory discourses. Michael Klein and I (2019) discuss the brief safety narratives that appear in these short reporting forms and are later adapted for clinical study reports as listing specific information in a defined order and hence existing outside humanistic or social science principles of narrative. Readers of pharmacovigilance calls to action are expected to understand the complex web of regulatory requirements and medical ethics that would ground decision-making—these reports present information for expert interpretation and judgement rather than making an argument.

Clinical study reports also refer to a broader matrix of documents and guidelines that suggest sites for argument and action. ICH E6 (R2) (2016) details the appropriate contents of a clinical protocol, or methods used to conduct a clinical trial, and the other documents and activities that must be completed before starting a study. These documents call investigators to actions described later in study reports. Although not technically regulations, ethical codes of conduct for clinical trials, such as the Declaration of Helsinki (2018), also make calls to action that are recounted in clinical study reports. Hamilton (2014), for instance, begins with the linkage of clinical study reporting and ethics (see Benau 2020, Wood & Foote, 2009). As study reports recount prior actions, the reader must trace those actions back to the original call or other rhetorical activity. Increased transparency among health authorities means that more clinical study reports and protocols will be publicly available (see Tomlin, 2008). These documents provide greater context compared with clinical trial results posted on government-mandated registries such as EudraCT or clinicaltrials.gov or even the clinical trial summaries for laypersons required in the European Union, providing a greater opportunity for technical communication interpretation and use (Gillow, 2015; Schindler, 2020; ICMJE, 2019 Regulation EU 536/2014).

Many studies presented in regulatory documents are later published, and publications may use clinical study data to make arguments or calls to action. That regulatory documents also require adequate references to the published biomedical and scientific literature (see ICH E3, 1995) creates a clear linkage between these discourses. Unsurprisingly, the standard scientific format described by Bazerman (1988) or Scott L Montgomery (2017) is broken down further in biomedical research contexts. The CONSORT guidance (2010) provides a consistent structure and format for publishing clinical trial results in peer-reviewed journals, which parallels ICH E3 (1995). CONSORT (2010) is intended to facilitate meta-analyses and other uses of data, especially those from randomized, controlled clinical studies and, unlike ICH E3, requests a benefit-to-risk assessment or statement based on the study data and existing published literature. The ICMJE Recommendations for the Conduct, Reporting, Editing, and Publication of Scholarly Work in Medical Journals (2019) go a step further, suggesting that authors "explore the implications of your findings for future research and for clinical practice or policy" (p. 17). ICMJE (2019), thus, encourages a call for specific, reasonable actions supported by the presented data.

The ICMJE Recommendations (2019) provide further insights into biomedical research values by describing the ethics of study conduct, authorship, peer review, and editing. The current ICMJE Recommendations (2019) refer to regulatory activities that support data transparency, such as data posting in trial registries, as a measure of the publishability of clinical studies. These circumstances link regulatory and publication functions and also support the assertion I made earlier that people documenting clinical trial results tend to understand their work as deeply related to content and clinical ethics. As with ICH E3 (1995) and CONSORT (2010), the ICMJE Recommendations (2019) specify that authors should use prose only for discussion and analysis or to highlight items of interest and ought not to repeat data across tables, figures, and text. Finally, the ICMJE (2019) refers readers to the specific guidelines and checklists, like CONSORT (2010), for quality publications developed by the EQUATOR network for various types of studies. ICMJE (2019) and Montgomery (2017) also note that biomedical journals publish expert opinion pieces, treatment guidelines, and editorials. These genres require explicit arguments or calls to action and therefore hew much more closely to the types of narratives Bazerman (1988) calls for. Overall, then, publications are likely to offer the type of rhetoric valued in technical communications scholarship; nevertheless, guidelines are important in deciphering the relationships between publication genres and other sources (like study reports) for the same data.

Clinical trial results can also be combined into large meta-analyses of data across studies, such as those performed by the Cochrane Collaboration, to identify trends that might not be evident within individual studies or clinical programs. Such analyses can be reported to regulatory bodies or published, where they can form an object of study. Christa Teston (2017), in *Bodies in Flux* offers an extended analysis of the Cochrane systematic review genre, beginning with the idea that these reviews function as a "stabilized-for-now set of guidelines" (p. 24) intended to aid in practical decision-making for healthcare professionals. Teston conducts an Toulmin analysis to examine Cochrane reviews about cancer therapies, locating sites where different claims originate and characterizing the outcomes of various interim medical and data handling decisions. Of particular interest to Teston is what she terms "evidential cutting" (2017, p. 24), or the selection of information that is suitable to include or exclude from a systematic review. The idea of work as *stabilized for now* that Teston suggests is a helpful way of understanding not only Cochrane reviews, but also regulatory dossiers, works that codify a current state of knowledge. Teston's analysis reveals some important features of the Cochrane review, which, like regulatory documents and biomedical publications, is judged by specific guidelines that limit rhetorical action.

Teston's work (2017), however, much like Kessler and Graham's (2018), betrays a lack of transferability to the source context under examination. For instance, Teston coins an acronym (CSR for Cochrane systematic review) which in biomedical research, including the Cochrane Collaboration, already refers to the clinical study report, and hence may be obfuscating for Cochrane's core expert audiences. Since the Cochrane Collaboration often relies on clinical study reports to do its work, their systematic reviews tend to be referred to as "Cochrane reviews" (Cochrane, 2020–2021). This might seem like a minor semantic point given the obvious merit of Teston's book, but it is a material barrier to accessing audiences familiar with either regulatory discourses or the Cochrane Collaboration. In other words, this semantic activity, as with Kessler and Graham's (2018) acronym for product labels, might lead an expert to assume that the author's conclusions are flawed. The notion of "evidential cutting" (p. 24) is also problematic because systematic reviews require the inclusion of all relevant evidence—what Teston refers to as cutting is a means discerning information and data that can be combined responsibly (or not) to form an evidence base. By characterizing this as "cutting" (p. 24) "evidential" (p. 24) material, Teston might be seen as misunderstanding how scientific evidence functions. Here once again, Stormer's models of articulation (2004) and recursivity (2013) might be helpful in interpreting Cochrane reviews against other types of biomedical documentation and in understanding how the rhetorical performances inherent in producing these works might intersect. Recourse to guidelines for activities like weighing and categorizing types of evidence might also have been helpful not only for building an understanding but also for seeing how these discourse communities use language. Thus, while Teston reviews the PRISMA guideline for publications of systematic meta-analyses and considers how evidence-based medicine experts do their work, her language use creates a distinct type of rhetorical performance.

I noted previously that Bazerman (1988) criticizes publications, like those described by ICMJE (2019), as ignoring the needs of readers accustomed to following linear arguments. However, by reading broadly across guidance documents, it becomes evident that structured formats such as CONSORT (2010) and ICH-mandated documentation are essential frameworks for biomedical epistemology, particularly for regulated activities such as drug development. If we further consider the aims of the CONSORT guidance (2010), ICMJE Recommendations (2019), or ICH E3 (1995), such as promoting meta-analyses (Cochrane, 2020–2021) to help protect public health by identifying trends across studies and general use, then a picture emerges of discourse communities built on an expectation of exchange that is strongly benefitted by format consistency, and specific modes of *taxis*, that allow reading across rather than within documents. These modes of reading also promote recursivity by clearly identifying current understanding as continually subject to future revision. Montgomery (2017) comments on the value of these structures, especially for international exchanges in English and also in allowing researchers to discard invalid work or data without dislodging larger frames of reference. Together, ICMJE, CONSORT, and ICH build a picture of individual clinical study reports and publications as the building blocks of both current and future knowledge, and while these documents are not rhetorically null, they do, in fact, demonstrate the presence of a large body of written genres that *intentionally* do not comply with humanistic sensibilities for argument, claim-making, and explanation because of the nature (and intended sites) of persuasion and calls to action in biomedical inquiry and practice, especially in regulated contexts.

■ Making Sense of Biomedical Inquiry

So, where does this discussion leave technical communication scholarship? The value of technical communication insights into regulatory documentation may be expanding as the role of professional regulatory writers becomes more intellectual. Tomlin (2008) surveyed regulatory medical writers, who linked increased transparency of clinical data over the preceding decade with more intellectual job demands. Increased transparency also created a need to address new audiences. The intellectual role of the medical writer has therefore continued to expand into new areas, as Danny Benau (2020) and Clemow and colleagues (2018) report, and these new areas include increasing numbers of laypersons. When considering technical communication insights, Slack and colleagues' (1993) discussion of power relations is a useful way of understanding the tension between the writer as thinking subject and intellectual contributor rather than a so-called extra pair of hands. Furthermore, technical communicators are already experts at addressing complex information to general audiences and likely will have valuable insights to offer regulatory writers seeking to address laypersons.

I previously suggested that technical communication expertise could enhance the design of regulatory forms like the patient lay summary of clinical study results (Consultation, 2018; DeTora, 2018; Gillow, 2015; Schindler, 2020). Early examples of lay summaries looked very much like clinical study report synopses. Yet lay summaries should combine text and visual elements, like infographics, to present clinical study data effectively for a general readership (Consultation, 2018; EU Regulation 536/2014; Gillow, 2015; Schindler, 2020). In his discussion of lay summaries, Thomas Schindler (2016), an industry expert, notes that plain language cannot fully capture the subtleties of complex scientific content. His subsequent work (2020) situates the comic book as an essential mode of communicating clinical data with certain lay audiences, like children. The theoretical concept of simultaneous mobilization (through Groensteen [1999/2011]), I suggested earlier then, might provide a very direct theoretical framework for the practical work of managing lay summary contents. This framework could then be adapted, by recourse to rhetorical articulation and recursivity to other settings in which information must be derived from a complex scientific format and then presented to general readers. Hybrid, prosthetic modes of thinking are a strength of technical and professional communication; thus, technical communicators are in a unique position to manage these activities, particularly when as medical writer Claire Gillow (2015) indicates, firm regulatory guidance is lacking and creative thinking is needed. Technical communication also offers models for framing explanations, like the link between statistical analyses and individual data offered by Cuppan and Bernhardt (2012) that may be helpful to individuals making medical decisions, even if regulatory audiences would question some particulars.

This is not to say that the field of technical communication has nothing to learn. Expert practices and guidelines in biomedicine remain an underexamined discourse for technical communicators, and one that offers many possibilities outside the direct regulatory context Bernhardt and Cuppan (2012) describe. The existing highly structured and prescribed formats in biomedicine can offer technical and professional communicators an opportunity to concentrate on creative thinking and problem-solving in the articulation of data to broad audiences. The vast array of guidelines, which has only been treated here in a superficial manner, also presents a challenge for technical communication and a new area of inquiry with the potential for real-world impacts. A few earlier examples hint that existing work, while promising, could benefit from a deeper dive into the vast meta-discourse of biomedical writing, a body of knowledge that renders visible many insider discourses. The special strengths of humanistic modes of thinking and rhetorical approaches to textual evidence should enable technical communicators to add real value to these discussions even as they expand their own knowledge and experience.

■ References

Angeli, Elizabeth L. (2019). *Rhetorical work in emergency medical services*. Routledge.
Battisti, Wendy P., Wager, Elisabeth, Baltzer, Lise, Bridges, Dan, Cairns, Angela, Carswell, Christopher I, Citrome, Leslie, Gurr, James A, Mooney, La Verne, Moore, B. Jane, Peña, Teresa, Sanes-Miller, Carol, Veitch, Keith, Woolley, Karen & Yarker, Yvonne E. (2015). Good publication practice for communicating company-sponsored medical research: GPP3. *Annals of Internal Medicine, 163*, 461–464. https://doi.org/10.7326/M15-0288.
Bazerman, Charles. (1988). *Shaping written knowledge: The genre and activity of the experimental article in science*. University of Wisconsin Press.
Benau, Danny. (2020). An overview of medical and regulatory writing. In L. DeTora (Ed.), *Regulatory writing: An overview* (2nd ed., pp. 1–14). Regulatory Affairs Professionals Society.
Bernhardt, Stephen A. (2003). Improving document review practices in pharmaceutical companies. *Journal of Business and Technical Communication, 17*(4), 439–473.
Bernick, Philip, Bernhardt, Stephen A. & Cuppan, Gregory P. (2008). The genre of the clinical study report in drug development. In B. Heifferon & S. Brown (Eds.), *Rhetoric of healthcare: Essays toward a new disciplinary inquiry* (pp. 115–132). Hampton Press.
Clemow, David B., Wagner, Bertil, Marshallsay, Christopher, Benau, Danny, L'Heureux, Darryl, Brown, David H., Dasgupta, Devjani G., Girten, Eileen, Hubbard, Frank, Gawrylewski, Helle M., Ebina, Hiroko, Stoltenborg, Janet, York, J. P., Green, Kim, Wood, Linda F., Toth, Lisa, Mihm, Michael, Katz, Nancy R., Vasconcelos, Nina-Maria, Sakiyama, Norihisa, … Aoyama, Yumiko. (2018). Medical Writing Competency Model—Section 2: Knowledge, Skills, Abilities, and Behaviors. *Therapeutic innovation & regulatory science, 52*(1), 78–88. https://doi.org/10.1177/2168479017723680.
Cochrane Library (2020–2021). Cochrane Database of Systematic Reviews. https://www.cochranelibrary.com/cdsr/about-cdsr
Consultation Document. (2018). *Summary of clinical trial results for laypersons. recommendations of the expert group on clinical trials for the implementation of regulation (EU) no. 536/2014 on clinical trials on medicinal products for human use*. Version 2. The European Commission. https://ec.europa.eu/health/sites/default/files/files/eudralex/vol-10/2017_01_26_summaries_of_ct_results_for_laypersons.pdf.
Content and Format of an NDA (New Drug Application). US 21 C.F.R § 314.50 (1985).
Committee on Publication Ethics (COPE) Council. (2017). *COPE Ethical guidelines for peer reviewers — English*. Version 2. https://doi.org/10.24318/cope.2019.1.9.

Cuppan, Gregory. P & Bernhardt, Stephen A. (2012). Missed opportunities in the review and revision of clinical study reports. *Journal of Business and Technical Communication*, 26(2), 131–170.
DeTora, Lisa. (2018). Principles of technical communication and design can enrich writing practice in regulated contexts. *Communication Design Quarterly*, 6(1), 11–19. https://doi.org/10.1145/3230970.3230974.
DeTora, Lisa. (2020a). Competing mentalities: Situating scientific content literacy within technical communication pedagogy. In Michael J. Klein (Ed.), *Effective teaching of technical communication* (pp. 271–285). The WAC Clearinghouse; University Press of Colorado. https://doi.org/10.37514/TPC-B.2020.1121.2.14.
DeTora, Lisa. (Ed.). (2020b). *Regulatory writing: An overview* (2nd ed.). Regulatory Affairs Professionals Society.
DeTora, Lisa & Klein, Michael J. (2019). Invention questions for intercultural understanding: Situating regulatory medical narratives as narrative forms. *Journal of Technical Writing and Communication*, 50(2), 167–186. https://doi.org/10.1177/0047281620906134.
European Medical Writers Association. (n.d.). Web Site. https://www.emwa.org.
Food and Drugs, 21 C.F.R. § 200-369.21, 600-680.3, 800-898.14.
Gillow, Claire L. (2015). Layperson summaries of clinical trial results: Useful resources in the vacuum of regulatory guidance. *EMWA Journal*, 24(4), 205–209.
Graham, S. Scott., Kessler, Molly M., Kim, Sang-Yeon, Ahn, Seokhoon & Card, Daniel. (2018). Assessing perspectivalism in patient participation: An evaluation of FDA patient and consumer representative programs. *Rhetoric of Health and Medicine*, 1(1), 58–89. https://doi.org/10.5744/rhm.2018.1006.
Groensteen, Thierry. (2011). *System of comics*. (Ann Miller, Trans.). University of Mississippi Press. (Original work published 1999)
Gross, Alan. (2019). *The scientific sublime*. Oxford University Press.
Hamilton, Sam. (2014). Effective authoring of clinical study reports: A companion guide. *Medical Writing*, 23(2), 86–92.
International Committee of Medical Journal Editors. (2019). *Recommendations for the conduct, reporting, editing, and publication of scholarly work in medical journals*. http://www.icmje.org/recommendations/.
International Council for Harmonisation of Technical Requirements for Pharmaceuticals for Human Use (ICH) E2A-E2F Expert Working Groups. (1994 to 2014). ICH E2A-E2F. Pharmacovigilance. https://www.ich.org/page/efficacy-guidelines.
ICH E3 Expert Working Group. (1995). *ICH E3 content and structure of clinical study reports*. https://www.ich.org/page/efficacy-guidelines.
ICH E6 (R2) Expert Working Group. (2016). *Integrated addendum to ICH E6(R1): guideline for good clinical practice E6(R2)* https://www.ich.org/page/efficacy-guidelines.
ICH E8 Expert Working Group. (1997). *ICH E8 general considerations for clinical trials*. Retrieved from: https://www.ich.org/page/efficacy-guidelines.
ICH E9 Expert Working Group. (1998). *ICH E9 statistical principles for clinical trials*. https://www.ich.org/page/efficacy-guidelines.
ICH E10 Expert Working Group. (2000). *ICH E10 choice of control group and related issues in clinical trials*. https://www.ich.org/page/efficacy-guidelines.
ICH M4 (R4) Expert Working Group. (2016). *Organisation of the common technical document for the registration of pharmaceuticals for human use M4* https://www.ich.org/products/ctd.html.

ICH M4E (R2) Expert Working Group (2016). *Revision of M4E guideline on enhancing the format and structure of benefit-risk information in ICH efficacy—M4E(R2)*. https://www.ich.org/products/ctd.html.

Kessler, Molly M. & Graham, S. Scott. (2018). Terminal node problems: ANT 2.0 and prescription drug labels. *Technical Communication Quarterly, 27*(2), 121–136. https://doi.org/10.1080/10572252.2018.1425482.

Kessler, Molly M. & Graham, S. Scott. (2020). The path intended and the path taken: A rejoinder to Dr. O'Connell. *Technical Communication Quarterly, 29*(1), 93–95.

Lakoff, George & Johnson, Mark. (2003). *Metaphors we live by*. Chicago University Press.

Melonçon, Lisa. (2017). Patient experience design: Expanding usability methodologies for healthcare. *Communication Design Quarterly, 5*(2), 19–28.

Miller, Carolyn R. (1979). A humanistic rationale for technical writing. *College English, 40*(6), 610–617. https://www.jstor.org/stable/375964.

Miller, Carolyn R. (1984). Genre as social action. *Quarterly Journal of Speech, 70*(2), 151–167.

Montgomery, Scott L. (2017). *The Chicago guide to communicating science* (2nd ed.). Chicago University Press.

National Commission for the Protection of Human Subjects of Biomedical and Behavioral Research. (1978). *The Belmont Report: Ethical principles and guidelines for the protection of human subjects of research*. The Commission. https://videocast.nih.gov/pdf/ohrp_belmont_report.pdf.

O'Connell, Cathleen. (2020). Reconsidering an essential premise in Kessler, M. M. & Graham, S. S. (2018). Terminal Node Problems: ANT 2.0 and Prescription Drug Labels. *Technical Communication Quarterly, 27*(2), 121–136. *Technical Communication Quarterly, 29*(1), 90–92, https://doi.org/10.1080/10572252.2019.1692909.

Regulation (EU) 536/2014. *Clinical trials on medicinal products for human use, and repealing directive 2001/20/EC*. European Parliament, Council of the European Union. https://ec.europa.eu/health/sites/default/files/files/eudralex/vol-1/reg_2014_536/reg_2014_536_en.pdf.

Renear, Allen H. & Palmer, Carole A. (2009). Strategic reading, ontologies, and the future of scientific publishing. *Science, 325*(5942), 828–832. https://doi.org/10.1126/science.1157784.

Regulatory Affairs Professional Society. (2021). Fundamentals of Regulatory Affairs. https://www.raps.org/publications-resources/fundamentals.

Schindler, Thomas. (2016). The joys of the impossible—The writing of lay summaries of clinical study results. *ClinicalTrialsArena*, December. http://www.clinicaltrialsarena.com/news/data/the-joys-of-the-impossible-the-writing-of-lay-summaries-of-clinical-study-results-5697962.

Schindler, Thomas. (2020). Lay summaries of clinical study results. In L. DeTora (Ed.), *Regulatory writing: An overview* (2nd ed., pp. 265–276). Regulatory Affairs Professionals Society.

Schulz, Kenneth F., Altman, Douglas G., Moher, David & CONSORT Group (2010). CONSORT 2010 statement: Updated guidelines for reporting parallel group randomised trials. *BMJ* (Clinical research ed.), 340, c332. https://doi.org/10.1136/bmj.c332.

Slack, Jennifer D., Miller, David, J. & Doak, Jeffrey. (1993). The technical communicator as author: Meaning, power, authority. *Journal of Business and Technical Communication, 7*(1), 12–36.

Stormer, Nathan. (2004). Articulation: A working paper on rhetoric and *taxis*. *Quarterly Journal of Speech*, *90*(3), 257–284.
Stormer, Nathan. (2013). Recursivity: A working paper on rhetoric and *mnesis*. *Quarterly Journal of Speech*, *99*(1), 27–50.
Teston, Christa. (2017). *Bodies in flux: Scientific methods for negotiating medical uncertainty.* University of Chicago Press.
Tomlin, Rita C. (2008). Online FDA regulations: Implications for medical writers. *Technical Communication Quarterly*, *17*(3), 289–310.
Winchester, Chris (2017). AMWA-EMWA-ISMPP joint position statement on the role of professional medical writers. *Medical Writing*, *26*(1), 7–8.
Wood, Linda F. & Foote, Mary Anne. (Eds.). (2009). *Targeted regulatory writing techniques: Clinical documents for drugs and biologics.* Springer.
World Medical Association. (2018). *Declaration of Helsinki—Ethical principles for medical research involving human subjects.* https://www.wma.net/policies-post/wma-declaration-of-helsinki-ethical-principles-for-medical-research-involving-human-subjects/.

Chapter 9: A Critique of Disability and Accessibility Research in Technical Communication Through the Models of Emancipatory Disability Research Paradigm and Participatory Scholarship

Sushil K. Oswal
UNIVERSITY OF WASHINGTON

Zsuzsanna B. Palmer
GRAND VALLEY STATE UNIVERSITY

Abstract: The last two decades have seen an increase in articles about disability and accessibility within technical and professional communication (TPC) scholarship. As disabled users make up a sizeable portion of all users that our field aims to serve, this development is certainly a welcome change. However, as this chapter points out, disability and accessibility scholarship within the field has fallen short of keeping up with recent developments in the field of disability studies. Through a critique of three articles within the TPC field, the first half of this chapter highlights areas in our scholarship that need improvement in order to not only keep up with developments in other fields but also to better address the needs of this specific group of users. The second half of the chapter then introduces participatory design and participatory action research from the perspective of emancipatory research paradigm as two approaches to interface and product design, research, and pedagogy and shows how these approaches have the potential to propel TPC scholarship towards being more inclusive and mindful of users with disability. The chapter concludes with two substantive examples that foster participatory design and participatory action research as a way to illustrate the practical application of these approaches in research and pedagogy. The seven-point heuristics introduced in this chapter can be employed as an independent tool for assessing the value of disability studies-centered research and pedagogy in a variety of settings.

Keywords: accessibility, critical social model of disability, emancipatory research paradigm, participatory design, participatory action research

As technical communication experts are primarily focused on facilitating communication in complex environments among people, the frameworks that inform this work have to incorporate theories that not only illuminate the needs of people with all abilities but also influence the design process of new technology and research for all users. Ten years after the Americans with Disabilities Act became law, an awareness emerged in the early 2000s that discussions of disability need to be incorporated into research and pedagogical considerations in the field with special attention to how disability studies might inform the teaching of medical and scientific writing (Wilson, 2000). In the same timeframe, following the publication of the Web Content Accessibility Guidelines (WCAG) in 1999, concerns of website accessibility moved to the foreground of practitioner and academic discourse in technical communication, with a clear purpose of bridging the gap between usability and accessibility (Slatin & Rush, 2002; Theofanos & Redish, 2003). John Slatin and Sharron Rush focused on the instrumental aspects of making webpages accessible, whereas Mary Frances Theofanos and Janice Redish's study was one of the earliest efforts in TPC at examining the accessibility problems faced by blind and low-vision users when visiting websites. While Slatin had first-hand experience of web accessibility issues as a blind user, Theofanos and Redish drew on observations of blind users as practitioners in the field. Similarly, Jeff Carter and Mike Markel (2001) introduced the problems faced by disabled users and offered some practical advice to web developers when creating websites.

These initial articles were followed by an increasing number of publications in the last two decades about incorporating disability and accessibility into our professional discourse (Knight & Oswal, 2018; Konrad, 2018; Melonçon, 2014; Melonçon, 2018; Moeller, 2014; Oswal, 2014, 2018; Oswal & Melonçon, 2014; Walters, 2010). The earliest of these articles (Walters, 2010) discusses the introduction of the concept of universal design and accessibility into the author's technical communication classroom. The approach described in this article made the students aware of accessibility issues and assistive technology pitting accessibility against multimodality without the involvement of disabled users. While disability studies theory is introduced in this article, the classroom activities included in this course did not specifically engage the basic tenets of disability studies particularly through any direct involvement with disabled users. Sushil Oswal (2014) wrote a position paper that argues for participatory design by expounding its benefits to designers and technical communicators. As the title of Marie Moeller's (2014) piece indicates, this article engages critical disability studies for the purpose of deconstructing the concept of normalcy on medical advocacy websites.

Working from the disappointing results of a national survey of online instructors, Sushil Oswal and Lisa Melonçon (2014) advocate for a greater attention to accessibility in the ever-increasing number of online technical communication courses. This article also includes practical suggestions for technical communication instructors on how to design accessible online courses. The guest editorial by Oswal (2018) offered a detailed overview of disability studies literature and explained the relevance to the teaching of business and professional communication. An example application of these concepts is presented by Melonçon (2018), where she describes how instructors can orient themselves to disability and accessibility in a professional communication classroom. A second editorial by Melinda Knight and Sushil Oswal (2018) emphasizes the need for accessibility research to move beyond the classroom to business and professional communication workplaces from the perspective of disability studies. An example of such research is Annika Konrad's (2018) study of several blind professionals' practices in the workplace.

While many of these articles have aimed to make technical and professional communication pedagogy more inclusive, several other publications are based on now outdated models of disability even though they might have served an instrumental purpose (Theofanos & Redish, 2003, 2005; Wilferth & Hart, 2005). In addition, some publications fail to go beyond the illustration of accessibility issues, only focus on the technical aspect of remediation, and separate the author's own pedagogical practices and disabled users' experiences from general directions outlined for others to follow (Dolmage, 2009; Palmeri, 2006).

The purpose of this chapter is to assess the current state of disability/accessibility research in technical and professional communication in comparison with the theoretical advances in disability studies research. Through the detailed analysis of three often-cited research- and pedagogy-oriented articles related to disability and accessibility in our field, this chapter will provide a deeper understanding of where technical communication research stands in relation to the most up-to-date theories in disability studies. This analysis will establish where our field makes meaningful contributions to equity and inclusion for all users, and where it falls short and needs to adopt a different approach to research and pedagogy. The chapter then will propose our approach to disability and accessibility research employing participatory design and participatory action research approaches which give disabled users a key role in the research and design processes. While our first example will offer our vision of participatory design, our second example will be based on our own classrooms and will show how participatory action research approach can be combined with inclusive pedagogy.

■ An Analysis of Select Technical Communication Articles

To provide insight into the state of technical and professional communication research focusing on disability, we selected three often-cited articles from the last two decades and analyzed them through the use of a new, seven-point heuristics model. We introduce this seven-point heuristics model for analyzing disability and accessibility-related scholarship in TPC because it helps to evaluate TPC work from the perspective of disability studies. This heuristic is grounded in the basic premise of emancipatory research paradigm in disability studies that asserts that no research can represent disabled people without their direct participation in all aspects and stages of research. In the disability studies field, when we talk about disability, the functional aspect of a person's impairments is limited to describing how this impairment keeps that person from performing a particular function within an ableist environment. On the other hand, disability is a much broader concept because it includes the physical and social environment within which a disabled person interacts with their environment on a daily basis, but it also covers such additional aspects as personal attitudes toward impairment and disability, social stigma, and a baggage of considering disability from the medical establishment's mechanical view of the disabled body. We can only gain valid insights into these aspects through participatory action research with disabled users and learning about their experiences through first-hand interactions. Just as we cannot have feminist research without a full participation of women, we cannot claim to have the right to speak for disabled users without giving their voices a predominant place in our research activity. We define this distinction between the functional aspects of impairment and the broader environment within which disabled people interact with society because approaches to teaching accessibility within technical communication still largely favor the functional approach which focuses on the disabled person's impairment and leaves out the fuller user experiences of people with disabilities. We are rethinking participatory design to help the field do participatory designs better but also to conduct better research.

The articles we take up for our analysis illustrate the present problems with TPC scholarship, particularly in how it represents disabled users, their accessibility needs, and their role in the production of TPC knowledge about disability. The purpose of our analysis was to determine how inclusive these articles were in light of recent disability studies research in order to identify the assumptions behind these articles as well as their strengths in becoming agents of social change. For this reason, we have developed these seven heuristics that guide our analysis of these illustrative samples:

1. Does the article address the functional aspect of disability only, or does it encompass the disability experience as a whole?
2. Where do the articles move the disabled users, consumers, students, workers, and educators from the margins to the center, or where do they allow the disabled to take center stage and have a literal voice in the design discourse?
3. Where do the articles simply evoke the topic of disability as a trendy topic, or where do they suggest concrete steps to counter ableism, inaccessible designs, and exclusionary pedagogies?
4. Do the articles give any meaningful examples where the authors have modified their own pedagogies, designs, and policies?
5. Is (Are) the author(s) willing to share their embodied experiences of disability directly or indirectly?

6. Does the article challenge the larger physical, social, political, cultural, economic, or institutional contexts and structures that, in the first place, create the need for the article's topic to be researched?
7. What contribution does the article make to "participatory accessibility"?

The first article analyzed is "Disability Studies, Cultural Analysis, and the Critical Practice of Technical Communication Pedagogy" by Jason Palmeri (2006). This is a commonly known and often-cited article in technical communication as it was one of the first publications that applied theories from disability studies to critically examine how the discourse and professional practices within our field contribute to a clear division between normal and deviant bodies. In this article, Palmeri takes a critical look at safety communications and usability and provides examples of texts and rhetorical moves where these subdisciplines further reinforce deep divisions in society. He shows that in many ways, discourse in safety communications and usability either subscribes to the medical model of disability that aims to rehabilitate people so that they can become like the "normate"—to borrow a term from Rosemarie Garland-Thomson—user, or it capitalizes on the charity model where the emotions of seeing people with disability are used to motivate society to take remedial action (Garland-Thomson, 1997, p. 8; Kleege, 2011; Longmore, 2015; Mattlin, 1991). While these observations were certainly effective in providing a critical view of these professional practices, a closer look at Palmeri's article shows that it does not go far enough in its criticism of the field and in its application of this critical stance to his own work.

While Palmeri's (2006) article does challenge the underlying assumptions within technical communication that further society's reinforcement of an environment and institutions favoring its able-bodied members, the strength of this article is in the act of critiquing and calling attention to an issue in the form of questions, not necessarily in providing a solution to the problem. In fact, the recommendations Palmeri includes are provided as a series of questions that instructors (in general) could incorporate into their curricula as part of their students' critical inquiry. These questions, as Palmeri states, could guide students' critical interaction with the professional discourse and could further their understanding about usability and accessibility, the functional aspect of disability, while personally experiencing assistive technology such as screen readers. Ideally, as explained in the article, this approach to teaching technical communication would allow students to arrive at a stage where they are ready to challenge the underlying norms of the whole discourse community, the norms that underpin the ideology of normalcy. However, Palmeri does not provide concrete examples of his own actual pedagogical projects, course descriptions, or lesson plans, and thus his call to action remains on the level of hypothetical suggestions as opposed to easy-to-implement and thoroughly tested pedagogical tools with an actual impact.

Our analysis of Palmeri's article has also revealed that while the author is conscious and open about his temporarily able-bodied condition and thus shows awareness of the subjectivity and embodied nature of disability, the article discusses disability overwhelmingly as an abstract concept without incorporating the lived experiences and voices of people with disabilities. This becomes evident especially where the article suggests using participatory design for product development for the field in general, yet when student projects in this area are outlined, nowhere in the discussion is it mentioned to possibly include disabled users in the design and testing phase of accessibility classroom projects. This is one of the ways in which this article remains on the level of awareness-raising about the issues surrounding disability. It functions well as a place to start critical discussions about technical communication's role in reinforcing societal norms that favor the more powerful actors of society, but it fails to fully embrace the value that participatory design could add to academic and professional projects focusing on equal access and chipping away some of this power imbalance between designers and disabled users (Kesby et al., 2007).

The second article, "Accessibility Scans and Institutional Activity: An Activity Theory Analysis" by Clay Spinuzzi (2007), is a great example of approaching the topic of accessibility from a functional point of view. Several screenshots of automated accessibility scan results demonstrate the practical aspect of website accessibility. These screenshots are accompanied by a detailed explanation of the additional need for interpretative scans by human actors to catch accessibility violations that are not detectable by machines. But Spinuzzi's article goes further; it argues that "accessibility is a rhetorical enterprise" (p.190) because a consensus is necessary among all the different stakeholders to achieve it. Understanding the interplay between the division of labor, actors, tools, community, rules, and objectives can help us discern just how complex of a process it is whereby a website can be declared accessible. According to the article, the complexity resulting from the interaction within and between activity systems makes the outcome of website accessibility difficult to achieve. Why is website accessibility still a goal that needs to be achieved? The article cites two main reasons for this: compliance with regulations and improving the user experience of disabled users. Though each of these reasons makes the work of creating access worthwhile, the article's main conclusion emphasizes that accessibility is a "moving target" (p. 198) not only because regulations and technologies change but also because it is difficult to prepare for individual variation in the training and application of adaptive technologies at the level of the end user.

While this article exemplifies great care and significant investment of the author's time to make a large collection of websites accessible, declaring accessibility a moving target sends a somewhat different message than the activities described. Accessibility regulations and adaptive technologies certainly change, but so do other types of regulations and technologies used by the "normate" user. In fact, many social and technological factors influence just how much any individual is able to benefit from digital technologies. Accessing information from different types of devices, geographical areas, and networks can significantly impact the user experience in ways that cannot always be assessed ahead of time. In addition, technology and digital literacy skills of users are also factors that affect the user experience and cannot always be anticipated. Thus, by singling out accessibility as a moving target because of the lack of information on how exactly each disabled user will interact with a website, the article implicitly suggests that it is more difficult to ensure a positive user experience for this specific group of users. In this sense, while disabled users are a central concern in this article, they are marginalized from mainstream because their knowledge and application of adaptive tools cannot be anticipated. The idea of incorporating users with disabilities into the design process, thus arriving at a more nuanced understanding of their interactions with websites, does not get mentioned as a solution for this issue.

Further, Spinuzzi's (2007) article, with its truly descriptive focus on activities and regulations as they exist in our society, does not allow for any type of critique of the status quo. Accessibility regulations and institutional policies are fully accepted at face value; none of the activities described go beyond compliance with these rules. As for the user experience, the relationship between adaptive tools and mainstream technologies is never questioned; in fact, examples of website design trends that make the use of adaptive technologies difficult are mentioned, but not critiqued. Overall, while the article provides a detailed view of all the factors involved in making a website accessible on a functional level, no tangible improvements for the disabled user result from such an approach. As men—both colonial and native—speak on behalf of the Sati woman in Gayatri Chakravorty Spivak's (1988) "Can the Subaltern Speak?" throughout this article, Spinuzzi and activity theory speak for the disabled users. Despite all the focus on the activity scans in this article—whose images, by the way, are altogether inaccessible to blind users—the disabled users themselves remain absent from Spinuzzi's article.

The third article, "A Version of Access" (2016), written by Casey Boyle and Nathaniel A. Rivers, approaches accessibility from a philosophical point of view. Accessibility, in the article, is posed as a type of motivation for creating different versions of texts and thus is described as a way to encourage difference. The premise of the article is an occasion when the authors created an audio version of their article for online publication in order to make this article accessible. While creating this audio version, the authors started to add features to it, such as music to signal the beginning and the end of segments that did not have equivalencies in the written text. The article then explores the value of these versions and argues that the differences between the original and nonequal versions open up new avenues for accessible design.

In order to establish versioning as a neutral process, the authors include architectural drawings of a building and argue that the various entrances to the same building, among these, doors at the top of wheelchair ramps, expose the entrants through each door to a different version of the building. While this analogy works well in theory, it does not take into account that wheelchair ramps are often added as afterthoughts to the sides or backs of older buildings and thus many times lead to obscure parts of buildings before they connect to the main area. A person entering a building through the back door certainly does not get the same impression of the space inside as someone who goes into the building through the main entrance. Thus, arguing for the validity of nonequal versions or compositions, some that are created to make information accessible, to a certain extent promotes the creation of nonequal versions and thus denies the right of people with disabilities to equal embodied experiences.

After setting the scene with the building analogy, the article unpacks this new approach of accessibility the authors call nonequal design through framing it around the binary of consumptive access versus rhetorical access. It makes a similar argument to Spinuzzi's (2007) article as it recites how the constantly changing rules, abilities, and technologies impact the task of creating accessibility and thus transform it into a rhetorical concern. Here, Boyle and Rivers (2016) argue that understanding accessibility from this rhetorical perspective will result in "prioritizing multiplicity as standard" (p. 36) and thus will not privilege any version as original. This would eventually lead to, the authors state, transforming the environment so that disability is not erased but valued. Further, this type of approach will lead to accessibility serving as a motivator to create generative difference. The article concludes by describing three main principles for nonequal design: syncopation, medium specificity, and versioning.

Our analysis questions to evaluate this article helped us to reveal that its strength lies in the authors' following of their own advice. While the article does not reveal the disability status of either of these authors, it suggests the creation of a different social order where everything would be multi-versioned. This vision grew out of an attempt to make texts accessible, and the article suggests several ways in which approaching this work with the nonequal design perspective might bring about social change where texts no longer need to be made accessible but will already evolve as several versions with their own specific rhetorical strengths and affordances. While the nonequal design approach seems to share the same principles as AccessFirst design (Hendren, 2014), which promotes creating products already accessible, it differs from this approach by supporting the creation of different versions as opposed to a specific version that is born accessible.

Further, Boyle and Rivers (2016) explain their theory to the reader without including any voices of people with disabilities. The reader is left wondering what people with their embodied experience of disability might think about versioning, and whether this approach would satisfy them. The only way to really know if this theory has any practical relevance and thus would make a difference in people's lives would be to include people with disabilities in the nonequal design process and then research whether it results in better outcomes. If it does not, the theory, however eloquent, will remain at the theory level without any real potential to bring about real social change that improves the embodied experience of disabled people. In closing this section, we invite the scholarly community to use our seven-point heuristic model as an open-ended analytical tool for evaluating disability studies-focused accessibility scholarship from the perspective of emancipatory and participatory research paradigms. This heuristic model is not only a practical tool for assessing disability scholarship in TPC, but it also can be employed as a pedagogical approach in graduate courses as a framework for teaching the underlying, fundamental tenets of disability studies.

In the second half of this chapter, we first introduce participatory design and participatory action research theories as ways to give a central place to disabled users in the TPC research and pedagogy. We frame these participatory research models within the disability studies research paradigm, which dictates that all research about disability should be emancipatory and applicable to the lived experiences of disabled people. We then incorporate two examples that show how these theoretical frameworks have great potential for making TPC research and pedagogy far more inclusive and participatory of disabled people. These two examples show how extending the definition of participatory design to be participatory action research solves some of the problems we pointed out in our critique of TPC literature above.

Proposing Participatory Design and Action Research for the 21st Century

Participatory design refers to design processes that involve users as co-designers and co-creators of product and design concepts. This methodology is rooted in the belief that users possess unique knowledge about their bodies and contexts of use which designers might not share, as it integrates the "genuine decision-making power of the co-designers and the incorporation of their values in the design process and its outcome" (Van der Velden & Mörtberg, 2015, p. 42). Through the involvement of users, participatory design engages the dialectics of "tradition and transcendence" to narrow the distance between what is and what could be (Ehn, 1989; Oswal, 2014). While participatory design methods have deep roots in the Scandinavian work methods research, these design methods have been developed for specific situations in different parts of the world, and vary in purposes and outcomes (Ehn, 2017). The maturation of these methods in the design field over the past four decades has led researchers to define the basic understandings of the field. According to Jesper Simonsen and Toni Robertson (2012), participatory design is "a process of investigating, understanding, reflecting upon, establishing, developing, and supporting mutual learning between multiple participants in collective 'reflection-in-action'" (p. 2). Besides establishing participatory design as a set of practices that aims to equalize power between designers and users, Finn Kensing and Joan Greenbaum (2012) propound four other principles to guide participatory design: 1) situation-based actions, 2) mutual learning among designers and users, 3) sharing of knowledge about tools and techniques, and 4) openness to alternative visions about technology. Since participatory design practices can entail work among designers, researchers, and participants with significant power differential, researchers and theorists in this field have more recently tried to address the questions of ethics to protect vulnerable participants and participant interests (Christiansen, 2014; Frauenberger et al., 2015; Kelly, 2019) and under the label of user-centered design in technical communication (Salvo, 2001).

We endorse participatory design activity between designers and disabled users as a viable proposition for conceptualizing accessible and useable products, processes, and spaces because participatory design research is not about, or on behalf of, disabled users; it is disabled users taking the front seat on the drafting board with professional designers to employ their distinct know-how about disability which originates from their bodily differences and diverse contexts of purpose and use. In the case of "context of use," design work with disabled users differs significantly from design work with other users because most participatory studies do not focus on this aspect of design. Disabled people bring viewpoints of their own of being in and with the material and social world which shape, at least in part, their human desires, needs, and expectations. Disabled bodies traverse through these worlds at a different pace, in diverse ways, and for succinct purposes to fulfill these needs, desires, and worldly goals which might appear odd, out of place, or even undesirable to a nondisabled eye and a presumably fit body. But participatory design as a process does not have to only apply to product or interface design; it can also be applied to research designs as it has been applied in the contexts of participatory action research (Priestley et al., 2010). We see research designs involving disabled participants and experts to explore scholarly questions relating to disability, or nondisability, as a far more robust model of scholarly inquiry than the research conducted by nondisabled academics. Projects not using a participatory research design model result in products and processes emerging out of only second-hand knowledge of disability—and heavily ridden with ableist assumptions about materiality and presumptions about the disabled body. Most importantly, discounting participatory action research will also lead to ableist research foci which are often devoid of an interest in the value of disabled life and of disability being a way of being in the normate, socio-material worlds.

Because disability in most parts remains invisible in human societies—despite its presence everywhere—and because disabled people's lived experiences are incomparable sources of knowledge about the human body, we as TPC professionals, researchers, and pedagogues with our own lived experiences of disability believe that participatory research designs are essential for our field to remove its veil of disability ignorance and experience the value of disability first hand. Our ableist academic values have so far denied a place to disability in the university beyond the disability service offices and testing rooms. Even after half a century of Section 504—which gave disabled children a right to secondary school education in the United States—and more than a quarter century of the Americans with Disabilities Act—which allowed disabled students to be considered for college admission—our programmatic and curricular designs are awash with ableist notions of knowledge, bodies, and human life while disability waits at our classroom doors yet to be admitted to the scholarly spaces. Even when we let disabled students into our classrooms, our exclusionary curricular designs and content—both pedagogically and physically—treat them as occasional guests and expect them to leave their disability outside because we have not yet learned to create a place for disability in our highbrowed academic disciplines. Worse yet, our research paradigms cling on to the pretense that everybody has been carved out to map perfectly on Galton's bell curve and only these bodies are a fit subject for our scholarly inquiries (Cowan, 1972; Devlin et al., 1995; Fendler & Muzaffar, 2008). In the next section, we present an example of a participatory research design that situates disability in the center while critiquing the status quo in the design of the U.S. academy. Our example also introduces a research method that makes a focal space available to a junior, disabled researcher to articulate her agenda in her own voice.

An Example of Participatory Design of a Research Project

According to the critical social model, disability is not simply a condition defined by an impairment or an individual's functioning level but is also the product of the interaction between individuals and their physical surroundings, institutional structures, and social environments (Kruse & Oswal, 2018; O'Day & Killeen, 2002). Emancipatory research designs have "proven their power to describe and clarify the interdependence of human interaction, cultural attitudes, institutional processes, and public policies" (O'Day & Killeen, 2002, p.9). On the other hand, lived interactions of disability with technology, spaces, and people are complex, and disability-focused user experience (UX) studies can encapsulate fresh insights into how disabled users adapt human bodies, senses, and minds and how they can develop novel, and often individualized, techniques to perform mundane, as well as complex, tasks. These types of studies can also teach us how our widespread, ableist actions and attitudes limit human potential to participate in the everyday life of the academy and of this world.

Emancipatory research guided by the critical social model of disability and participatory design also affects the nature of questions researchers ask and the analysis they perform on the data. For instance, a traditional researcher would ask, "How does your bipolar illness keep you from participating in your classes?" and hold the mental disability of the student responsible for their learning difficulties. The same question framed within the emancipatory research paradigm might ask, "How do your professors' attitudes about disability, their approach to the delivery of course content, the classroom structure, and the level of accommodations affect your learning?" thus shifting the burden of blame away from the student's mental or physical impairment and pointing it back toward the societal and environmental factors—the design of the institutional physical and social infrastructure, the ableist university policies, the exclusionary curriculum that perceives human difference as deviant, and the deeply-entrenched, normate pedagogies.

All of these societal and environmental factors in academia do not take into consideration a variety of bodies as the members of their learning community in the overall conceptualization and planning of the higher education enterprise. The research activity—which in contemporary societies predominantly takes place in the university—is also steeped in these exclusions and views disabled bodies as aberrant. For these reasons, the proponents of emancipatory research from the critical social model of disability assert that "Disabled people have come to see research as a violation of their experience, as irrelevant to their needs and as failing to improve their material circumstances and quality of life" (Oliver, 1992, p. 105).

Academics conducting research studies with disabled people often disassociate themselves from their participants once they have gathered data, and any later contact is generally for the formality of validating their results. The multiple steps of data analysis, the writing of the study, and the dissemination process are fully controlled by the academic and professional needs of the researcher. The voices of the disabled participants at the writing stage not only become subservient to the demands of the conventions of the genre chosen for dissemination, publication venue, and the dissemination process itself but also get removed from the original context and purposes for which the participants might have invested their time and energy. The claim from almost three decades ago—"research has been and essentially still is, an activity carried out by those who have power upon those who do not"—still holds water for most research designs in the academy (Oliver, 1992, p. 110).

The research design by Allison Kruse and Sushil Oswal (2018) described here is an example of participatory research design which focuses on the lived experiences of an undergraduate technical communication student with a bipolar disorder diagnosis and a professor with a sensory disability. In this participatorily designed scholarly work, Kruse not only presents an account of an ableist university campus through an autoethnography, but also goes on to subtly bracket the ableisms disabled students often internalize in the elite environment of the university. Such internalization by disabled students refrains them from questioning the problematics of their existence in a space especially reserved for learning and critique.

The basics of this participatory study design are ordinary. Kruse is exposed to a minor discussion of disability and accessibility in one of the four courses she studies with Oswal, a professor with a sensory disability. Toward her senior year, Kruse expresses interest in conducting a term-long independent research project relating to disability and accessibility under Oswal's direction. The research project is defined by Kruse's academic interests and soon moves in the direction of more emphasis on disability studies and the access conditions in the academy. This study of published research also begins the process of disability disclosure for Kruse with her professor, and this is the point when the student becomes the informed participant and expert of lived experiences with bipolar disorder. The research project at this point takes a turn toward an additional component of Kruse's composing of an autoethnography of the academic accommodations of a person with bipolar disorder diagnosis, and this project now becomes the focus of a joint research project for publication where Kruse becomes the lead author due to her expertise with the lived experiences of a mental disability in the academy. Equally significant is the role she begins to play in the design of the manuscript for publication, which by no means resembles the shape of a typical scholarly article as the readers of this chapter might discover in their perusal of the end result (Kruse & Oswal, 2018) of this participatory collaboration.

We selected this research article published in an open-access, European journal as an example of participatory research design to highlight the productive potential of this type of research where an established researcher co-designs a study of academic ableism with a disabled student, walks her through the research and publication process, and participates in analyzing and reporting the results. At some point in this process, the two become co-authors in the professional sense of the word and, through Kruse's participatory autoethnography, construct a visionary design of university which not only performs its fiduciary duty under The Americans with Disabilities Act to educate all students that enter through its gates but also sees it as an inseparable component of the educational ethics.

Kruse and Oswal (2018) categorically avoid making a legal claim—as stated in their abstract and again stressed in their introduction and analysis later in the article. As the synopsis of the circumstances of this collaboration narrated above reveals, the two co-authors, who recruited each other in different ways for becoming participants in this project and collaboratively compiling the implications of this autoethnographic study by the primary author, flip-flopped in determining whether or not their position statement should emphasize the legal over the ethical. The two authors deliberated over the issue together and weighed the purpose of their project again. The legal aspect was eventually pushed back because they determined that their goal was not to ask for more legal accommodations but to make the academy less ableist. The nature of mental disabilities required acceptance, not legal redress. Thus, at the point where Kruse began to rewrite her analysis of the autoethnography for a journal audience, Oswal's role as an expert in this independent research study had fallen by the wayside because Kruse's expertise of writing about the lived experiences of a mental disability had come to occupy that space.

By the time Kruse and Oswal finished deciding how to write the implications of the study and make a proposal to design the academic environments for a more disabled-friendly mental and physical space, they were two disability activists taking scholarly decisions and applying their individual expertise in the lived experiences of disability for a shared task. The social relations of research production also had moved to another space, and the professor was now a co-activist of sorts against the barriers for students with mental disabilities. Their joint article—which began as an autoethnography of a student with a bipolar disorder diagnosis—moved closer to the theme of "changers for change" (Lather, 1987; Oliver, 1992). The emancipatory paradigm of research and the participant design methodologies had built "trust and respect" among the collaborators, and the resulting "reciprocity" facilitated "a politics of the possible" between the two members of the academy to confront social oppression (Oliver, 1992, p. 107 & p. 110).

On the research design side, this study tries to bridge the researcher-participant gap in conceiving, conducting, and writing up this scholarly project. We see that by making the Kruse autoethnography a centerpiece of this scholarly article, Kruse and Oswal foreground what would otherwise have been a marginalized "participant voice" in a more traditionally structured scholarly article in the form of third-person descriptions and scattered quotations from the participant narrative. By conceptualizing, designing, and composing the article as co-authors, they try to dismantle the researcher-participant and instructor-student hierarchy and present an alternative research design for studying the academy. In fact, for the purposes of our chapter, their relationship is strictly that of two scholars collaborating on a project where Kruse is the lead author and her narrative voice defines the purpose and structure of this study. Had there been an opportunity available, they might have disseminated their results through other means—a conference paper or a website blog—before publishing this work in the *Social Inclusion* journal. Further on, as we worked on this chapter, Kruse reviewed this section about the article development process to provide her feedback.

While we do not want to construct another hierarchy by indulging in the discourse of empowerment in this context, the outcome of the Kruse and Oswal collaboration is an activist experience of two disabled members of the academy—one as the user of its services and the other as an employee—who have participated in a collaborative act of social action employing the emancipatory, participatory research design and the scholarly genre of an article. Just as a visual artist with little knowledge of web design might become a participant in the development of a website with a web art designer to get their work recognized but might end up becoming a web art designer themself, Kruse and Oswal participated in this project and participatorily designed this research study to realize the potential of their different expertise about disability, disability studies, and scholarly work (Alexander, 2010). Further on, Kruse had used her autoethnography as a form of narrative inquiry meant for reflection, analysis, and interpretation from a personal and local context to a wider institutional or socio-cultural frame and gained a voice to critique the academy (Berger & Quinney, 2005; Chang, 2008; Ellis et al., 2011).

Defining Participatory Action Research for the Technical Communication Classroom

Before turning to the discussion of our second example, we also want to define and differentiate participatory design research concepts from participatory action research—the latter being quite relevant for our TPC pedagogy to prepare students in the basics of accessible design through action research with disabled users (Foth & Axup, 2006). Here, we will highlight the most important aspects of participatory action research, which has been successfully employed in the global south for health, socioeconomic, and pedagogical purposes (Etmanski & Pant, 2007; Tanabe et al., 2018; Wallerstein et al., 2017).

Participatory approaches have also been employed by grassroots groups for community-based action research, particularly in the majority world, as a response to the university-based researchers who tend to look down upon underprivileged participants and small-scale non-governmental organizations (NGOs) of this nature (Brown & Tandon, 1983; Hall, 1982/2002; Kothari, 2001; Parpart, 1995). Disabled people in the global north share many of the characteristics and exclusions with these majority world groups of the global south, and participatory action research is an attractive option for them to advance their socioeconomic agenda because the disabled are among the poorest of the poor amidst an ocean of middle-class consumerism and the wasteful opulence of the rich. Instead of all the talk about social justice, TPC classes can employ participatory action research to work with and to learn from disabled participants. Often when we talk about social justice, we are talking about someone delivering social justice to someone else—in this context, a disabled person—thus creating the giver/recipient binary. Social justice approaches help their advocates accrue social capital for themselves, build careers, and practice professional and social power through their words in an arena of activity where they are, in fact, perpetuating structural inequities at the cost of further marginalizing the recipient. Disabled people rejected this position many decades ago and hence the slogans "Nothing about us without us" and, more recently, "Nothing without us" (Charlton, 1998; Crowther, 2007).

From the perspective of disabled users, participatory design practices have room for defining and redefining the fundamental concepts of designs; processes; products; the imagined and real contexts of use; and relationships among designers, researchers, and disabled participants, the last being of utmost relevance to bodies with a difference. We find participatory action research well-suited for usability and accessibility-centered pedagogy in the human-centered design and technical communication courses to immerse our students in work with a rarely explored customer category. As compared to other "do good" approaches like service learning and social justice, participant action research does not only engage disabled users in the technical communication activity, it lets them occupy a central space in all aspects of the inquiry whether it is aimed at theory building or is tackling a practical problem. Bob Dick and Davydd J. Greenwood (2015) stress that "for action researchers a key concept is a dual commitment to both participation and action. Action research is done with, rather than on, the participants" (p. 194). Participatory action research cracks the binary of theory and method due to its firm commitment to a cycle of research and reflection aimed at refining methods and building theory that could help participants solve their problems. Quoting Dick and Greenwood again, "the core of action research is the constant confrontation of reflection and action, theory and method, theory and practice aimed at producing understanding and effective action" (2015, p. 195). Participatory action research is particularly relevant for the disabled participants because they have been marginalized in the academy since time immemorial and academic research—whether it is Galton's scientific ideas of normalcy or the medical establishment's castigation of the disabled body—has played a key role in this marginalization (Fendler & Muzaffar, 2008; Oliver, 1996; Priestley, 1999).

An Example of Participatory Action Research in the Classroom

For implementing the inclusive pedagogy agenda in the TPC classroom, we advocate for participatory action research-oriented curriculum that engages undergraduate students in inclusive data gathering, data analysis, writing up of results, and presentation of results to a live audience of peers or clients. The projects in such a curriculum would directly involve disabled consumers' and employees' day-to-day user experiences with technology, information designs, websites, and, of course, print documents (Davis, 2000; McFarlane & Hansen, 2007). We describe this pedagogical approach through an example from our own classes. Both of us teach accessibility concepts in our TPC web design assignments, and we usually assign readings from published research and "how-to" articles by practitioners to familiarize students with the accessibility problems as well as to instruct them to design accessible pages. We share an instance of the participation action research that, in fact, happened on the initiative of a student and which went beyond traditional involvement of disabled users as cursory testers. While we cannot share direct excerpts from the work of this student group because a member of this group took this participatory action research initiative rather spontaneously, we provide a detailed description of how the pedagogy of such participatory action research can be orchestrated. We might also disclose that our course under discussion was covered by an Institutional Review Board (IRB) approval for an international teaching collaboration among three instructors—two of whom are the authors of this chapter—but our application did not specifically include interactions with disabled participants—a protected class under the U.S. federal government's guidelines for research with such "subjects" as well as those of our universities. Consequently, our research approval at this time allows somewhat limited use of student work in our publications. (For more details about the purpose and nature of this international, intercultural collaboration, see Koris et al., 2019; Oswal & Palmer, 2018; Palmer et al., 2020). We, nevertheless, chose this example of participatory action research pedagogy because it was successful in achieving the desired results, required limited planning on behalf of the group, and affected the whole class' overall understanding about disabled user experience, accessibility issues, and the value of participatory design research itself.

Disabled Users as Experts and Equals

In the aforementioned teaching project, our students work in groups on web design and web accessibility projects in a client-provider relationship using low-tech tools, such as email and the Moodle learning management system, for their collaboration. The purpose of this assignment unit is to help students learn:

- what accessibility barriers web users with diverse disabilities face;
- what disability laws exist to ensure accessibility and their inherent limitations;
- what WCAG 2.0 and Section 508 accessibility guidelines are and how they are often implemented;
- how to conduct a website accessibility test employing an automated checker or a screen reader and collect pertinent data;
- how to interpret the data from these test results, including the skills for reading the reports produced by the automated checker software;
- and, of course, how to package the results from the data analysis for a live presentation as well as a written report.

In Sushil's program's gateway course, "Technical Communication in the Workplace," which is generally populated by information technology, computer science, and technical communication majors, students evaluate the website drafts designed by Zsuzsanna's business communication students earlier in the same semester. Then, Zsuzsanna's student groups revise their websites' designs using the accessibility test reports composed by Sushil's groups. Sushil's students write these reports after having conducted machine tests on these website drafts employing automated tools like *WAVE* and *AChecker* along with a variety of color contrast checkers of the group's choice to evaluate how well the websites meet the WCAG standards, WCAG AA being the desired level of accessibility. These groups' testing procedures can also involve testing of the web pages with Microsoft's Narrator or Freedom Scientific's JAWS-for-Windows screen reader by student teams. The students informally interviewed their instructor—who is an experienced screen reader user—to learn how he employed assistive technology to interact with web pages and what personal preferences he had for various features of a web page.

Although students are interacting with an experienced, disabled web user and have the opportunity to see the context of use from a technologically literate instructor's perspective, these interactions are happening within the unseen boundary of instructor-student relationship in a classroom setting. In a recent iteration of this course collaboration, however, a member of one of the student groups decided to observe and interview a fellow employee with cerebral palsy who not only used a screen reader but also used it differently. The employee under discussion had some residual sight but was dependent on the screen reader for reading and writing online. Their additional disabilities mixed with their residual sight—which gave them a good sense of the sort of ease and comfort sighted web users experience when online—made this user a highly vocal critic of the web design community.

Not only did this participant's comments give this group some powerful insights into screen reader use and web accessibility problems, but they also made students conscious of how this employee spoke about the poor quality of online designs with a sense of privilege and entitlement. The students' interest and trust in this participant's knowledge and the suggestions they received for improving the design of the website this group was testing markedly benefited this group's report. Later, this group leveraged this participant knowledge to support their recommendations when they video-conferenced with their web designer partner group to brief them on their report. During the class presentation of their group's website evaluation, the group included slides about this disabled tester's feedback. Amidst their presentation, the lead student interjected an aside: "I wish that we had Jim [a pseudonym] participating in this presentation to help us understand the web accessibility barriers he confronts on a daily basis and what accessibility features he will like to see in these web pages." We agree, and, as many European researchers affirm this sentiment, disabled participants should be involved in all stages of research (Iversen & Leong, 2012; Van der Velden & Mörtberg, 2015). Whereas this student's impromptu remark during the presentation suggests that he has come to realize the purpose and meaning of self-representation, his earlier conversation with Sushil about the spontaneous steps he had taken to observe, interview, and record this participant's testing-oriented, action research on web pages expressed a sense of awe in receiving feedback from a typical disabled user. During this conversation, the student also compared the results of his own test on these web pages with Microsoft's Narrator to those of Jim and explained how he had made so many assumptions about disabled users which were not accurate at all. Not only was this experience of participatory action research transformative for this student, his discussion also affected other students' attitudes toward the learning about disability and accessibility in this course. The participatory experience also served as an additional motivation for this student to take this project further, and he later converted the accessibility guidelines he wrote for this class assignment into a short article for *Intercom* (Marquardt, 2019).

On the instructor end, we're trying to incorporate such a participatory action research component in this web testing unit and are running into the usual hurdles—institutional rules about not disclosing the identity of disabled students—some of which are essential to protect student privacy and disability stigmatization. Students on their own are, however, free to find contacts for such testing on and off campus. For example, in the past, in another course of ours, students found participants for such action research through their connections with the student government. Sushil is also looking for IRB-approved research models which would permit ongoing participatory action research involving student groups. If we move to such a model, it would expand the scope of this project and further complicate the pedagogy of this testing unit, but our students would also have the opportunity to learn about the intricacies of IRB-approved research with participants from protected classes. The greatest advantage of incorporating participatory action research in this unit is that the accessible design pedagogy will give a front seat to the disabled users by including their voices directly. The students will learn how to include disabled participants in their action research as well as to value their viewpoint as experts in their area of disability and, sometimes, as assistive technology experts in their own right as long-term users. As we know from the long history of participatory design research, the participants don't only provide us with insights into user preferences but also open up a window into their economic, cultural, and aesthetic values (Iversen et al., 2010; Iversen & Leong, 2012; Schuler & Namioka, 1993; Voß et al., 2007). Participatory action research pedagogy can get along with the research models advocated by disability studies scholars under the rubric of emancipatory paradigm, and disability values-led participatory designs can serve all users a great deal better (Barnes, 2009 Morris, 1992; Oliver, 1997). If we do our due diligence to recruit participants with assistive technology experience and learn to mediate participatory design activity with humility, we might also see the emergence of novel designs that get out of the old rut of "features and more features" without delivering additional affordances for the capabilities of diverse humanity. By overcoming our ivory tower arrogance and ableism toward disability, not only can these participants become our co-creators and co-problem-solvers in conceptualizing more complete designs but we also might fill the gawking gaps in our academic training about disability. We need not remind our readers that the best of our human-centered designs at this time serve less than 80 percent of humanity, because these designs do not meet the needs of at least 20 percent of the human population with an array of disabilities. These users pay for these designs like all other consumers, invest their time in learning the use of these designs, and yet cannot achieve even their basic functional purposes due to the built-in design flaws in our technologies, web pages, and information. We iterate that these flaws exist because of our ignorance about disabled bodies, the value systems these users embody, and how they employ our design products to their purposes.

■ Conclusion

At this point, we stop to ask ourselves this rhetorical question: "Should participatory approaches be an essential aspect of the design cycle if we desire to develop accessible and usable interfaces, interactions, and products that provide disabled users with the same type of user experience that nondisabled users have come to expect?" and we answer it with a resounding "Yes." Our chapter interrogates the approaches and attitudes that posit unlimited authority in technological determinism and expert knowledge to solve disability and accessibility problems. The outcome of this interrogation is that neither our TPC pedagogy nor our research practices appear accessible from the perspective of disabled users. Instead, we advance the participatory design and participatory action research approaches, which engage disabled users right up front, for gaining primary insights into what disabled bodies desire and how they function with people, technologies, and communication. The implication of such methodological change for the TPC field will require a paradigm shift about how we perceive disability, interact with disabled users, and conceive the concept of accessibility itself. It would mean that our field will have to make room for disabled bodies in our classrooms, research projects, and field practice because disabled people are almost invisible in these spaces. It would also mean that we will have to seek out opportunities to actively learn from the user experiences and user expertise about accessibility that these bodies will bring with themselves to our discipline (Oswal, 2019). By ceding some of our expert power to these participants, our field would become more inclusive and more complete. Considering the limited knowledge most designers, developers, and researchers possess about disability, assistive technology, and, above all, disabled people, without conducting participatory design work with active involvement from disabled users, experts, and potential users, we can't even pretend to have done a reasonable needs assessment for determining, at least, threshold-level design characteristics for accessibility.

We further advocate that TPC professionals adopt a design regime driven by a participatory and reiterative user testing cycle in a variety of user contexts and environments over the life cycle of processes and products so that the initial design features do not get lost at later stages. Additionally, we argue that we need a new framework for assessing information and communication designs which goes beyond following the WCAG 2.0 checklist and would benchmark accessibility progress relative to the autonomy and ease bestowed upon disabled users in achieving their professional and personal ends (Leuthold et al., 2008). We also ask designers, developers, and technical communicators to question the introduction of inaccessible, trendy technologies that, in fact, serve only a small percentage of even nondisabled users. We end this chapter by repeating the affirmative note: Yes, active participation by disabled users in conceptualizing, implementing, and testing designs can serve as a lynchpin to make accessible products and processes a reality at the end of the production cycle. Scholarly work relating to such design projects can also contribute novel and constructive knowledge to our field. The adoption of disabled-centered participatory action research for our pedagogy will not only prepare our students for a more just and equity-oriented practice but also lessen ableist attitudes in the academy and in their future workplaces.

■ Acknowledgments

We thank Allison Kruse for reviewing this chapter and providing us with her feedback on the Kruse and Oswal section. We also take this opportunity to thank Joanna Schreiber and Lisa Meloncon for their editorial comments on the earlier drafts of this chapter. Sushil thanks his youngest, Hitender, for offering material and moral support throughout the development of this project.

■ References

Alexander, Amanda. (2010). *Collaboratively developing a web site with artists in Cajamarca, Peru: A participatory action research study* [Unpublished doctoral dissertation]. The Ohio State University.

Barnes, Colin. (2009). An ethical agenda in disability research: Rhetoric or reality? In Donna M. Mertens & Pauline E. Ginsberg (Eds.) *The handbook of social research ethics* (pp. 458–473). Sage.

Berger, Ronald J. & Quinney, Richard. (Eds.). (2005). *Storytelling sociology: Narrative as social inquiry*. Lynn Reinner.

Boyle, Caseu & Rivers, Nathaniel A. (2016). A version of access. *Technical Communication Quarterly*, 25(1), 29–47. https://doi.org/https://doi.org/10.1080/10572252.2016.1113702.

Brown, L. David & Tandon, Rajesh. (1983). Ideology and political economy in inquiry: Action research and participatory research. *The Journal of Applied Behavioural Science*, 19, 277–294.

Carter, Jeff & Markel, Mike. (2001). Web accessibility for people with disabilities: An introduction for web developers. *IEEE Transactions on Professional Communication*, 44(4), 225–233.

Chang, Heewon. (2008). *Autoethnography as method*. Left Coast Press.

Charlton, James I. (1998). *Nothing about us without us: Disability oppression and empowerment*. University of California Press.

Christiansen, Ellen. (2014). From "ethics of the eye" to "ethics of the hand" by collaborative prototyping. *Journal of Information, Communication and Ethics in Society*, 12(1), 3–9.

Cowan, Ruth S. (1972). Francis Galton's statistical ideas: The influence of eugenics. *Isis*, 63(4), 509–528.

Crowther, Neil. (2007). Nothing without us or nothing about us? *Disability & Society*, 22(7), 791–794.

Davis, John M. (2000). Disability studies as ethnographic research and text: Research strategies and roles for promoting social change? *Disability & Society*, 15(2), 191–206.

Devlin, Bernie, Fienberg, Stephen E., Resnick, Daniel P. & Roeder, Kathryn. (1995). Galton redux: Eugenics, intelligence, race, and society. *Journal of the American Statistical Association*, 9, 1483–1488.

Dick, Bob & Greenwood, Davydd. (2015). Theory and method: Why action research does not separate them. *Action Research*, 13(2), 194–197.

Dolmage, Jay. (2009). Disability, usability, universal design. In S. Miller-Cochran & R. Rodrigo (Eds.), *Rhetorically rethinking usability: Theories, practices, and methodologies* (pp. 167–190). Hampton Press.

Ehn, Pelle. (1989). *Work-oriented design of computer artifacts* (2nd ed.). Lawrence Erlbaum Associates.

Ehn, Pelle. (2017). Scandinavian design: On participation and skill. In D. Schuler & A. Namioka (Eds.), *Participatory design: Principles and practices* (pp. 41–77). CRC Press.

Ellis, C., Adams, T. & Bochner, A. (2011). Autoethnography: An overview. *Historical Social Research*, 36, 273–290.

Etmanski, Catherine & Pant, Mandakini. (2007). Teaching participatory research through reflexivity and relationship: Reflections on an international collaborative curriculum project between the Society for Participatory Research in Asia (PRIA) and the University of Victoria (UVic). *Action Research*, 5(3), 275–292.

Fendler, Lynn & Muzaffar, Irfan. (2008). The history of the bell curve: Sorting and the idea of normal. *Educational Theory*, 58(1), 63–82.

Foth, Marcus & Axup, Jeff. (2006). Participatory design and action research: Identical twins or synergetic pair?. In Gianna Jacucci, Finn Kensing, L. Wagner & Jeanette. Blomberg (Eds.), *Proceedings of Participatory Design Conference 2006: Expanding Boundaries In Design* (pp. 93–96). ACM.

Frauenberger, Christopher, Good, Judith, Fitzpatrick, Geraldine & Iversen, Ole S. (2015). In pursuit of rigour and accountability in participatory design. *International Journal of Human-Computer Studies*, 74, 93–106.

Garland-Thomson, Rosemary. (1997). *Extraordinary bodies: Figuring physical disability in American culture and literature*. Columbia University Press.

Hall, Budd. (2002). Breaking the monopoly of knowledge: Research methods, participation and development. In Rajesh Tandon (Ed.), *Participatory research: Revisiting the roots* (pp. 9–21). Mosaic Books. (Original work published 1982)

Hendren, Sara. (2014, October 16). All technology is assistive: Six design rules on disability. *Wired*. https://www.wired.com/2014/10/all-technology-is-assistive.

Iversen, Ole S., Halskov, Kim & Leong, Tuck W. (2010). Rekindling values in participatory design. In *Proceedings of the 11th Biennial Participatory Design Conference* (pp. 91–100). ACM.

Iversen, Ole S. & Leong, T.uckW. (2012). Values-led participatory design: Mediating the emergence of values. In *Proceedings of the 7th Nordic Conference on Human-Computer Interaction: Making Sense through Design* (pp. 468–477). ACM.

Kelly, Janet. (2019). Towards ethical principles for participatory design practice. *CoDesign* 15(4), 1–16. https://doi.org/10.1080/15710882.2018.1502324.

Kensing, Finn & Greenbaum, Joan. (2012). Heritage: Having a say. In *Routledge international handbook of participatory design* (pp. 21–36). Routledge.

Kesby, Mike, Kindon, Sarah & Pain, Rachel. (2007). Participation as a form of power: Retheorizing empowerment and spatializing participatory action research. In Sarah Kindon, Rachel Pain & Mike Kesby (Eds.), *Participatory action research approaches and methods* (pp. 19–25) Routledge.

Kleege, Georgina. (2011). "What keeps me in the ghetto?" *Parallel Lines*, pilot edition. https://web.archive.org/web/20180104183519/http://www.parallellinesjournal.com/article-what-keeps.html.

Knight, Melinda & Oswal, Sushil K. (2018). Disability and Accessibility in the Workplace: Some Exemplars and a Research Agenda for Business and Professional Communication. *Business and Professional Communication Quarterly*, 81(4), 395–398.

Konrad, Annika. (2018). Reimagining work: Normative commonplaces and their effects on accessibility in workplaces. *Business and Professional Communication Quarterly*, 81(1), 123–141.

Koris, Rita, Oswal, Sushil K. & Palmer, Zsuzsanna B. (2019). Internationalizing the communication classroom via technology and curricular strategy: Pedagogical takeaways from a three-way online collaboration project. In Paige K. Turner, Soumia Bardhan, Tracey Quigley Holden & Eddah M. Mutua (Eds.), *Internationalizing the communication curriculum in an age of globalization: Why, what, and how* (pp. 235–243). Routledge; Taylor & Francis.

Kothari, Uma. (2001). Power, knowledge and social control in participatory development. In Bill. Cooke & Uma Kothari (Eds.), *Participation: The new tyranny?* (pp. 139–151). Zed Books.

Kruse, Allison K. & Oswal, Sushil K. (2018). Barriers to higher education for students with bipolar disorder: A critical social model perspective. *Social Inclusion, 6*(4), 194–206. https://doi.org//10.17645/si.v6i4.1682.

Lather, Patti. (1987). Research as praxis. *Harvard Educational Review, 56*, 257–273.

Leuthold, Stefan, Bargas-Avila, Javier A. & Opwis, Klaus. (2008). Beyond web content accessibility guidelines: Design of enhanced text user interfaces for blind internet users. *International Journal of Human-Computer Studies, 66*(4), 257–270.

Longmore, Paul K. (2015). *Telethons: Spectacle, disability, and the business of charity*. Oxford University Press.

Marquardt, S. (2019, September). Recommendations for accessibility & usability. *Intercom 66*(5) p. 30–31.

Mattlin, B. (1991, September 1). An open letter to Jerry Lewis: The disabled need dignity, not pity. *Los Angeles Times*.

McFarlane, Hazel & Hansen, Nancy E. (2007). Inclusive methodologies: Including disabled people in participatory action research in Scotland and Canada. In Sarah Kindon, Rachel Pain & Mike Kesby (Eds.), *Participatory action research approaches and methods* (pp. 88–94). Routledge.

Melonçon, Lisa. (Ed.). (2014). *Rhetorical accessability: At the intersection of technical communication and disability studies*. Routledge.

Melonçon, Lisa. (2018). Orienting access in our business and professional communication classrooms. *Business and Professional Communication Quarterly, 81*(1), 34–51.

Moeller, Marie. (2014). Pushing boundaries of normalcy: Employing critical disability studies in analyzing medical advocacy websites. *Communication Design Quarterly, 2*(4), 52–80.

Morris, Janine. (1992). Personal and political: A feminist perspective on researching physical disability. *Disability, Handicap, and Society, 7*(2), 157–166.

O'Day, Bonnie & Killeen, Mary. (2002). Research on the lives of persons with disabilities: The emerging importance of qualitative research methodologies. *Journal of Disability Policy Studies, 13*(1), 9–15.

Oliver, Michael. (1992). Changing the social relations of research production. *Disability, Handicap, and Society, 7*(2), 101–114.

Oliver, Michael. (1996). *Understanding disability: From theory to practice*. St. Martin's Press.

Oliver, Michael. (1997). Emancipatory research: Realistic goal or impossible dream? In Colin Barnes & Geof Mercer (Eds.), *Doing disability research* (pp. 15–31). The Disability Press.

Oswal, Sushil K. (2014). Participatory design: Barriers and possibilities. *Communication Design Quarterly Review, 2*(3), 14–19.

Oswal, Sushil K. (2018). Can Workplaces, Classrooms, and Pedagogies Be Disabling? *Business and Professional Communication Quarterly, 81*(1). 3–19.

Oswal, Sushil K. (2019). Breaking the exclusionary boundary between user experience and access: Steps toward making UX inclusive of users with disabilities. In *Proceedings of the 37th ACM International Conference on the Design of Communication* (pp. 1–8). ACM. https://doi.org/10.1145/3328020.3353957.

Oswal, Sushil K. & Melonçon, Lisa. (2014). Paying attention to accessibility when designing online courses in technical and professional communication. *Journal of Business and Technical Communication*, *28*(3), 271–300.

Oswal, Sushil K. & Palmer, Zsuzsanna B. (2018). Can diversity be intersectional? Inclusive business planning and accessible web design internationally on two continents and three campuses. In L. A. Whittle (Ed.), *2018 Proceedings of the Association for Business Communication conference* (p. 1–23). Association for Business Communication. https://www.businesscommunication.org/d/do/1650.

Palmer, Zsuzsanna B., Oswal, Sushil K. & Koris, Rita. (2020, October 23). Reimagining business planning, accessibility, and web design instruction: A stacked interdisciplinary collaboration across national boundaries. *Journal of Technical Writing and Communication*. https://doi.org/10.1177/0047281620966990.

Palmeri, Jason. (2006). Disability studies, cultural analysis, and the critical practice of technical communication pedagogy. *Technical Communication Quarterly*, *15*(1), 49–65. https://doi.org//10.1207/s15427625tcq1501_5.

Parpart, Jane L. (1995). Deconstructing the development "expert": Gender, development and the "vulnerable groups." In Marianne H. Marchand & Jane L. Parpart (Eds.), *Feminism, postmodernism, development* (pp. 221–243). Routledge.

Priestley, Mark. (1999). *Disability politics and community care*. Jessica Kingsley.

Priestley, Mark, Waddington, Lisa & Bessozi, Carlotta. (2010). Towards an agenda for disability research in Europe: Learning from disabled people's organisations. *Disability & Society*, *25*(6), 731–746. http://doi.org/10.1080/09687599.2010.505749.

Salvo, Michael J. (2001). Ethics of engagement: User-centered design and rhetorical methodology. *Technical Communication Quarterly*, *10*(3), 273–290.

Schuler, Douglas & Namioka, Aki. (Eds.). (1993). *Participatory design: Principles and practices*. CRC Press.

Simonsen, Jesper & Robertson, Toni. (Eds.). (2012). *Routledge international handbook of participatory design*. Routledge.

Slatin, John M. & Rush, Sharron. (2002). *Maximum accessibility: Making your web site more usable for everyone*. Addison-Wesley Longman Publishing.

Spinuzzi, Clay. (2007). Accessibility scans and institutional activity: An activity theory analysis. *College English*, *70*(2), 189–201.

Spivak, Gayatri C. (1988). Can the subaltern speak? In Cary Nelson & Lawrence Grossberg (Eds.), *Marxism and the interpretation of culture* (pp. 271–313). University of Illinois Press.

Tanabe, Mihoko, Pearce, Emma & Krause, Sandra K. (2018). "Nothing about us, without us": Conducting participatory action research among and with persons with disabilities in humanitarian settings. *Action Research*, *16*(3), 280–298.

Theofanos, Mary F. & Redish, Janice G. (2003). Bridging the gap: Between accessibility and usability. *Interactions*, *10*(6), 36–51.

Theofanos, Mary F. & Redish, Janice G. (2005). Helping low-vision and other users with web sites that meet their needs: Is one site for all feasible? *Technical Communication*, *52*(1), 9–20.

Van der Velden, Marcel & Mörtberg, Christina. (2015). Participatory design and design for values. In Marcel J. Van Den Hoven, Pieter E. Vermaas & Ibo van de Poel (Eds.) *Handbook of ethics, values, and technological design: Sources, theory, values and application domains* (pp. 41–66). Springer.

Voss, Alex, Hartswood, Mark, Procter, Rob, Rouncefield, Mark, Slack, Roger & Büscher, Monika. (Eds.). (2007). *Configuring user-designer relations: Interdisciplinary perspectives*. Springer Verlag.

Wallerstein, Nina, Giatti, Leando, Bógus, Cláudia, Akerman, Marco, Jacobi, Pedro, de Toledo, Renata, Mendes, Rosilda, Acioli, Sonia., Bluehorse-Anderson, Margaret, Frazier, Shelley & Jones, Marita. (2017). Shared participatory research principles and methodologies: Perspectives from the USA and Brazil—45 years after Paulo Freire's "Pedagogy of the Oppressed." *Societies*, *7*(2), NP. https://doi.org//10.3390/soc7020006.

Walters, Shannon. (2010). Toward an accessible pedagogy: Dis/ability, multimodality, and universal design in the technical communication classroom. *Technical Communication Quarterly*, *19*(4), 427–454.

Wilferth, Joe & Hart, Charles. (2005). Designing in the dark: Toward informed technical design for the visually impaired. *Computers and Composition Online*. http://cconlinejournal.org/wilferthhart/wilferthhart.htm.

Wilson, James C. (2000). Making disability visible: How disability studies might transform the medical and science writing classroom. *Technical Communication Quarterly*, *9*(2), 149–161.

Chapter 10: Localize, Adapt, Reflect: A Review of Recent Research in Transnational and Intercultural TPC

Nancy Small
University of Wyoming

Abstract: Technical and professional communication (TPC) has been a border-crossing field since its inception, and as globalization continues to create new avenues for research and practice, now is an opportune time to review what kinds of intercultural and transnational projects are being pursued as well as to consider how to be ethical agents in these projects. After relating the fraught process of defining "transnational" and "intercultural," this chapter describes a meta-analysis of articles published in major TPC journals during a five-year window (2014–2018). The analysis categorizes different types of projects and seeks out advice emerging from scholars' experiences. The study reveals a wide range of transnational research settings which resist being easily delimited and determines that space in journal articles to reflect on cross-cultural complexities is scarce. Limited reflections from scholars in cross-cultural projects indicate that working in intercultural and transnational spaces requires persistent localization, ongoing adaptation, and a reflective, reflexive mindset. Taken together, these lessons point to ongoing (re)positionality at the center of successful intercultural work. Based on the results of this review, the author recommends the field develop a formal statement of ethics for transnational and intercultural research. That ethic should be human-centered and mindful of social justice principles.

Keywords: research, transnational, intercultural, positionality, ethics

By its very motivation and nature, technical and professional communication (TPC) has always been a border-crossing field and practice because it sits at the intersection of technical content and application of communication principles. The teaching of technical writing emerged from hybrid spaces in engineering programs of the early twentieth century (Connors, 1982), and the ongoing "role of the technical communication practitioner stems from the need for members from two distinct professions to connect" (Amare, 2002, p. 128). Beyond being a site where disciplines meet, TPC serves as a "high encompassing culture" bridging the sciences and humanities (Amare, 2002, p. 129). Technical communication professionals also cross divides in expertise and experience, between subject matter experts and varied audiences (Rice-Bailey, 2016). Spanning differences of language, perspective, and practice is at the heart of what we do, but what foundations have we developed in working across borders most effectively and ethically? And what innovations are going on in cross-cultural projects?

This chapter surveys the current state of a particular kind of TPC border crossing: transnational and intercultural. My purposes are to highlight the diverse sites and locations of TPC work and to critically examine our disciplinary discourses regarding the challenges of complex intercultural spaces. Despite TPC activities being situated in a wide array of locations, we have limited outlets through which to share our insights and lessons learned about the complexities of carefully and ethically navigating those situations. As TPC continues to evolve and grow, we grapple with defining and describing the notably far-reaching sites and goals of our discipline. I take up some of that grappling here through a survey of recent transnational projects, asking about the varieties of border-crossing they do as well as the lessons cultivated from research situated in complex spaces. Emerging through my study are ongoing struggles and limited successes in defining and describing the terms of "(inter)cultural" work despite TPC activities being located in a fascinating array of such situations.

Now into our second century as a discipline, forces of globalization continue to open new spaces, drive new questions and innovative practices, and provide new opportunities in learning to operate in culturally diverse situations. Scholarship has kept pace with these changes, particularly in the last two decades. For example, Barry Thatcher (2001) disrupted traditional notions of validity in intercultural research. J. Blake Scott and Bernadette Longo's (2006) special issue of *Technical Communication Quarterly* (*TCQ*) considered the complications of the field's "cultural turn" by "expanding methods for talking about the influences of sociocultural contexts [and by] foregrounding new critical perspectives on intercultural communication" (p. 4). Another *TCQ* special issue, edited by Huiling Ding and Gerald Savage (2013), pushed back against the traditional "nation-centric mindset" via a collection of articles on transnational communication processes and products. In that issue, Steven Fraiberg (2013) called for a "less bounded" and "less static" approach to methods and practices in global contexts. As his study demonstrated, more flexibility is needed to contextualize and untangle meaning when multiple culturally embedded symbol systems are at play (pp. 23–24). Guiseppe Getto's (2015) introduction to a special issue in *Rhetoric, Professional Communication, and Globalization* further applied the tricky concept of "culture" by tracing it along multiple axes. He framed technical communicators as "capacity builders" whose daily tasks make them "purveyors of a large variety of professional cultures" (p. 1). Barry Thatcher and Kirk St.Amant's (2011) edited collection spoke to the growth of TPC taught across borders and advised faculty on course and program level development. The first-hand storytelling by TPC practitioners in Han Yu and Gerald Savage's (2013) *Negotiating Cultural Encounters* invited readers to witness real-life complexities that intercultural tensions create in the workplace. Angela Haas and Michelle Eble's (2018) *Key Theoretical Frameworks: Teaching Technical Communication in the Twenty-First Century* interlocked social justice with the very nature of technical communication being intercultural and potentially global (pp. 10–11). Issues of quality and methods when working across borders are foundational and perpetual to TPC's disciplinary identity.

Despite being part of TPC's foundation, present, and future, transnational and intercultural projects continue to be precarious affairs. For example, St.Amant (2017) described how development in Indian and Chinese markets consequently drives demand for online TPC courses to be delivered to overseas audiences. He advised course developers to consider infrastructure "friction points"—specific hardware, software, and bandwidth factors likely to impact course functionality for international users. In other words, St.Amant pointed out how, without careful consideration, an online course designed for global reach might not function within the real-world situation of the varied end users. Such design-user mismatches are not limited to educational endeavors. I have witnessed the potentially fraught nature of transnational and intercultural projects myself. For six years, I was on the faculty at an international branch campus of a USAmerican university in Qatar. During that time, I saw first-hand clumsy and privileged, yet well-intentioned, visiting researchers—most often USAmerican, western European, and White—desiring to use Middle Eastern spaces as locations of *outsider* knowledge making. In other words, these misguided attempts amounted to intellectual colonization and perpetuation of western domination. Such troubles extended to teaching, as my colleague and I have described (Rudd, 2018; Small, 2017).

As TPC continues to expand and evolve, particularly along with global developments, we must take stock of how we are designing and discussing our projects. Through discipline-wide reflection and conversation, we can better understand the state of our activities and identify principles of better practice that will help us avoid the "good intentions" trap (Gorski, 2008). We must continue to cultivate our discipline's cross-cultural ethics in support of designing and facilitating more socially just projects. In an effort to explore the recent range of and approaches to transnational and intercultural work in TPC and to consider our commitment to building better practices, I designed a literature review motivated by the following questions:

1. In what ways do TPC scholars work within or across transnational and intercultural spaces?
2. What lessons are TPC scholars sharing about their experiences in these spaces?
3. How can these individual lessons be gathered and organized in order to inform others about better practices in their own transnational and intercultural projects?

My chapter proceeds by defining key terms related to my inquiry, explaining my review method, and presenting results organized in response to my motivating questions. The primary outcome of my study is that transnational and intercultural research involves complex and multi-layered positionalities, and I conclude with that discussion as well as point towards future research.

■ Definitions

To begin answering my research questions, I first had to consider what the terms "transnational" and "intercultural" mean. In general, transnational work moves across borders that are geopolitical, cultural (including national, ethnic, dis/ability, gender, and socioeconomic), linguistic, disciplinary, organizational, temporal, modal, or a combination of these and others. Borders can be real, requiring documentation or cultural ambassadors to facilitate passage, or they can be imagined, assigned, or performed, such as those identity borders that diversify our individual and local community experiences. Movement across borders can be singular or multiple (for example, see Rose & Racadio, 2017, p. 8). The term "transnational" is contested because it centers on "nations" as a category of identity and because of its association with economic imbalance. Transnationalism is often explored in terms of "elites" and "migrant laborers" and, therefore, is associated with racism and socioeconomic disadvantage (Croucher, 2012, p. 18). Although transnational work is associated with economic forces of globalization, it can be understood in much broader ways. It links to thinking *beyond* national histories and singular perspectives on economic, social, and cultural flows. Transnationalism invokes movement, while the terms "multinationalism" or "multiculturalism" typically indicate diversity within the same site. Transnational work often is intercultural, but intercultural work is not necessarily transnational.

While establishing a working definition of "transnational" was relatively straight-forward, defining "intercultural" was a different story because pinpointing the meaning of "intercultural" involves the struggle of determining what counts as "culture." Although over 150 different definitions of culture exist, efforts at establishing a unified, shared definition fail. Instead, any singular definition risks monolithically essentializing, erroneously stabilizing, and failing to address the roles of ideology and power (Baldwin et al., 2006). In TPC, St.Amant (2013) broadly defined culture as "an organizational system, or a *worldview*" prescribing acceptable behaviors and therefore creating "the *rhetoric*—or the communication practices and style—its members use when interacting" (pp. 481–482). Culture can also be understood in processual terms of flows and border-working, such as moving across, transcending, and disrupting socially and politically constructed divisions (Ding & Savage, 2013, pp. 2–3). As Natasha Jones (2014) reflects, "Culture can be dynamic and fluid, even hard to define and identify" as well as "found in the most unexpected places" (p. 15).

Attempts to hem in "culture" as an element of communication are, by nature, partial and open to critique; therefore, TPC scholars often define culture indirectly through contextual factors. For example, Longo (1998) suggested that technical communicators and researchers conceive of cultural studies as being situated within histories "constructed at a certain place and time" (p. 64) and often focused on the functions and silences of everyday objects and practices (p. 67). Getto and St.Amant (2014) framed culture in terms of expectations, design, and user experience. Proposing their process of persona development, they suggested a culturally complex persona should include demographic elements, attitudinal and behavioral indicators, and contextual data. Therefore, reversing their framing, a persona can represent a culture via aspects of identity, activity, and context (location, as well as history, social, political, and economic situations). Getto (2015) addressed the problem of complexity by examining culture from multiple perspectives: local, meso, and global (pp. 4–5). He demonstrated that any cultural perspective is automatically prismatic, as well as dependent upon the position from which you view it. Rather than define culture through a series of binary terms, Getto operationalized it along axial intersections: local-technological, local-cultural, global-technological, and global-cultural. Through applying his framework to a specific communication situation, we can understand "culture" as reified through tensions regarding local preferences and expressions of collective identity and in terms of broader contexts and networks that influence norms and practices. For Getto, "culture" is an integrated system of influences that co-create a particular site of inquiry. As the axes shift, the situation changes; some tensions are amplified while others are quieted.

In the wakes of TPC's epistemological turns towards social construction and social justice, a focus on communication between or among cultures invites more nuanced, critical, and complex study of the sociopolitical influences of the field (Scott & Longo, 2006). Haas and Eble (2018) asserted that "all technical communication contexts are multi- and inter-cultural" (p. 8), offering as an example ubiquitously globalized flows and distributions of communication (e.g., a "local" company may have international/multi-national stakeholders). They also established that intercultural communication is not limited to crossing geopolitical borders (Haas & Eble, 2018, p. 10). However, approaching a definition of culture through the lens of communicative competence is no less challenging and only reinforces the "field's reluctance to specify what intercultural competence means" (Yu, 2012, p. 170). Although Yu (2012) landed on a working definition of intercultural competence as "the ability to communicate appropriately and effectively in international and cross-cultural technical communication situations based on one's sensitivity, awareness, and skills" (p. 171), the nature of what constitutes a "culture" in her work as well as in Haas and Eble's introduction remains unspecified.

Scholars in decolonizing and critical cultural studies have emphasized the dangers of objectifying and isolating "culture" as an "object of study" (Powell et al., 2014). Powell et al. (2014) argued for an understanding of culture as *"relational and constellated,"* based on "encounters people have with one another within and across particular systems of shared belief" (p. 5). The "constellation" perspective emphasizes multiple practices of meaning-making and "allows for multiply-situated subjects to connect to multiple discourses at the same time" and for relationships among actors and discourses "to shift and change without holding a subject captive" (Powell et al., p. 5). Shawn Wilson's (2008) paradigm for indigenous research addressed the dangers of treating cultures (and encultured people) as objects by promoting relationality, reciprocity, and respect in intercultural interactions.

Also resisting conceptions of culture that encase and petrify, Ding and Savage (2013) asked TPC scholars to adapt an "alternative conceptualization of cultures and the 'intercultural' that moves beyond the nation-centric mindset to investigate alternative approaches to straightforward applications of cultural heuristics and cultural dimensions" (p. 1). Although Ding and Savage did not specifically cite Geert Hofstede's (2001) ubiquitous *Culture's Consequences* as emblematic of a limited nation-centric heuristic point of view, other scholars have critiqued the reductive motivation to simplify and predict human identities and behaviors (Agboka, 2014, p. 299) as well as the outdated nature of Hofstede's study and its use in transnational and intercultural research (e.g., McSweeney, 2002; Osland et al., 2000;).

Efforts to define "culture" as the central feature of "intercultural communication" have not brought me to a satisfying solution. Therefore, I offer the following definition solely for the purposes of moving forward on inquiring about the range of ways TPC scholars work in transnational and intercultural spaces: *Culture* is a situated, shared, and constantly shifting set of values, norms, symbols, and processes that motivate (re)creation of group or collective identity. Implicit in culture are real and imagined borders and borderlands inherent in the construction of "insiders," "outsiders," "in-betweeners," "crossers," and "returners." Although my literature survey narrows to focus on articles in a subset of border-crossing situations, intercultural communication can happen without any travel at all—with the people in our shared work and living spaces (see, for example, N. Jones, 2014). All are "contact zones," or "social spaces where cultures meet, clash, and grapple with each other, often in contexts of highly asymmetrical relations of power" (Pratt, 1991, p. 34). Even for researchers and practitioners who remain in their home places, sharing outcomes of their transnational activities can help us develop awareness of and sensitivity to issues of intercultural interactions in our own organizations and projects.

■ Methods

My research curiosities regarding transnational and intercultural TPC activities, lessons learned, and better practices invited a broad survey across the field to consider the diversity of projects published via our scholarly outlets. Because journals are published with more frequency than book-length works, I chose to design my inquiry as a meta-analysis of articles published in seven TPC outlets: *Communication Design Quarterly* (*CDQ*); *IEEE Transactions on Professional Communication* (*IEEE*); *Journal of Business and Technical Communication* (*JBTC*); *Journal of Technical Writing and Communication* (*JTWC*); *Rhetoric, Professional Communication and Globalization* (*RPCG*); *Technical Communication* (*TC*); and *Technical Com-*

munication Quarterly (*TCQ*). The initial corpus consisted of 609 original articles in 126 issues published between 2014 and 2018. The five-year window encompassed the most recent research at the time, and although five years ultimately is an arbitrary cut-off point, it yielded a sufficiently large starting corpus. Most of these journals published four issues per year, with the exceptions being *RPCG*, which published one per year, and *CDQ*, which had an extra issue in 2016. I looked at original research articles but not book reviews or editorial commentaries. *IEEE* has a category for "teaching cases," and I included those because other journals publish similar materials as original articles. My inquiry did not include non-academic or book-length sources. Table 10.1 lists the volumes and numbers of articles per year.

Table 10.1. Corpus for journal analysis, publication years 2014–2018

Journal	Volume #s	# of Articles
CDQ	2–6	108
IEEE	57–61	111
JBTC	28–32	75
JTWC	44–48	100
RPCG	6–10	25
TC	61–65	96
TCQ	23–27	94
Totals	**126**	**609**

A first pass through the corpus involved reading abstracts and, if necessary, skimming the article for a better understanding of its focus, looking for projects that directly or indirectly engaged cultural differences and/or moved across borders. My analytical process started with pre-existing expected categories based on a general intercultural communication understanding of identities used to explore bordered groups (e.g., national, regional, ethnic/racial, linguistic, age, etc.). However, I also took a grounded-theory-inspired approach of being open to emergent themes. Through this process, I identified 143 articles, or 23 percent of the total corpus, as centered on at least one border-crossing factor. The first part of my results and discussion surveys these outcomes, which fell into the following categories:

- Disciplinary
- Academic/Practitioner/Public
- Temporal
- Digital or Technological
- Economic
- Generational
- Dis/Ability

- Linguistic (Translation)
- Cultural/Theoretical
- National

Although many articles crossed a combination of categories, my discussion provides examples according to which category emerged as primary in the study. In a second pass, I studied methods sections of transnational/intercultural studies for cues that the project included human participants as opposed to working only with texts, theories, or pre-existing data. Of the 143 articles in my first sample, 33 (23%) met this criterion. Because I was interested in TPC research activities beyond academy walls, I considered articles on local pedagogical practices and curriculum design outside the scope of this subset.

For the purposes of answering research questions two and three, I made a final pass focused specifically on articles where researchers reflected and shared "lessons learned" about transnational projects with human participants. Figure 10.1 summarizes my sorting process. The requirement for reflection further narrowed the sample of 33 down to just seven articles. In the following results and discussion section, I begin by describing the wide and varied ways TPC scholars work interculturally.

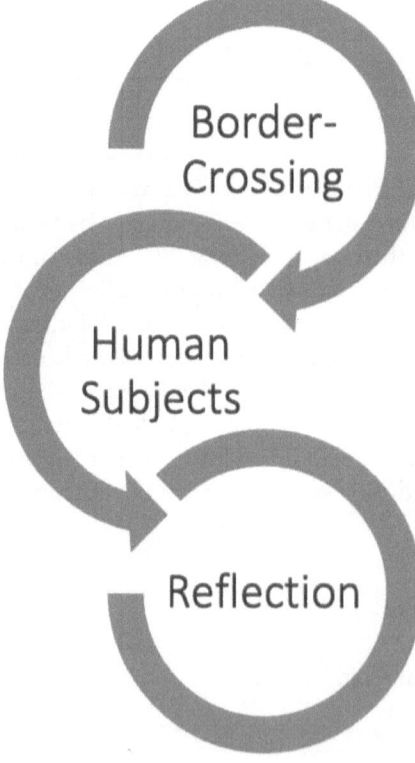

Figure 10.1. Sorting process.

Results and Discussion

Types of Transnational and Intercultural Research

The diversity of TPC's border-crossing activities was exciting and encouraging because the range of complex cultures within which our work is situated demonstrates the far reach of the field's curiosity and influence. Recent examples included Chad Wickman's (2015) connections between theoretical physics and TPC, and Susan Popham's (2014) argument for "multi-disciplinary identities" spanning social work and science. Another frequent topic—bridging the academic-practitioner and public divides—was intercultural communication awareness and skills. For example, Liberty Kohn's (2015) review of university-workplace partnerships and Russel Hirst's (2016) reflection on an academic partnership with the nuclear industry exemplified how professional identity—and in the case of Hirst, also disciplinary identity—continues to be a site of border crossing. Tatiana Batova's (2018a) scholarship was situated at the crossroads of academia and industry through a broad literature review combined with student feedback in service of developing a curriculum more effective at teaching USAmerican students about globalization and TPC. Bridging both disciplinary and academic-industry divides, Hirst's (2016) teaching case detailed how he set up a student intern project at a nuclear manufacturer and offered advice about how to make such partnerships run smoothly.

Including but also extending beyond disciplinary and public divides, TPC research moves across time and spaces. For example, Chelsea Milbourne (2016) argued that eighteenth-century science displays shaped audiences' reception of and expectations regarding both social and material worlds. John Ramey (2014) traversed time and regional culture by writing about an eighteenth-century Creole lawyer living in San Domingo and a technical manual for slave ownership. Other articles brought nineteenth-century Chinese business communication (Sinclair & Blachford, 2015) and Song Dynasty medical texts (Zhang, 2016) into contemporary conversation. Not surprising considering TPC's focus on communication technologies, extension from analog into digital and cyber contexts remains at the forefront in TPC publications. For example, recent articles by Jennifer Sano-Franchini (2017), Douglas Walls (2017), Josephine Walwema (2016), Wang and Gu (2015), and Jo Mackiewicz (2014) analyzed websites and social media through a variety of cultural perspectives, including Dutch, German, and Chinese lenses.

Recent publications also confirm TPC's concerns about communicating across socioeconomic classes, generations, and dis/ability statuses. Emma Rose's (2016) study of how people function in *"resource-constrained contexts"* (p. 433) considered how bus riders who were homeless or experiencing shelter precarity navigated public transportation in Seattle, Washington. Kim Campbell et al.'s (2017) inquiry into plain language in the US included a comparison of blue-collar, pink-collar, and white-collar style preferences across economic (as well as linguistic and ethnic/racial) borders. Two examples foregrounding questions of cross-generation TPC were Rhonda Stanton's (2017) challenge of age-based stereotypes in the workplace and N. Lamar Reinsch and Jonathan Gardner's (2014) study of communication abilities as a factor in promotion for baby boomers and Gen-Xers. Articles by Liz Hutter and Hutter Lawrence (2018) as well as Sushil Oswal (2014) offered insights from the deaf community and from disability studies to suggest how researchers and practitioners can improve usability testing and participatory design.

The next categories foreground two types of translation: linguistic conversion of meaning from one symbolic system to another and the translation required when using theory developed in one cultural context to explain activities in another cultural context. Researching across national, regional, and ethnic cultures inspires questions about developing better theories and practices of translation. For example, Laura Gonzales and Heather Turner (2017) described how translators in the US worked with a variety of technical communication tools, including information design and usability observations, as integral to their daily work. Additional studies in the US asked how health and medical systems can be designed to better meet the needs of immigrants with varying English language skills (Koerber & Graham, 2017; Rose et al., 2017). Articles coded for the "cultural/theoretical" category proved intercultural in their analytical moves, applying a theory from one broad culture (usually western/European/USAmerican) to a different cultural context. For example, Ding's (2018) article defined and contextualized "whistle-blowing" through the U.S. legal system yet applied the concept to a case study of "a 76-year-old veteran physician who retired from the China People's Liberation Army (PLA) General Hospital" (p. 38). Similarly, Jung-Yoon Yum and Se-Hoon Jeong (2015) applied western theories of crisis communication and attribution to an experiment using undergraduates in a South Korean university. What might be the implications of working across these kinds of borders, where cultural assumptions embedded in the theories shape interpretations of local practices?

My last category of transnational and intercultural research tacks in to studies that spoke to my own experiences crossing national borders. In his 2017 article, Fraiberg, listed as a professor at a U.S. university, returned to Israel to continue studying TPC, this time focusing on entrepreneurship and rhetoric. His project demonstrated the important role of translation in transnational work, as he provided his readers both with literal conversion of Hebrew conversations into English and with rich contextualization and interpretation of his data. Rebecca Walton et al. (2016) conducted a study in an international humanitarian organization, which began with 25 online video interviews with people from 19 countries and included an additional 95 interviews over "two-week research visits to six countries" (p. 89). Their conclusions broadly called for fieldwork as a means of understanding textual production—in other words, they argue that TPC research is better when it does not rely on textual analysis alone. In another study, set in Vietnam, Rebecca Walton and Sarah Beth Hopton (2018) interviewed local participants to better understand non-western perspectives on the use of Agent Orange. Their project argued that community-based research must attend to the local situation and not necessarily generalize from strategies used in western contexts.

In answer to my first research question—"In what ways do TPC scholars work within or across transnational and intercultural spaces?"—journal publications represented an expanding range of TPC border crossings. However, although compelling, these results were neither simple nor straightforward. As the next section details, surveying these types of activities reaffirmed the complex diversity of how intercultural research can unfold.

■ Questions Emerging from the Results

Even as my literature review began to yield answers, it generated new possibilities and conundrums concerning cross-cultural project design. Some studies published in US-based journals were located in other countries and were conducted and authored by researchers at universities in those same countries. For example, Jenni Virtaluoto and colleagues (2016) were all at a Finnish university and conducted their research at Finnish locations. Yvonne Cleary (2016), affiliated with an Irish university, conducted her inquiry into communities of practice in Ireland. Rodney Jones (2014) published on food labels in Hong Kong while working at a university there. Of course, the place where a researcher works does not dictate the person's cultural identity (although it may shape that identity indeed), but authorial information in TPC publications does not offer enough detail to know how familiar, adapted, or assimilated the researchers are to their contexts. Publishing international research in U.S. journals—which are also subscribed to by people around the world—means that all our work is transnational and intercultural, as scholarship circulates globally.

Not knowing a researcher/author's self-ascribed identify is not necessarily a problem, but when values, attitudes, and perceptions of a localized group are discussed, readers may wonder about how the writer's positioning as a local and cultural insider and/or outsider shapes their interpretations. For example, Nikita Basov and Vera Minina (2018), affiliated with a Russian university, used interview data from another study to analyze collaborations in Portugal in their article on professional networks (also a discipline-spanning project). The study is interesting and well written but leads to questions about how the researchers addressed a complexity of multiple cultural identities and positionalities in their data interpretation. Xiaobo Wang and Baotong Gu (2015) provided rich explanations of cultural values in their study of the social media platform WeChat in China but did not identify their own positioning or sources of expertise for their interpretive insights.

As another example, James Kiwanuka-Tondo and Keon Pettiway (2016), researchers from U.S. universities, wrote a SWOT analysis of how complex climate change data in Kenya is designed for a variety of audiences, including sector officers, media outlets, politicians and policy makers from the sub-region, and end users ("farmers, women, youths, traders, and fishermen"; p. 80). That analysis was an exemplar for localized research projects by unpacking the range of needs to which the study responded. However, little information was provided on how the researchers' backgrounds prepared them for their analysis. One of them, Kiwanuka-Tondo, spent about five weeks in Kenya gathering documents (p. 77), but no biographical information is provided about the author other than his affiliation with North Carolina State University, leaving readers unsure of his positionality as insider/outsider. Kiwanuka-Tondo and Pettiway's goal for their analysis was to propose more effective information design methods for authors working with "complex climate science information" (p. 78), a process which should involve deep knowledge about the audiences, contexts, and uses of the science information. Indeed, the authors do focus on audience (pp. 80–82); however, the "study did not investigate the production, consumption, or reception of messages or interfaces" of the documents they studied (Kiwanuka-Tondo & Pettiway, 2016, p. 82). If the authors did not study the reception of the climate data through primary research with actual audience members, then their analysis seems to be based on assumptions about the specific audiences, especially those that are local.

Other studies did not provide sufficient information to know how cultural differences might be taken into account. For example, a survey conducted with engineering students about their experience with a writing center did not include a demographic breakdown of the students (Weissbach & Pflueger, 2018). Although this study spanned across disciplinary borders—engineering and writing studies—information about the cultural and even educational backgrounds of the tutors and their student clients may have added further nuance to evaluating the effectiveness of feedback. An assumption seems to be that a shared U.S. university culture supersedes the impact of individual student backgrounds. A survey of technical editors (Kreth & Bowen, 2017) included responses from international participants but either did not gather information about or did not explain how those international/transnational spaces might have affected editorial processes or products. A study of cultural conflict in student teams (Wang, 2018) did not include information on the students' ethnic, national, or other cultural identities even though student backgrounds could have impacted conflict and negotiation styles.

Another question this review brought up is this: what counts as a "crossing"? For example, do live video interactions allowing researchers to stay in their own home spaces while communicating with people around the world count as transnational? Stefania Passera et al.'s (2017) project is a case in point. They conducted a mixed-methods survey and experiment engaging 122 business professionals in 24 countries; however, the researchers remained in their Finland location while using live video (a "webinar") to interact with their participants. At first glance, the project does seem to be transnational. After all, it connected people across borders. However, because the researchers stayed in their home country, they did not have access to the rich and varied contexts within which their participants worked. Having lived abroad, I experienced the significant difference between being an online visitor to another place, a tourist only on the ground for a few days or weeks, a long-term visitor to the country living and working in that host environment, and a resident with more permanent notions of belonging. In other words, technology offers the chance to connect with transnational (and intercultural) places, but remains limited in the depth and breadth of that connection.

Border crossings are not binary either, and transnational/intercultural meeting spaces become hybrid sites of work and research. Projects located in MNCs, or multinational corporations (e.g., Shin et al., 2015; Batova, 2018b; Yin et al., 2015), create even further complexity. In these studies, participants might come from a range of different national and/or regional cultural perspectives yet may have also adapted to their host nation where the MNC is located. The MNC itself may become a hybrid space with a broader general culture—and specific organizational culture—of its own. Transnational education projects are equally complicated. Robert Davison et al. (2017) sought better practices in establishing online international student teams. Their project spanned three sets of countries: Hong Kong and the US, Hong Kong and the UK, and Hong Kong and Singapore. The Hong Kong-Singapore location was even more culturally complex because the Singapore team included Norwegian students on an exchange program. Beyond that, teams identified by national location included students identified by another nationality. For example, the UK team included a Greek student, and a team in Singapore included a Norwegian exchange student (Davison, 2017, p. 323). Davison et al. identified challenges for student teams, particularly around forming relationships and trust, and discussed setting up teams from an instructor's point of view. But the informative study did not address yet another layer of cultural complexity: face-to-face "local" teams had collaborated transnationally via virtual technologies, meaning their activities were set in a hybrid space, both physical (for the local teams) *and* online (for the team-to-team meetings). While the article provided excellent recommendations about ice breakers, trust, and time, it did not discuss how the limitations and constraints of the virtual space created a culture of its own. In sum, some of the challenges—such as those associated with tone and word choice—could have factored into the feel and functioning of the text-based digital environment.

Although lack of specific information about intercultural factors and silence regarding their potential impacts on the research process might be frustrating, I remain sympathetic to these researcher/writer situations. Studies set in border-crossing contexts—which can cover a wide variety of intercultural locations—can be exceptionally complex, and authors simply do not have the space in a standard research publication to address everything. Addressing influential identities, positionality, and contexts is challenging, and based on the general trend of omission, reviewers and editors either do not want to read that information or do not think to ask about its inclusion. In my own experience, such metacommentary and reflexive work is generally cut from our manuscripts, deemed outside of the write-up's scope. Similarly, publications present scholarly work as a cleanly conceived and executed process. Although Peter Smagorinsky (2008) implores us to make the methods section the "epicenter" of our scholarly write-ups, publication processes more often whitewash challenges and struggles. Editors, reviewers, and readers are perceived as only wanting to see the version of the methods that "worked," as a model of rigor and thoroughness. Yet researchers know that our work is never cleanly conceived or executed. Reflections on a study's limitations are common and often required, but broader sharing about actual struggles in the planning, data collection, interpretation, and writing processes—crucial to ethical and rigorous research practices—proves uncommon.

■ Reflections and "Lessons Learned"

After identifying complicated outcomes for my first research question, my second question led to a deeper look at the subset of national border-crossing projects to better understand the lessons scholars have learned working in complex spaces. Of the articles emerging from the specific focus on transnational work with human participants, three offered brief reflection while four were more substantive in their metacommentary. Sean Williams et al. (2016) studied entrepreneurship through narrative interviews with professionals in China, Spain, and the US. Translation was not an issue because each of the authors was fluent in the language of one of the three countries, and the authors noted that their positionings as cultural insiders supported their interpretations of the data. Their other main reflection was that they refrained from considering the cultures they were studying as homogenized monoliths (Williams et al., 2016, p. 382). Bin Ai and Lifei Wang (2017) co-authored a case study about "Jack," a Chinese-Australian immigrant. Both authors identified themselves as Chinese cultural insiders, which boosted the ethos of their contextualized interpretations. Additionally, Ai shared that he was an immigrant himself, lending him special insight into Jack's world: "[Ai's] layered and shifting identities enable [him] to reflect upon his identity work in [Jack's] transnational business world" (Ai & Wang, 2017, p. 205). Ai also reflected on a change in research methods—his diary became a data source (p. 205)—and shared some of those diary entries, allowing readers to witness the researcher's (Ai's) approaches as he built a relationship with Jack (p. 207).

While reflections in these articles boosted author ethos by identifying close cultural and interpersonal connections, Andrew Mara (2017) illustrated how transnational projects can be challenging for outsider researchers. Mara reported on a user-experience design project at a Kenyan university and provided insight into how local contexts affected the project's design. His article shared some study complexities, including an "awakening" to understand that Kenyan cancer statistics were compiled in ways other than what had been assumed, the "surprising differences" in how different audiences read similar information, and the decision to "take greater care" in understanding his participants' perceptions (Mara, 2017, p. 51). Andrew Mara was involved in a second Kenya-based study along with Miriam Mara, and outcomes of Mara (2017) may have helped revise the research design for Mara and Mara (2018), discussed below.

Before getting to the heart of TPC scholars' advice on transnational research, I pause here for a bit of a cheater's move. Because my process of narrowing the scholarship brought me to a sample size of only four articles with substantive transnational research reflection, I added two more that were part of my corpus but technically did not meet my criteria because they were located in US-based spaces. N. Jones (2014) wrote about her experiences observing a nonprofit agency, and she pointedly argued that, in TPC, organizations should be considered as "cultures." Her reframing of workplace environments as spaces of shared symbolic systems in which "culture is a lot more subtle than most people realize" (N. Jones, 2014, p. 15) extended the importance of my inquiry back to "non-transnational" spaces. As Godwin Agboka (2014) confirmed, "academic research is always *cultural*, in many respects, and is always laden with political, power, and social justice concerns" (p. 299). My second addition is by Emma Rose and Robert Racadio (2017). They provided a fuller retrospective critique based on their study working with immigrant populations who needed information about health care in Seattle, Washington. Although their work was physically located in the US, they crossed intercultural borders via their participants' national identities and first languages. Their rich detail on the research team's background allowed readers to better understand the intercultural dynamics affecting the study design. Their discussion of the effects of back translation, challenges of translation precision, scenario design, and facilitation style demonstrated an array of ways the study could have been redesigned in response to intercultural differences among the team and participants.

I am also making an organizational shift from the previous category-based approach to a now thematic review. In answer to my second research question about "lessons learned," three themes emerged from the substantive reflections found in Agboka (2014), N. Jones (2014), Longo (2014), Mara and Mara (2018), Rose and Racadio (2017), and Walton et al. (2015). Those themes are lessons learned about localization, adaptation, and reflexivity. Together, they reveal a multilayered and dynamic TPC researcher positionality. The scope of this chapter precludes a thorough explication of all insights these scholars shared about their experiences, so the following paragraphs only provide highlights. If you are interested in intercultural and transnational research, all six original articles are worth careful reading.

The first central theme emerging from lessons learned about transnational and intercultural research is the imperative to localize the project by centering the host country contexts as well as the needs of the host country stakeholders. The groundwork has already been laid connecting "participatory localization" methods to intercultural work and social justice (Agboka, 2013). Walton et al. (2015) spend seven pages of their article detailing how they localized their Rwandan project at every phase from design to dissemination of the results. Engaging a local translator as a co-investigator provided crucial expertise, and their reflection foregrounds the potential articulatory functions translators play as cultural ambassadors, contextualizers, and data analyzers (Walton et al., 2015, pp. 49–50). An emphasis on working with—rather than observing—host participants and others at the project site requires researchers to step outside of the typical confines of the researcher role. For example, Rose and Racadio (2017) wished they had spent more time "training and mentoring" the staff at their Seattle community health center (pp. 21–22). Sharing their knowledge on how to observe user experiences would have served everyone in the center and would have contributed to a bigger positive impact. Longo (2014) described an almost four-year project that sought to locate then relocate a collaboration between her team in the US and partners in the Democratic Republic of Congo. Longo's article, an experience report-style case study, tells the story of her sustained attempts to provide host country organizations with technologies to address their needs. She writes about the challenges of "'making sense' of the research situation and relationships" and how "significant differences of worldview" ultimately doomed the endeavor (Longo, 2014, pp. 208–209). She reflects that relationship-building—in other words, the time and money to travel—was needed to bridge those differences. In all these reflection-heavy articles, localization is a product of relationship-building, whether through a cultural ambassador such as a translator, through time spent in the research location, or through being fully embedded in the research context.

Relationships take time to build, and cross-cultural relationships can require significant commitment because they bridge differences in worldviews and values. For example, Mara and Mara (2018) spent 11 months over a five-year period learning the language, observing local contexts, growing a network of contexts, developing localized subject matter expertise, studying literary works, and learning about the local culture. Many researchers do not have the resources for such investments, however (see Rose & Racadio, 2017, pp. 21–22). Relationships can be complicated by the insider/outsider status of the researcher (Agboka, 2014, p. 307), can change when new stakeholders or participants enter the project, and can have lasting consequences. Relationships are risky, inviting emotional ties and empathy, and when authentic and successful, can lead to more substantive understanding in the project and a "warming sense of acceptance" (N. Jones, 2014, p. 37). However, when relationships fail to develop, projects can feel disjointed and lead to troubling questions of whether or not the work will have any benefit to host participants (see Longo, 2014, pp. 212–213). "Relational accountabilities" are tied to ethical issues of respect and reciprocity, and indicate more than bonds between humans (Wilson, 2008). Relationality is multidimensional, interweaving histories and ancestors, lands and locations, rituals and practices, and more. Scholars and practitioners working in transnational and intercultural spaces engage in much more than consent form signing and data gathering, and learning to build relationships, to act ethically (contextualized within cultures, too), and to enact respect and reciprocity can leave us feeling unsure of ourselves.

A second theme in recent scholars' lessons learned is a tolerance for uncertainty and a willingness to adapt to circumstances in host spaces. The unpredictable nature of transnational and intercultural research means that it has been called "messy," meaning "unpredictable, mutable, contingent, serendipitous, complex, and challenging" (Walton et al., 2015, p. 45). Uncertainty may lead a researcher to feel her "position as the researcher and author of [her] research is unstable and decentered" (Longo, 2014, p. 208). Adaptation may involve a revision of basic research questions and methods, when time and relationship-building with participants reveal your original questions to be "heavily biased" and based on "hasty generalization," applying western assumptions about communication practices to the host site (Agboka, 2014, p. 309). Adapting can mean changes to interview questions and method protocols (N. Jones, 2014, p. 31; Rose & Racadio, 2017, p. 17) and/or the "power distribution [being] flipped" and a complete revision of post-research results dissemination in the host location (Walton et al., 2015, p. 61). Rather than *in-situ* amendments, adaptation may be an anticipated part of the project-planning process, as Mara and Mara (2018) demonstrated when they kept their "interview protocol and survey deliberately broad" (p. 100). They also allowed time at their Kenyan locations to get to know the location, then drew interview participants from "personal connections, community member suggestions, and in-person requests at health care facilities" (Mara & Mara, 2018, p. 102). In other words, they prioritized relationship-building as a means of directing their data gathering.

Recent TPC scholarship on engaging uncertainty through analytical frameworks extends the usefulness of "agile thinking" from a usability methodology to being a means of repositioning and reconceptualizing uncertainty as generative (Walsh & Walker, 2016). However, within the confines of Institutional Review Board (IRB) approvals, limited time and travel resources, and the academic publishing pressure cooker, uncertainty and surprises are viewed less as opportunities for repositioning and creative problem-solving and more as anxiety- and frustration-producing "messy challenges." Another area of TPC scholarship addressing uncertainty is "metis" intelligence, similar to Anzaldúa's (1987) *mestiza* consciousness. Metis intelligence, or metis thinking, is an agile and flexible approach to settings and situations that "are complicated, messy, chaotic, changeable, and ambiguous" (Pope-Ruark, 2014, p. 337). Metis thinking can unbind researchers and practitioners from self-imposed constraints regarding what is possible. A metis-based praxis "rounds out the profile of a civically engaged rhetor" (Pope-Ruark, 2014, p. 336). Therefore, this mindset is a strong match for transnational and intercultural research because it is a means of localizing and adapting.

A third theme emerging from the most substantively reflective articles is the need to be reflexive in intercultural spaces. Reflexivity begins in reflection (of one's own positioning, power, and privilege) but then goes further to consider broader contextual factors as well as how the researcher's chosen theoretical and methodological lenses shape all stages of a research project (N. Jones, 2014, p. 25). Reflexive thinking underpins the first two themes, localization and adaptation, as it pushes scholars to become aware of their own positions within their intercultural research contexts and how their own identities as scholars as well as insiders/outsiders affect the organizations and relationships within which they work. For example, N. Jones (2014) described her own grappling with identity and "othering" of her participants (p. 27). Reflexive thinking can reveal researchers' blind spots and assumptions of how their own cultures and educations have served as a form of indoctrination working against better practices (Agboka, 2014, p. 308). Through reflexive thinking, scholars become more aware of how local cultural logics beyond their control shape their transnational and intercultural locations (Mara & Mara, 2018, p. 96–97). For Agboka (2014), reflexive thinking invited "unlearning" a colonial mindset (p. 304). For Longo (2014), responsiveness to those logics and realities became a "matter of personal ethics more than professional responsibility or participatory design" (p. 214). Agboka (2014) calls on TPC scholars to use reflexivity as part of confronting harmful colonizing practices. He says we must constantly "question our own assumptions . . . be critical of our own approaches; question our insider posture . . . ; and be humble in our contacts with participants" (p. 299). Humility and willingness to cede control—as products of reflective and reflexive thinking—are persistently associated with lessons learned in transnational and intercultural spaces (N. Jones, 2014, p. 37; Longo, 2014, p. 212; Walton et al., 2015, p. 63).

■ Positionality Writ Large and Ongoing

Taken together, stories of "lessons learned" regarding localization, adaptability, and reflexivity constellate into a narrative of *positionality writ large* and *positionality as an ongoing process*. "Positionality refers to the stance or positioning of the researcher in relation to the social and political context of the study—the community, the organization or the participant group" and begins with locating the researcher along a continuum of insider-outsider identities (Rowe, 2014, p. 627). This view of positionality grounds it in the relative privilege that the researcher has in relation to the project participants and stakeholders. In other words, "One person's position is usually in relation to other people's positions, is shaped by history, and is highly contextualized" (Jones et al., 2016, p. 220). The relational nature of positionality can reveal power imbalances and systemic unearned advantages (see Walton, et. al., 2019). However, positionality is also multidimensional and dynamic. In studying the reflections of authors working in intercultural and transnational spaces, terms of "positionality" become even more important and complex.

The "highly contextualized" nature of positionality means it is more than a matter of how researcher(s) are stationed in comparison to their colleagues, participants, and stakeholders. Contextualization means that a researcher's positionality is a complex function of interwoven relationships to host organizations (e.g., N. Jones, 2014); to local hierarchies (Agboka, 2014, pp. 307–308); to local symbolic systems and norms (e.g., requiring translation and cultural ambassadors); to local legal systems (Mara & Mara, 2018, p. 100); to local social, economic, and political contexts (Longo, 2014, p. 207, 213); and more. Outside of the project's location, positionality extends to larger epistemological and methodological relationships: the researchers to their theories, methods, and goals. An ongoing reflexive practice and agile attitude mean positionality is an ongoing *re*-positioning as a project proceeds, as relationships develop, and as power dynamics emerge. As recent TPC scholarship has argued, we must continually (re)localize our work—which includes repositioning ourselves—specifically in relation to how we continue to (re)define users, communities, and diversity (Shivers-McNair & San Diego, 2017).

As an ongoing process of planning and adapting, positionality continues after data collection concludes, throughout interpretation and processes of representation (e.g., write-up and dissemination of results). As TPC scholars and professionals move through this process, positionality yet again shifts. Researchers and practitioners move from primary accountability being with their local participants to it being with their reviewers, editors, and other audiences. That shift in positionality and accountability—from our host locations to the series of publication gatekeepers—may be one reason why not much reflection on the influence of intercultural complexity is included in our collective scholarly work. Whereas ongoing reflexivity and (re)positioning are intense parts of the planning and data gathering project phases, the writing and publication phases typically streamline the focus to include only a description of what "worked" and not how the process demanded adaptation. Revealing our uncertainties and agilities should reinforce—rather than risk—representing ourselves as methodical and rigorous.

An agile attitude towards shifting positionality aligns with our field's history of continually reexamining itself (see St.Amant & Melonçon, 2016, pp. 271–272). Researcher positionality goes hand-in-hand with TPC's orientation towards praxis and social justice. But how? If positionality is at the heart of our work and if border crossing permeates much of what we do, then what gravitational force is at the center of our individual and collective positions? How do we avoid simply stumbling across shifting sands? The answer, as Walton (2016) asserts, is that we must ground what we do in a "human-centered" principle. That principle requires a persistent, reflexive (re)consideration of our positionings and how they are intertwined with power and privilege. By considering how the "3Ps" of positionality, privilege, and power shape TPC, we can "examine macrolevel concepts that can impact social capital and agency" (Jones et al., 2016, p. 220). In other words, a localized, adaptable, and reflexive (re)positioning in all of our culturally infused research spaces should motivate our efforts to justly address issues of power and hierarchy.

■ Conclusions

This literature review set out to explore three questions about the nature of transnational and intercultural research in TPC, about the lessons border crossers are learning, and about how individual reflections can be gathered and organized to help other researchers and practitioners make better informed, more ethical, and more socially just decisions about their own projects. The project has revealed that transnational and/or intercultural research is a healthy part of both our foundations and a source of innovative methods and knowledge making. My project has illustrated that recent TPC border crossing happens in a myriad of interesting and overlapping ways, from inquiries that jump across time to those that bridge generational, linguistic, and embodied perspectives. Even defining "intercultural" remains complex and generates as many questions as it does answers. Specifically focusing on transnational research with human participants, recent publications in TPC reveal a range of globalized projects, yet only a limited number of the scholars offer influential reflections and advice concerning the challenges of working across national and cultural borders. Their lessons center on the importance of localization, of being flexible or agile, and of constantly learning (and unlearning) through reflective and reflexive thinking.

We now arrive at my last research question, about how we can organize and share transnational and intercultural research advice to promote better practices in the field. First, authors must be encouraged to share the ways their projects met challenges, adapted, and resulted in rich reflections. Beyond the solidly written methods section, authors must be offered (and must take up) the space for sharing such metacommentary. But even if we have a growing movement to do that sharing, TPC as a field of research and practice should do more. We have strong statements on ethics from the Association of Teachers of Technical Writing (n.d.), the National Council of Teachers of English (2015), and the Society for Technical Communication (1998). However, none includes guidance on better practices for transnational and/or intercultural projects, perhaps because they assume that ethics transcend borders. However, because cultures have their own attendant norms and systems, that assumption will not necessarily hold, and because of the expanding ways we work, a statement of transnational and intercultural ethics would be timely and useful for both researchers and practitioners in the field. The ethic should consider methods of increasing inclusion, building relationships, sharing power, decolonizing practices, and pursuing more just practices at all phases of the research process: planning, gathering, analyzing, and representing outcomes. Kirk St.Amant and Lisa Melonçon (2016) "encourage researchers to think more broadly about what it is that TPC does while also thinking more narrowly about how individual research projects contribute to the larger whole" (p. 274). Creating a shared statement of transnational and intercultural research ethics would make progress towards their call, and, although addressing only one aspect of what we do (research across borders), ethic-building conversations that focus on research topics, practices, and praxes also can generate reflection that may yield additional ideas for unifying common grounds regarding the field.

The scholars and articles discussed here offer a starting point, but I also acknowledge that defining my study's scope and choosing to narrow my corpus in the way I did introduced limitations. If I were to change my definition of "intercultural" based on N. Jones' (2014) call to address organizations as cultural groups, then many more of the 609 articles in my original corpus would have counted in the sorting process. By choosing to leave out pedagogy-focused articles, I may have missed other thoughtful advice (e.g., Ballentine, 2015), and by focusing primarily on projects involving human participants, I likely missed additional frameworks and guidance (e.g., St.Amant, 2015). By limiting my scope to articles, I also have not delved into edited collections or other manuscripts. Moving towards an ethic would require casting a broader net as well as engaging in deeper discussions and collaborations with representatives from across TPC.

Through developing an ethic, we can continue the conversation of amorphous concepts such as "culture," "borders," and "transnational." The process of discussing, proposing, testing, creating, and recreating a shared ethic would reveal and amplify questions about TPC priorities and realities. It would support the continued development and evolving skill sets demanded by the field (Shalamova et al., 2018) and should contribute broader disciplinary commitments to socially just ways of researching, collaborating, and generally being in the world (Walton et al., 2019). We can also use a research ethic to inform ongoing innovation in research methods. As technologies continue to transform possibilities for researching across places and spaces, thinking about the implications of our processes remains crucial. Finally, researchers who are new to moving across borders would benefit from both the ethic and the shared reflections it would inspire. An ethic should hold the TPC field to high standards and support excellence in the teaching and mentoring of future transnational and intercultural researchers. However, it should also guide researchers to action: "research is not seen as worthy or ethical if it does not help to improve the reality of the research participants," and the best research changes the researchers themselves (Wilson, 2008, p. 37). Striving for better—more informed, more critically examined—practices will indeed be "messy" and complex but will serve to strengthen TPC as a field.

■ References

Agboka, Godwin Y. (2013). Participatory localization: A social justice approach to navigating unenfranchised/disenfranchised cultural sites. *Technical Communication Quarterly*, 22(1), 28–49.

Agboka, Godwin Y. (2014). Decolonial methodologies: Social justice perspectives in intercultural technical communication research. *Journal of Technical Writing and Communication*, *44*(3), 297–327.

Ai, Bin & Wang, Lifei. (2017). Transnational business communication and identity work in Australia. *IEEE Transactions on Professional Communication*, *60*(2), 201–213.

Amare, Nicole. (2002). The culture(s) of the technical communicator. *IEEE Transactions on Professional Communication*, *45*(2), 128–132.

Anzaldúa, Gloria. (1987). *Borderlands: La frontera*. Aunt Lute.

Association of Teachers of Technical Writing. (n.d.). *Code of ethics*. Association of Teachers of Technical Writing. https://attw.org/about-attw/code-of-ethics/.

Baldwin, John R., Faulkner, Sandra L. & Hecht, Michael L. (2006). A moving target: The illusive definition of culture. In J. R. Baldwin, S. L. Faulkner, M. L. Hecht & S. L. Lindsley (Eds.), *Redefining culture* (pp. 27–50). Routledge.

Ballentine, Brian D. (2015). Creativity counts: Why study abroad matters to technical and professional communication. *Technical Communication Quarterly*, *24*(4), 291–305.

Basov, Nikita & Minina, Vera. (2018). Personal communication ties and organizational collaborations in networks of science, education, and business. *Journal of Business and Technical Communication*, *32*(3), 373–405.

Batova, Tatiana. (2018a). Global technical communication in 7.5 weeks online: Combining industry and academic perspectives. *IEEE Transactions on Professional Communication*, *61*(3), 311–329.

Batova, Tatiana. (2018b). Negotiating multilingual quality in component content-management environments. *IEEE Transactions on Professional Communication*, *61*(1), 77–100.

Campbell, Kim S., Amare, Nicole, Kane, Erin, Manning, Alan D. & Naidoo, Jefrey S. (2017). Plain-style preferences of US professionals. *IEEE Transactions on Professional Communication*, *60*(4), 401–411.

Cleary, Yvonne. (2016). Community of practice and professionalization perspectives on technical communication in Ireland. *IEEE Transactions on Professional Communication*, *59*(2), 126–139.

Connors, Robert J. (1982). The rise of technical writing instruction in America. *Journal of Technical Writing and Communication*, *12*(4), 329–352.

Croucher, Sheila. (2012). Americans abroad: A global diaspora? *Journal of Transnational American Studies*, *4*(2). https://escholarship.org/uc/item/07c2k96f.

Davison, Robert M., Panteli, Niki, Hardin, Andrew M. & Fuller, Mark A. (2017). Establishing effective global virtual student teams. *IEEE Transactions on Professional Communication*, *60*(3), 317–329.

Ding, Huiling. (2018). Cross-cultural whistle-blowing in an emerging outbreak: Revealing health risks through tactic communication and rhetorical hijacking. *Communication Design Quarterly Review*, *6*(1), 35–44.

Ding, Huiling & Savage, Gerald. (2013). Guest editors' introduction: New directions in intercultural professional communication [Special issue]. *Technical Communication Quarterly*, *22*(1), 1–9.

Fraiberg, Steven. (2013). Reassembling technical communication: A framework for studying multilingual and multimodal practices in global contexts. *Technical Communication Quarterly*, *22*(1), 10–27.

Fraiberg, Steven. (2017). Start-up nation: Studying transnational entrepreneurial practices in Israel's start-up ecosystem. *Journal of Business and Technical Communication, 31*(3), 350–388.

Getto, Guiseppe. (2015). Editor's introduction. *Rhetoric, Professional Communication, and Globalization, 7*(1), 1–11.

Getto, Guiseppe & St.Amant, Kirk. (2014). Designing globally, working locally: Using personas to develop online communication products for international users. *Communication Design Quarterly Review, 3*(1), 24–46.

Gonzales, Laura & Turner, Heather N. (2017). Converging fields, expanding outcomes: Technical communication, translation, and design at a non-profit organization. *Technical Communication, 64*(2), 126–140.

Gorski, Paul C. (2008). Good intentions are not enough: A decolonizing intercultural education. *Intercultural Education, 19*(6), 515–525.

Haas, Angela & Eble, Michelle. (2018). Introduction: The social justice turn. In A. Haas & M. Eble (Eds.), *Key theoretical frameworks: Teaching technical communication in the twenty-first century* (pp. 3–19). Utah State University Press.

Hirst, Russel. (2016). Bonding with the nuclear industry: A technical communication professor and his students partner with Y-12 national security complex. *Journal of Technical Writing and Communication, 46*(2), 151–171.

Hofstede, Geert. (2001). *Culture's consequences: Comparing values, behaviors, institutions and organizations across nations*. Sage.

Hutter, Liz & Lawrence, Halcyon M. (2018). Promoting inclusive and accessible design in usability testing: A teaching case with users who are deaf. *Communication Design Quarterly Review, 6*(2), 21–30.

Jones, Natasha N. (2014). Methods and meanings: Reflections on reflexivity and flexibility in an intercultural ethnographic study of an activist organization. *Rhetoric, Professional Communication, and Globalization, 5*(1), 14–43.

Jones, Natasha N., Moore, Kristen R. & Walton, Rebecca. (2016). Disrupting the past to disrupt the future: An antenarrative of technical communication. *Technical Communication Quarterly, 25*(4), 211–229.

Jones, Rodney H. (2014). Unwriting food labels: Discursive challenges in the regulation of package claims. *Journal of Business and Technical Communication, 28*(4), 477–508.

Kiwanuka-Tondo, James & Pettiway, Keon M. (2016). Localizing complex scientific communication: A SWOT analysis and multi-sectoral approach of communicating climate change. *Communication Design Quarterly Review, 4*(4), 74–85.

Koerber, Amy & Graham, Hilary. (2017). Theorizing the value of English proficiency in cross-cultural rhetorics of health and medicine: A qualitative study. *Journal of Business and Technical Communication, 31*(1), 63–93.

Kohn, Liberty. (2015). How professional writing pedagogy and university-workplace partnerships can shape the mentoring of workplace writing. *Journal of Technical Writing and Communication, 45*(2), 166–188.

Kreth, Melinda L. & Bowen, Elizabeth. (2017). A descriptive survey of technical editors. *IEEE Transactions on Professional Communication, 60*(3), 238–255.

Longo, Bernadette. (1998). An approach for applying cultural study theory to technical writing research. *Technical Communication Quarterly, 7*(1), 53–73.

Longo, Bernadette. (2014). RU There? Cell phones, participatory design, and intercultural dialogue. *IEEE Transactions on Professional Communication, 57*(3), 204–215.

Mackiewicz, Jo. (2014). Motivating quality: The impact of amateur editors' suggestions on user-generated content at Epinions.com. *Journal of Business and Technical Communication, 28*(4), 419–446.

Mara, Andrew. (2017). Framework negotiation and UX design. *Communication Design Quarterly Review, 5*(3), 48–54.

Mara, Miriam & Mara, Andrew. (2018). Blending humanistic and rhetorical analysis to locate gendered dimensions of Kenyan medical practitioner attitudes about cancer. *Technical Communication Quarterly, 27*(1), 93–107.

McSweeney, Brendan. (2002). Hofstede's model of national cultural differences and their consequences: A triumph of faith—a failure of analysis. *Human Relations, 55*(1), 89–118.

Milbourne, Chelsea R. (2016). Disruption, spectacle, and gender in eighteenth-century technical communication. *Technical Communication Quarterly, 25*(2), 121–136.

National Council of Teachers of English. (2015, March 31). CCCC guidelines for the ethical conduct of research in composition studies. *Conference on College Communication and Composition.* https://ncte.org/statement/ethicalconduct/.

Osland, Joyce S., Bird, Allan. (2000). Beyond sophisticated stereotyping: Cultural sensemaking in context. *The Academy of Management Executive (1993–2005), 14*(1), 65–77.

Oswal, Sushil K. (2014). Participatory design: Barriers and possibilities. *Communication Design Quarterly Review, 2*(3), 14–19.

Passera, Stefania, Kankaanranta, Anne & Louhiala-Salminen, Lena. (2017). Diagrams in contracts: Fostering understanding in global business communication. *IEEE Transactions on Professional Communication, 60*(2), 118–146.

Pope-Ruark, Rebecca. (2014). A case for metic intelligence in technical and professional communication programs. *Technical Communication Quarterly, 23*(4), 323–340.

Popham, Susan L. (2014). Hybrid disciplinarity: Métis and ethos in juvenile mental health electronic records. *Journal of Technical Writing and Communication, 44*(3), 329–344.

Powell, Malea, Levy, Daisey, Riley-Mukavetz, Andrea, Brooks-Gillies, Marilee, Novotny, Maria & Fisch-Ferguson, Jennifer. (2014). Our story begins here: Constellating cultural rhetorics. *Enculturation: A Journal of Rhetoric, Writing, and Culture, 18.* http://www.enculturation.net/our-story-begins-here.

Pratt, Mary L. (1991). Arts of the contact zone. *Profession,* 33–40.

Ramey, John W. (2014). The coffee planter of Saint Domingo: A technical manual for the Caribbean slave owner. *Technical Communication Quarterly, 23*(2), 141–159.

Reinsch, N. Lamar, Jr. & Gardner, Jonathan A. (2014). Do communication abilities affect promotion decisions? Some data from the c-suite. *Journal of Business and Technical Communication, 28*(1), 31–57.

Rice-Bailey, Tammy. (2016). The role and value of technical communicators: Technical communicators and subject matter experts weigh in. *Technical Communication Quarterly, 25*(4), 230–243.

Rose, Emma J. (2016). Design as advocacy: Using a human-centered approach to investigate the needs of vulnerable populations. *Journal of Technical Writing and Communication, 46*(4), 427–445.

Rose, Emma J. & Racadio, Robert. (2017). Testing in translation: Conducting usability studies with transnational users. *Rhetoric, Professional Communication, and Globalization, 10*(1), 5–26.

Rose, Emma J., Racadio, Robert, Wong, Kalen, Nguyen, Shalley, Kim, Jee & Zahler, Abbie. (2017). Community-based user experience: Evaluating the usability of health insurance information with immigrant patients. *IEEE Transactions on Professional Communication, 60*(2), 214–231.

Rowe, Wendy E. (2014). Positionality. In D. Coghlan & M. Brydon-Miller (Eds.), *The Sage encyclopedia of action research* (pp. 627–628). Sage.

Rudd, Leanne Mysti. (2018). "It makes us even angrier than we already are": Listening rhetorically to students' responses to an honor code imported to a transnational university in the Middle East. *Journal of Global Literacies, Technologies, and Emerging Pedagogies, 4*(3), 655–674.

Sano-Franchini, Jennifer. (2017). What can Asian eyelids teach us about user experience design? A culturally-reflexive framework for UX/I design. *Rhetoric, Professional Communication and Globalization, 10*(1), 27–53.

Scott, J. Blake & Longo, Bernadette. (2006). Guest editors' introduction: Making the cultural turn [Special issue]. *Technical Communication Quarterly, 15*(1), 3–7.

Shalamova, Nadya, Rice-Bailey, Tammy & Wikoff, Katherine. (2018). Evolving skill sets and job pathways of technical communicators. *Communication Design Quarterly Review, 6*(3), 14–24.

Shin, Wonsun, Pang, Augustine & Kim, Hyo Jung. (2015). Building relationships through integrated online media: Global organizations' use of brand web sites, Facebook, and Twitter. *Journal of Business and Technical Communication, 29*(2), 184–220.

Shivers-McNair, Ann & San Diego, Clarissa. (2017). Localizing communities, goals, communication, and inclusion: A collaborative approach. *Technical Communication, 64*(2), 97–112.

Sinclair, Paul & Blachford, Dongyan. (2015). The Guide to Kuan Hua: Language and literacy in the 19th-century Chinese business environment. *Journal of Business and Technical Communication, 29*(4), 403–427.

Smagorinsky, Peter. (2008). The method section as conceptual epicenter in constructing social science research reports. *Written Communication, 25*(3), 389–411.

Small, Nancy. (2017). Risking our foundations: Honor, codes, and authoritarian spaces. In K. Gray, H. Bashir & S. Keck (Eds.), *Western higher education in Asia and the Middle East: Politics, economics, and pedagogy* (pp. 223–242). Lexington Books.

Society for Technical Communication. (1998). *Ethical principles.* https://www.stc.org/about-stc/ethical-principles/.

St.Amant, Kirk. (2013). What do technical communicators need to know about international environments? In J. Johnson-Eilola and S. A. Selber (Eds.), *Solving problems in technical communication* (pp. 479–496). University of Chicago Press.

St.Amant, Kirk. (2015). Aspects of access: Considerations for creating health and medical content for international audiences. *Communication Design Quarterly Review, 3*(3), 7–11.

St.Amant, Kirk. (2017). Of friction points and infrastructures: Rethinking the dynamics of offering online education in technical communication in global contexts. *Technical Communication Quarterly, 26*(3), 223–241.

St.Amant, Kirk. & Melonçon, Lisa. (2016). Addressing the incommensurable: A research-based perspective for considering issues of power and legitimacy in the field. *Journal of Technical Writing and Communication, 46*(3), 267–283.

Stanton, Rhonda. (2017). Communicating with employees: Resisting the stereotypes of generational cohorts in the workplace. *IEEE Transactions on Professional Communication*, *60*(3), 256–272.

Thatcher, Barry. (2001). Issues of validity in intercultural professional communication research. *Journal of Business and Technical Communication*, *15*(4), 458–489.

Thatcher, Barry & St.Amant, Kirk. (Eds.). (2011). *Teaching intercultural rhetoric and technical communication: Theories, curriculum, pedagogies, and practices*. Baywood.

Virtaluoto, Jenni, Sannino, Annalisa & Engeström, Yrjö. (2016). Surviving outsourcing and offshoring: Technical communication professionals in search of a future. *Journal of Business and Technical Communication*, *30*(4), 495–532.

Walls, Douglas M. (2017). The professional work of "unprofessional" tweets: Microblogging career situations in African American Hush Harbors. *Journal of Business and Technical Communication*, *31*(4), 391–416.

Walsh, Lynda & Walker, Kenneth C. (2016). Perspectives on uncertainty for technical communication scholars. *Technical Communication Quarterly*, *25*(2), 71–86.

Walton, Rebecca. (2016). Supporting human dignity and human rights: A call to adopt the first principle of human-centered design. *Journal of Technical Writing and Communication*, *46*(4), 402–426.

Walton, Rebecca & Hopton, Sarah Beth. (2018). "All Vietnamese men are brothers": Rhetorical strategies and community engagement practices used to support victims of Agent Orange. *Technical Communication*, *65*(3), 309–325.

Walton, Rebecca, Mays, Robin E. & Haselkorn, Mark. (2016). Enacting humanitarian culture: How technical communication facilitates successful humanitarian work. *Technical Communication*, *63*(2), 85–100.

Walton, Rebecca, Moore, Kristen R. & Jones, Natasha N. (2019). *Technical communication after the social justice turn*. Routledge.

Walton, Rebecca, Zraly, Maggie & Mugengana, Jean Pierre. (2015). Values and validity: Navigating messiness in a community-based research project in Rwanda. *Technical Communication Quarterly*, *24*(1), 45–69.

Walwema, Josephine. (2016). Tailoring information and communication design to diverse international and intercultural audiences: How culturally sensitive ICD improves online market penetration. *Technical Communication*, *63*(1), 38–52.

Wang, Junhua. (2018). Strategies for managing cultural conflict: Models review and their applications in business and technical communication. *Journal of Technical Writing and Communication*, *48*(3), 281–294.

Wang, Xiaobo & Gu, Baotong. (2015). The communication design of WeChat: Ideological as well as technical aspects of social media. *Communication Design Quarterly Review*, *4*(1), 23–35.

Weissbach, Robert S. & Pflueger, Ruth C. (2018). Collaborating with writing centers on interdisciplinary peer tutor training to improve writing support for engineering students. *IEEE Transactions on Professional Communication*, *61*(2), 206–220.

Wickman, Chad. (2015). Locating the semiotic power of writing in science. *Journal of Business and Technical Communication*, *29*(1), 61–92.

Williams, Sean D., Ammetller, Gisella, Rodríguez-Ardura, Inma & Li, Xiaoli. (2016). A narrative perspective on international entrepreneurship: Comparing stories from the United States, Spain, and China. *IEEE Transactions on Professional Communication*, *59*(4), 379–397.

Wilson, Shawn S. (2008). *Research is ceremony: Indigenous research methods.* Fernwood Publishing.

Yin, Juelin, Feng, Jieyun & Wang, Yuyan. (2015). Social media and multinational corporations' corporate social responsibility in China: The case of ConocoPhillips oil spill incident. *IEEE Transactions on Professional Communication, 58*(2), 135–153.

Yu, Han. (2012). Intercultural competence in technical communication: A working definition and review of assessment methods. *Technical Communication Quarterly, 21*(2), 168–186.

Yu, Han & Savage, Gerald. (Eds.). (2013). *Negotiating cultural encounters: Narrating intercultural engineering and technical communication* (Vol. 1). John Wiley & Sons.

Yum, Jung-Yoon & Jeong, Se-Hoon. (2015). Examining the public's responses to crisis communication from the perspective of three models of attribution. *Journal of Business and Technical Communication, 29*(2), 159–183.

Zhang, Yuejiao. (2016). Illustrating beauty and utility: Visual rhetoric in two medical texts written in China's Northern Song Dynasty, 960–1127. *Journal of Technical Writing and Communication, 46*(2), 172–205.

Contributors

Michael J. Albers is Professor at East Carolina University (ECU), where he teaches in the professional writing program. His primary teaching areas are editing, information design, and usability. Before earning his Ph.D., he worked for ten years as a technical communicator, writing software documentation and performing interface design. His research interests include designing information focused on answering real-world questions, presentation of complex information, and human-information interaction.

Stephen Carradini is Assistant Professor in the technical communication program at Arizona State University. His research interests include disciplinarity, social media in the workplace, and extra-institutional individuals. His work has been published in journals such as *IEEE Transactions on Professional Communication*, *Journal of Technical Writing and Communication*, and *New Media & Society*.

Lisa Detora is Associate Professor and director of STEM writing at Hofstra University and guest faculty in medical humanities at Hofstra Northwell Medical School. Her scholarship bridges regulatory documentation, biomedical writing, medical humanities, and rhetorics of health and medicine. Her teaching interests include comic book studies, disability studies, and scientific writing. She is also the editor of *Regulatory Writing: An Overview* (2020) and *Bodies in Transition in the Health Humanities* (Routledge 2020).

Sara Doan is Assistant Professor of Technical Communication at Kennesaw State University, where she teaches data visualization, visual and information design, and the rhetoric of health and medicine. Her research on instructor feedback has appeared in *IEEE Transactions on Technical Communication* and in the proceedings of IEEE ProComm. Sara's current research on data visualizations and infographics about COVID-19 has appeared in the *Journal of Business and Technical Communication*.

Brenton Faber is Professor in the departments of Humanities & Arts and Biomedical Engineering at Worcester Polytechnic Institute. He is also a practicing paramedic with the Potsdam Volunteer Rescue Squad in Potsdam, NY. His research interests include scientific and medical communication, epidemiology, and professional communication. He is a founding editor of the *Northern New York Medical Review* and has recently completed a book examining the function of intention in professional communication.

Michael J. Faris is Associate Professor of Technical Communication and Rhetoric at Texas Tech University, where he co-administers the first-year writing program. His research focuses on digital rhetorics and literacies.

Marjorie Rush Hovde is Associate Professor of Technical Communication who teaches a variety of technical communication courses. Marjorie's research interests include workplace technical communication dynamics, usability of user documentation, and technical communicators' technological literacy.

Lisa Meloncon is Professor of Technical Communication and interim department chair at the University of South Florida. Her research focuses on programmatic issues in technical and professional communication, research methodology and methods, and the rhetoric of health and medicine.

Sushil K. Oswal is Professor of Human Centered Design in the School of Interdisciplinary Arts and Sciences and Affiliate Professor of Disability Studies in the College of Arts and Sciences at the University of Washington. He is the founding editor of *the Western ABC Bulletin*. Besides teaching accessible design, UX, and disability courses, he consults in the areas of designing accessible self-service kiosks and inclusive academic spaces.

Zsuzsanna B. Palmer is Assistant Professor in the Department of Writing at Grand Valley State University in Michigan, where she teaches professional writing, writing for the web, and document design. Her research has been published in *Business and Professional Communication Quarterly*, *Journal of Business and Technical Communication*, and *Journal of Technical Writing and Communication*, as well as in several edited collections about the international aspects of teaching professional communication.

Ashley Patriarca is Associate Professor of English at West Chester University of Pennsylvania, where she teaches business and technical writing courses. Her work focuses on issues of risk communication, social media, usability, and pedagogy.

Rebecca Pope-Ruark earned a Ph.D. in Rhetoric and Professional Communication from Iowa State University and taught writing, rhetorical theory, publishing, and grant writing for 12 years at Elon University. She is currently a faculty teaching and learning specialist in the Center for Teaching and Learning at the Georgia Institute of Technology as well as a coach, consultant, and facilitator on the subjects of writing, productivity, and burnout. She is the author of *Agile Faculty: Practical Strategies for Research, Service, and Teaching*.

Jacob D. Rawlins is Associate Professor in the Linguistics Department at Brigham Young University in Provo, UT, where he teaches courses in editing, publishing, and grammar. His research interests include applied rhetoric, professional communication pedagogy, workplace myth building, and interactive data visualizations.

Joanna Schreiber is Associate Professor of Technical and Professional Communication at Georgia Southern University. Her research interests include project management, trends in professional and technical editing, workplace studies, and technical communication programs and pedagogies. Her work has been published in *Technical Communication*, *Technical Communication Quarterly*, *Programmatic Perspectives*, and *Journal of Technical Writing and Communication*.

Matthew R. Sharp is Associate Professor of Communication in the Humanities and Communication Department at Embry-Riddle Aeronautical University in Daytona Beach, FL. His research analyzes organizational activity systems and their mediating technologies from both cultural and rhetorical studies perspectives.

Nancy Small is Assistant Professor of English and Director of First Year Writing at the University of Wyoming. Her work has appeared in *Peitho: Journal of the Coalition of Feminist Scholars in the History of Rhetoric & Composition* and *Journal of Technical Writing and Communication*. She has written about her transnational experiences in two edited collections and has a forthcoming monograph on rhetorical feminism and transnational spaces.

Jennifer R. Veltsos is Interim Associate Vice President for Curriculum and Dean of Graduate Studies at Minnesota State University, Mankato. She has taught undergraduate courses in business communication, technical communication, visual rhetoric and document design, and research methods; at the graduate level, she has taught managerial communication, proposals, and instructional design. From 2017 to 2019, she was the director of the Center for Excellence in Teaching and Learning.

Greg Wilson is a scholar in rhetoric of science and technical communication who has worked in academia, government, and industry.

www.ingramcontent.com/pod-product-compliance
Lightning Source LLC
Chambersburg PA
CBHW020518080526
44583CB00013B/652